T0137457

# Intelligent Systems Reference Library

Volume 58

*Series editors*

Janusz Kacprzyk, Polish Academy of Sciences, Warsaw, Poland
e-mail: kacprzyk@ibspan.waw.pl

Lakhmi C. Jain, University of Canberra, Canberra, Australia
e-mail: Lakhmi.jain@unisa.edu.au

For further volumes:
http://www.springer.com/series/8578

*About this Series*

The aim of this series is to publish a Reference Library, including novel advances and developments in all aspects of Intelligent Systems in an easily accessible and well structured form. The series includes reference works, handbooks, compendia, textbooks, well-structured monographs, dictionaries, and encyclopedias. It contains well integrated knowledge and current information in the field of Intelligent Systems. The series covers the theory, applications, and design methods of Intelligent Systems. Virtually all disciplines such as engineering, computer science, avionics, business, e-commerce, environment, healthcare, physics and life science are included.

George A. Anastassiou · Iuliana F. Iatan

# Intelligent Routines II

## Solving Linear Algebra and Differential Geometry with Sage

George A. Anastassiou
Department of Mathematical Sciences
University of Memphis
Memphis
USA

Iuliana F. Iatan
Department of Mathematics and Computer Science
Technical University of Civil Engineering
Bucharest
Romania

ISSN 1868-4394 ISSN 1868-4408 (electronic)
ISBN 978-3-319-35037-0 ISBN 978-3-319-01967-3 (eBook)
DOI 10.1007/978-3-319-01967-3
Springer Cham Heidelberg New York Dordrecht London


Softcover reprint of the hardcover 1st edition 2014


Printed on acid-free paper

Springer is part of Springer Science+Business Media (www.springer.com)

*Mathematics: the art of redescribing to the simplest of logical structures, that is an art of rewriting finite steps of thinking.*

G. A. Anastassiou

*The measure of success for a person is the magnitude of his/her ability to convert negative conditions to positive ones and achieve goals.*

G. A. Anastassiou

*Labor omnia vincit improbus.*

Virgil, Georgics

*The greatest thoughts come from the heart.*

Vauvenargues, Reflections and Maxims

*Friends show me what I can do, foes teach me what I should do.*

Schiller, Friend and foe

*Experience is by industry achieved, and perfected by the swift course of time.*

Shakespeare, The Two Gentlemen of Verona

*Good name in man and woman, dear my lord, Is the immediate jewel of their souls. Who steals my purse steals trash; 't is something, nothing; 'T was mine, 't is his, and has been slave to thousands; But he that filches from me my good name Robs me of that which not enriches him, And makes me poor indeed.*

Shakespeare, Othello Act III, scene 3

*L'ami du genre humain n'est point du tout mon fait.*

Molière, Le Misanthrope, I, 1

*Honor is like an island, rugged and without shores; we can never re-enter it once we are on the outside.*

Boileau, Satires

*Modesty is an adornment, but you come further without it.*

German proverb

# Preface

Linear algebra can be regarded as a theory of the vector spaces, because a vector space is a set of objects or elements that can be added together and multiplied by numbers (the result remaining an element of the set), so that the ordinary rules of calculation are valid. An example of a vector space is the geometric vector space (the free vector space), presented in the first chapter of the book, which plays a central role in physics and technology and illustrates the importance of the vector spaces and linear algebra for all practical applications.

Besides the notions which operates mathematics, created by abstraction from environmental observation (for example, the geometric concepts) or quantitative and qualitative research of the natural phenomena (for example, the notion of number) in mathematics there are elements from other sciences. The notion of vector from physics has been studied and developed creating vector calculus, which became a useful tool for both mathematics and physics. All physical quantities are represented by vectors (for example, the force and velocity).

A vector indicates a translation in the three-dimensional space; therefore we study the basics of the three-dimensional Euclidean geometry: the points, the straight lines and the planes, were in the second chapter.

The linear transformations are studied in the third chapter, because they are compatible with the operations defined in a vector space and allow us to transfer algebraic situations and related problems in three-dimensional space.

Matrix operations clearly reflect their similarity to the operations with linear transformations; so the matrices can be used for the numerical representation of the linear transformations. The matrix representation of linear transformations is analogous to the representation of the vectors through n coordinates relative to a basis.

The eigenvalue problems (also treated in the third chapter) are of great importance in many branches of physics. They make it possible to find some coordinate systems in which changes take the simplest forms. For example, in mechanics the main moments of a solid body are found with the eigenvalues of a symmetric matrix representing the vector tensor. The situation is similar in continuous mechanics, where the body rotations and deformations in the main directions are found using the eigenvalues of a symmetric matrix. Eigenvalues have a central importance in quantum mechanics, where the measured values of the observable physical quantities appear as eigenvalues of operators. Also, the eigenvalues are useful in the study of differential equations and continuous dynamical systems that arise in areas such as physics and chemistry.

The study of the Euclidean vector space in the fourth chapter is required to obtain the orthonormal bases, whereas relative to these bases the calculations are considerably simplified. In a Euclidean vector space, scalar product can be used to define the length of vectors and the angle between them. In the investigation of the Euclidean vector spaces very useful are the linear transformations compatible with the scalar product, i.e. the orthogonal transformations. The orthogonal transformations in the Euclidean plane are: the rotations, the reflections or the compositions of rotations and reflections.

The theory of bilinear and quadratic form are described in the fifth chapter. These are used with analytic geometry to get the classification of the conics and of the quadrics, presented in the Chap. 8.

In Analytic Geometry we replace the definitions and the geometrical study of the curves and the surfaces, by the algebraic correspondence: a curve and a surface are defined by algebraic equations, and the study of the curve and the surface is reduced to the study of the equation corresponding to each one (see the seventh chapter).

The above are used in physics, in particular to describe physical systems subject to small vibrations. The coefficients of a bilinear form behave for certain transformations like the tensor coordinates. The tensors are useful in theory of elasticity (deformation of an elastic medium is described through the deformation tensor).

In the differential geometry, in the study of the geometric figures, we use the concepts and methods of the mathematical analysis, especially the differential calculus and the theory of differential equations, presented in the sixth chapter. The physical problems lead to inhomogeneous linear differential equations of order $n$ with constant coefficients.

In this book we apply extensively the software SAGE, which can be found free online http://www.sagemath.org/.

We give plenty of SAGE applications at each step of our exposition.

This book is usefull to all researchers and students in mathematics, statistics, physics, engineering and other applied sciences. To the best of our knowledge this is the first one.

The authors would like to thank Prof. Razvan Mezei of Lenoir-Rhyne University, North Carolina, USA for checking the final manuscript of our book.

Memphis, USA, May 13, 2013 George A. Anastassiou
Bucharest, Romania Iuliana F. Iatan

# Contents

# Symbols

# Chapter 1
# Vector Spaces

## 1.1 Geometric Vector Spaces

### *1.1.1 Free Vectors*

Besides the notions used in Mathematics, created by abstraction from environmental observation (for example the geometric concepts) or quantitative and qualitative research of the natural phenomena (for example the notion of number) in mathematics there are elements from other sciences. The notion of vector from physics has been studied and developed creating vector calculus, which became a useful tool for both mathematics and physics. All physical quantities are represented by vectors (for example the force, the velocity).

In examining the phenomena of nature, we can meet two kinds of quantities:

1. scalar quantities (the temperature, the length, the time, the volume, the density, the area) which can be characterized by a number (which is measured by a specific unit);
2. vector quantities (the force, the velocity, the acceleration) which to measure their characterization is not sufficient, it is necessary to know the direction and the sense in which they operate.

To represent a vector-oriented segment is used a method from mechanics.

One denotes by $E_3$ the three-dimensional space of the Euclidean geometry.

**Definition 1.1** (see [1], p. 108). We call **oriented segment** (or **bound vector**) an ordered pair of points $(A, B) \in E_3 \times E_3$ and we denote it by $\overrightarrow{AB}$ (see Fig. 1.1).

G. A. Anastassiou and I. F. Iatan, *Intelligent Routines II*,
Intelligent Systems Reference Library 58, DOI: 10.1007/978-3-319-01967-3_1,

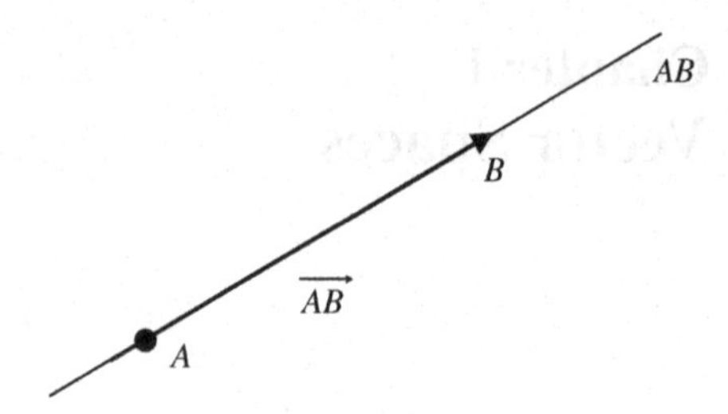


**Fig. 1.1** The representation of an oriented segment

THE CHARACTERISTICS OF AN ORIENTED SEGMENT

We consider the oriented segment $\overrightarrow{AB}$, for any two points $A, B \in E_3$.

- The points $A$ and $B$ are called the **origin** (the **starting point**) and respectively the **endpoint** (the **peak**) of the oriented segment. If $A = B$ then $\overrightarrow{AA}$ is the **null oriented segment**.
- If $A \neq B$ then the straight line determined by them is called the **support line** of $\overrightarrow{AB}$ and is denoted by $AB$.
- The **direction** of the oriented segment $\overrightarrow{AB}$ $(A \neq B)$ is the direction of the straight line $AB$.
- The **sense** on the support line, from $A$ to $B$ is called the sense of the oriented segment $\overrightarrow{AB}$.
- The distance between the points $A$ and $B$ means the **length** (the **norm** or **magnitude**) of the oriented segment $\overrightarrow{AB}$ and is denoted by $\left\| \overrightarrow{AB} \right\|$. If the origin of an oriented segment coincides with the endpoint (the null oriented segment) then the length of this segment is equal to 0.

**Definition 1.2** (see [1], p. 108). Two oriented segments $\overrightarrow{AB}$ and $\overrightarrow{CD}$, $A \neq B,\ C \neq D$ have the **same directions** if their support lines and are parallel or coincide.

**Definition 1.3** (see [2], p. 4 and [3], p. 4). Two oriented segments $\overrightarrow{AB}$ and $\overrightarrow{CD}$, $A \neq B,\ C \neq D$ with the same direction have the **same sense** if $B$ and $D$ belong to the same half-plane determined by a straight line $AC$ (see Fig 1.2).

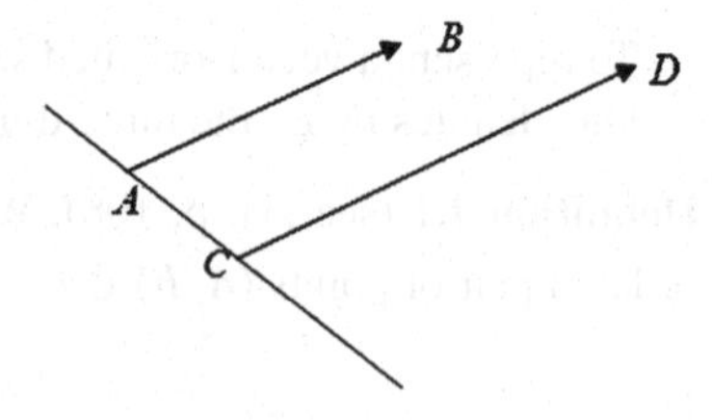


**Fig. 1.2** Example of two oriented segments, that have the same sense

**Definition 1.4** (see [1], p. 108). Two oriented segments $\overrightarrow{AB}$ and $\overrightarrow{CD}$, $A \neq B\ C \neq D$ one call **equipollent** if they have the same direction, the same sense and the same norm; if $\overrightarrow{AB}$ is equipollent with $\overrightarrow{CD}$ we shall write $\overrightarrow{AB} \sim \overrightarrow{CD}$.

**Theorem 1.5** (see [1], p. 108). The equipollent relation defined on the set of the oriented segments is an equivalence relation.

**Proof**

The equipollent relation is an equivalence relation since it is:

1. reflexive: $\overrightarrow{AB} \sim \overrightarrow{AB}$;
2. symmetric: $\overrightarrow{AB} \sim \overrightarrow{CD}$ involves $\overrightarrow{CD} \sim \overrightarrow{AB}$;
3. transitive: $\overrightarrow{AB} \sim \overrightarrow{CD}$ and $\overrightarrow{CD} \sim \overrightarrow{EF}$ involves $\overrightarrow{AB} \sim \overrightarrow{EF}$.

We can classify the vectors in the following way:

(a) *free vectors*, which have the arbitrary origin at any point in space, but whose direction, sense and length on space are prescribed;
(b) *bound vectors*, whose origin is prescribed;
(c) *sliding vectors* moving along the same support line, and their origin can be anywhere on the line.

**Definition 1.6** (see [4], p. 86). We call a **free vector** (**geometric vector**) characterized by an oriented segment $\overrightarrow{AB}$, the set of the oriented segments, equipollent with $\overrightarrow{AB}$:

$$\overline{AB} = \left\{ \overrightarrow{CD} | \overrightarrow{CD} \sim \overrightarrow{AB} \right\}.$$

Any oriented segment of this set is called the **representative** of the free vector $\overline{AB}$; therefore $\overrightarrow{CD} \in \overline{AB}$.

A free vector of the length:

- 1 is called **versor** (**unit vector**); generally one denotes by $\overline{e}$;
- 0 is called **null vector**; one denotes by $\overline{0}$.

**Definition 1.7** (see [4], p. 86). The length, direction and sense of a free nonzero vector means the **length**, **direction** and **sense** corresponding to the oriented segment that it represents.

The set of the geometric vectors from the space $E_3$ will be denote by $V_3$:

$$V_3 = \left\{ \overline{AB} | A, B \in E_3 \right\},$$

namely $V_3$ means the set of equivalence classes of the oriented segment $\overrightarrow{AB}$.

We can use the notations: $\|\overline{a}\|$ $\left\|\overline{AB}\right\|$ or $d\left(A, B\right)$ in order to designate the length of a free vector $\overline{a}$ or $\overline{AB}$.

**Definition 1.8** (see [1], p. 109). We say that two free vectors are **equal** and we write $\overline{a} = \overline{b}$ if their representatives are equipollent.

**Definition 1.9** (see [1], p. 109). Two free non-null vector $\overline{a}$ and $\overline{b}$ are **collinear** if they have the same direction (see Fig. 1.3).

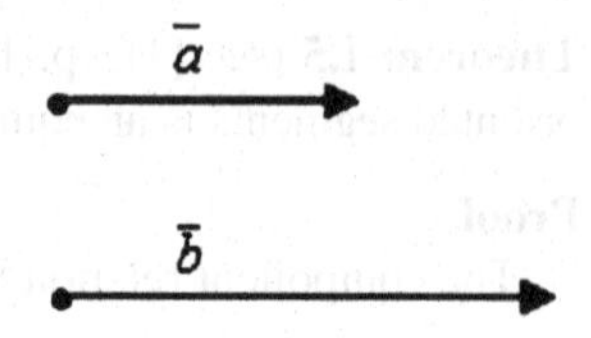


**Fig. 1.3** Example of collinear vectors

**Definition 1.10** (see [1], p. 109). Two collinear vectors which have the same length but they have opposite directions are called **opposite vectors**. The opposite of a free vector $\overline{a}$ is $-\overline{a}$ (see Fig. 1.4).

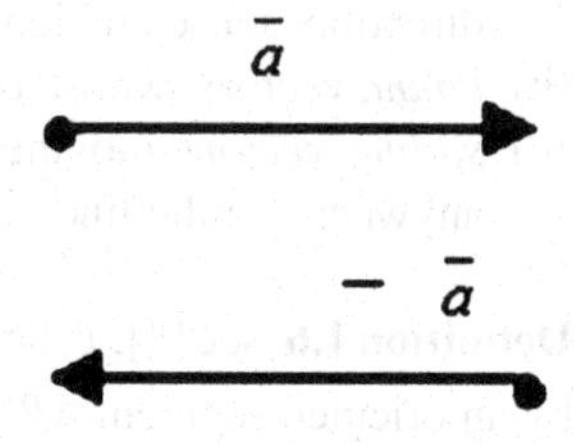


**Fig. 1.4** Example of an opposite vectors

**Definition 1.11** (see [1], p. 109). Three free vectors $\overline{a}$ , $\overline{b}$ , $\overline{c}$ are called **coplanar** if their support lines lie in the same plane (see Fig. 1.5).

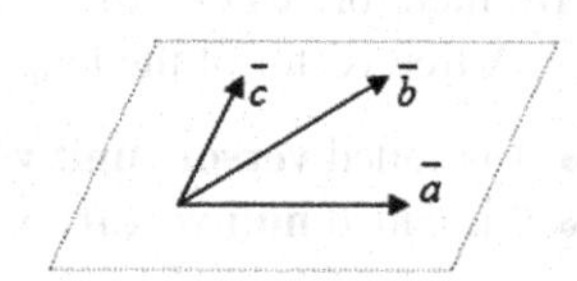


**Fig. 1.5** Example of some coplanar vectors

## 1.2 Operations with Free Vectors

We can define the following operations in the set $V_3$:

1. the addition of the free vectors;
2. the scalar multiplication;
3. the vector decomposition.

### *1.2.1 Addition of the Free Vectors*

An internal operation is defined on $V_3$ (the addition of the free vectors)

$$+ : V_3 \times V_3 \rightarrow V_3, \left(\overline{a}, \overline{b}\right) \rightarrow \overline{a} + \overline{b}.$$

Therefore, the sum of two or more vectors is also a vector, which can be obtained through the following methods:

(A) if vectors are parallel or collinear and

(a) they have the same sense, then the sum vector has the direction and the sense of the component vectors, and their length is equal to the sum of the lengths corresponding to the component vectors;
(b) they have opposite sense, then the sum vector has the common direction, the sense of the larger vector, and its magnitude is given by the differences of the two vector magnitudes.

(B) if the vectors have only a common origin, then their sum is determined using the parallelogram rule.

**Definition 1.12** (the **parallelogram rule**, see [1], p. 110). Let $\overline{a}, \overline{b} \in V_3$ be two free vectors, which have a common origin and $A \in E_3$ be an arbitrary fixed point. If $\overrightarrow{OA} \in \overline{a}$ ($\overrightarrow{OA}$ means the representative of the free vector $\overline{a}$ ) and $\overrightarrow{OC} \in \overline{b}$ then the free vector $\overline{c}$ represented by the oriented segment $\overrightarrow{OB}$ is called the **sum of the free vectors** $\overline{a}$ and $\overline{b}$; one writes $\overline{c} = \overline{a} + \overline{b}$ or $\overrightarrow{OB} = \overrightarrow{OA} + \overrightarrow{OC}$ (Fig. 1.6).

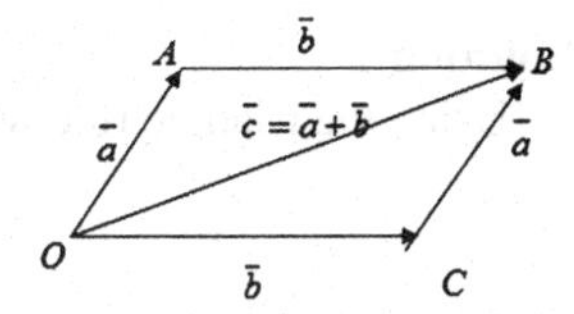


**Fig. 1.6** Illustration of the parallelogram rule

(C) if the vectors are arranged so as one extremity to be the origin of the other, to achieve their sum one applies the triangle rule.

**Definition 1.13** (the **triangle rule**, see [1], p. 110). Let $\overline{a}, \overline{b} \in V_3$ be two free vectors and $A \in E_3$ an arbitrary fixed point. If $\overrightarrow{AB} \in \overline{a}$ ($\overrightarrow{AB}$ is the representative of the free vector $\overline{a}$) and $\overrightarrow{BC} \in \overline{b}$ then the free vector $\overline{c}$ represented by the oriented segment $\overrightarrow{AC}$ is called the **sum of the free vectors** $\overline{a}$ and $\overline{b}$; one writes $\overline{c} = \overline{a} + \overline{b}$ or $\overrightarrow{AC} = \overrightarrow{AB} + \overrightarrow{BC}$ (Fig. 1.7).

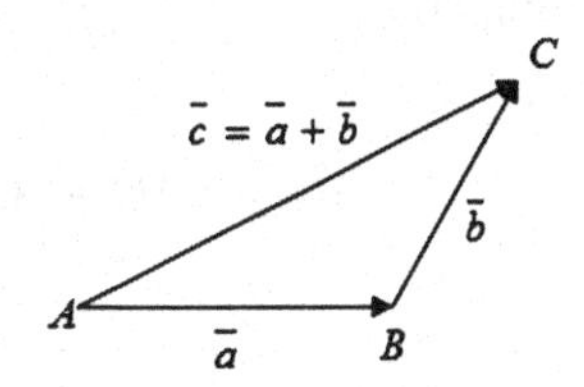


**Fig. 1.7** Illustration of the triangle rule

**Remark 1.14** (see [11], p. 118 and [10], p. 8). For many vectors, sitting in the same way, one applies the polygon rule, which is a generalization of the triangle rule; the sum vector is that which closes the polygon, joining the origin of the first component, with the last extremity.

The addition of the vectors is based on some experimental facts (composition of forces, velocities).

**Example 1.15** (see [5], p. 58). We suppose a segment $\overline{AB}$ and the points $M_1$ and $M_2$ that divide the segment into three equal parts. If $M$ is an arbitrary point outside the segment, express the vectors $\overline{MM_1}$ and $\overline{MM_2}$ depending on the vectors $\overline{MA} = \overline{a}$ and $\overline{MB} = \overline{b}$.

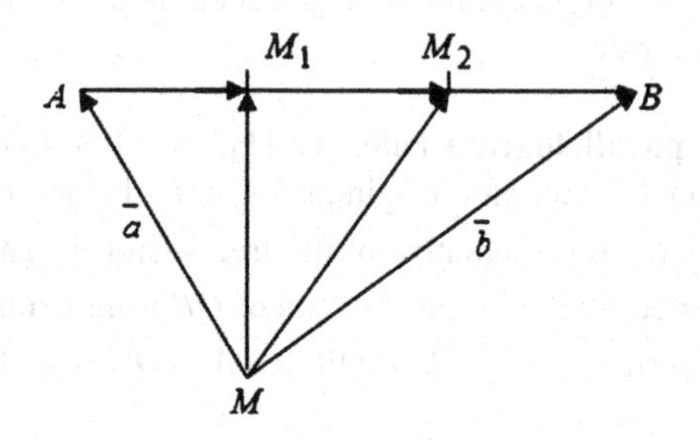


**Solution**

Using the triangle rule we have:

$$\overline{MM_1} = \overline{MA} + \overline{AM_1}.$$

Hence

$$\overline{AB} = \overline{AM} + \overline{MB} = -\overline{a} + \overline{b}.$$

We deduce

$$\overline{AM_1} = \frac{1}{3}\overline{AB} = \frac{\overline{b} - \overline{a}}{3}$$

and

$$\overline{MM_1} = \overline{a} + \frac{\overline{b} - \overline{a}}{3} = \frac{2\overline{a} + \overline{b}}{3}.$$

Similarly,

$$\overline{MM_2} = \overline{MA} + \overline{AM_2} = \overline{MA} + \frac{2}{3}\overline{AB} = \overline{a} + 2\frac{\overline{b} - \overline{a}}{3} = \frac{\overline{a} + 2\overline{b}}{3}.$$

A solution in Sage will be:

```
sage: a,b=var('a,b')
sage: AB=b-a;AM1=AB/3;AM2=2*AB/3
sage: MM1=a+AM1;MM2=a+AM2;MM1;MM2
2/3*a + 1/3*b
1/3*a + 2/3*b
sage: v1=arrow3d((-1,-1,-1),(-2,1,5),2)+text3d("M",(-1.2,-1.1,-1.1))
sage: v2=arrow3d((-1,-1,-1),(3,1,6),2)+text3d("A",(-2.4,1.1,5))
sage: v3=arrow3d((-2,1,5),(3,1,6),2)+text3d("B",(3,1.1,6))
sage: t=text3d("b",(2.3,-0.6,3.5))+text3d("a",(-2.4,0.7,3.1))
sage: p1=point3d((-1/3,1,16/3),size=20)+text3d("M1",(-0.4,1.1,5.6))
sage: p2=point3d((4/3,1,17/3),size=20)+text3d("M2",(1.4,1.1,6))
sage: (v2+v3+v1+t+p1+p2).show(aspect_ratio=1)
```

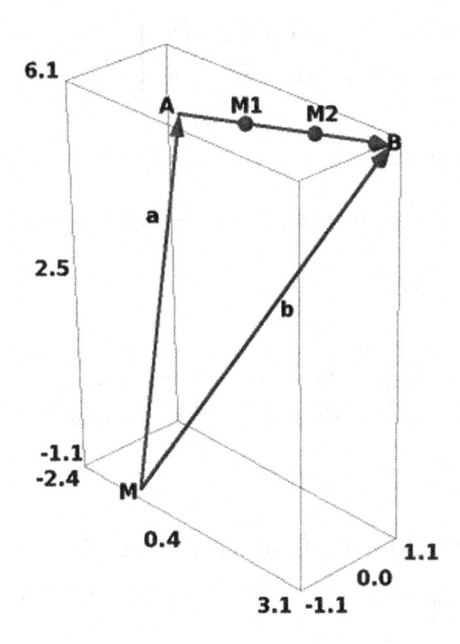


**Theorem 1.16** (see [6], p. 7). The addition of the free vectors determines an abelian group structure $(V_3, +)$ on the set of the free vectors.

**Proof**

We note that the addition of the free vectors is an internal well-defined algebrical operation, i.e. the free vector $\overline{c} = \overline{a} + \overline{b}$ doesn't depend on the choice point $A$ since from $\overline{AB} = \overline{A'B'}$ and $\overline{BC} = \overline{B'C'}$ it results $\overline{AC} = \overline{A'C'}$.

We check the properties of (Fig. 1.8):

1. *associativity*:

$$\overline{a} + \left(\overline{b} + \overline{c}\right) = \left(\overline{a} + \overline{b}\right) + \overline{c}, (\forall)\, \overline{a}, \overline{b}, \overline{c} \in V_3.$$

Let $O$ be a fixed point in space and $\overline{OA} = \overline{a}, \overline{AB} = \overline{b},\ \overline{BC} = \overline{c}$.

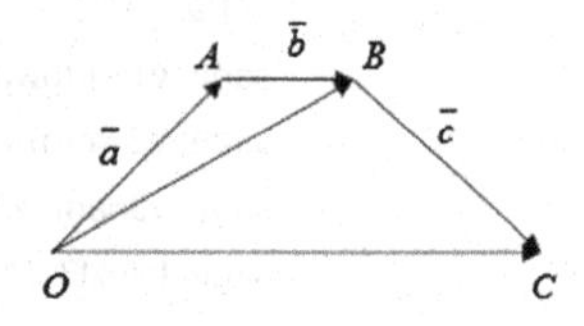

**Fig. 1.8** Illustration of the associative property

$$\left.\begin{array}{l} \left(\overline{a} + \overline{b}\right) + \overline{c} = \left(\overline{OA} + \overline{AB}\right) + \overline{BC} = \overline{OB} + \overline{BC} = \overline{OC} \\ \overline{a} + \left(\overline{b} + \overline{c}\right) = \overline{OA} + \left(\overline{AB} + \overline{BC}\right) = \overline{OA} + \overline{AC} = \overline{OC} \end{array}\right\} \Longrightarrow \overline{a} + \left(\overline{b} + \overline{c}\right) = \left(\overline{a} + \overline{b}\right) + \overline{c}.$$

2. $\overline{0}$ is the *neutral element*:

$$(\exists)\, \overline{0} \in V_3 \text{ such that } \overline{a} + \overline{0} = \overline{0} + \overline{a} = \overline{a}, (\forall)\, \overline{a} \in V_3.$$

Let $\overline{OA} = \overline{a},\ \overline{OO} = \overline{0},\ \overline{AA} = \overline{0}$.
We have:

$$\begin{aligned} \overline{OO} + \overline{OA} = \overline{OA} &\Longleftrightarrow \overline{0} + \overline{a} = \overline{a}, \\ \overline{OA} + \overline{AA} = \overline{OA} &\Longleftrightarrow \overline{a} + \overline{0} = \overline{a}. \end{aligned}$$

3. *simetrizable element*:

$$(\forall)\, \overline{a} \in V_3, (\exists) - \overline{a} \in V_3 \text{ such that } \overline{a} + (-\overline{a}) = (-\overline{a}) + \overline{a} = \overline{0}.$$

Let $\overline{a} = \overline{AB},\ -\overline{a} = \overline{BA}$.
We obtain

$$\begin{aligned} \overline{a} + (-\overline{a}) &= \overline{AB} + \overline{BA} = \overline{AA} = \overline{0}, \\ (-\overline{a}) + \overline{a} &= \overline{BA} + \overline{AB} = \overline{BB} = \overline{0}. \end{aligned}$$

□

**Remark 1.17** (see [4], p. 88). The existence of an opposite for a free vector allows the substraction definition of the free vectors $\overline{a}, \overline{b} \in V_3$ (Fig. 1.9):

$$\overline{a} - \overline{b} = \overline{a} + \left(-\overline{b}\right).$$

4. *commutativity*:

$$\overline{a}+\overline{b}=\overline{b}+\overline{a},\ (\forall)\,\overline{a},\overline{b}\in V_3.$$

Let $\overline{a}=\overline{OA}, \overline{b}=\overline{AB}$.

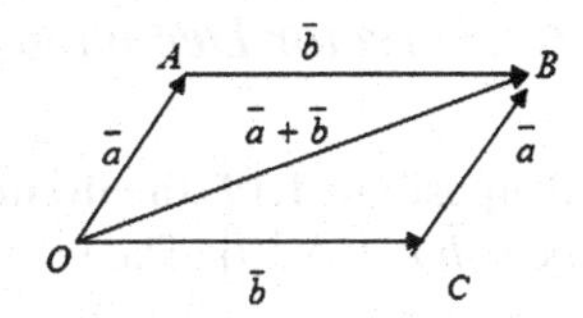


**Fig. 1.9** Illustration of the commutative property

We note that $OABC$ is a parallelogram.

Then we have:

$$\overline{OA}+\overline{AB}=\overline{OB},$$
$$\overline{OC}+\overline{CB}=\overline{OB}.$$

## *1.2.2 Scalar Multiplication*

We now define an external operation (scalar multiplication)

$$\bullet : \mathbb{R}\times V_3\to V_3,\ (t,\overline{a})\to t\overline{a},$$

- $\mathbb{R}$ being the set of the real numbers,
- $t\overline{a}=\overline{0}$ if $t=0$ or $\overline{a}=\overline{0}$;
- the free vector $t\overline{a}$ has:
  - the same direction with $\overline{a}$,
  - the same sense with $\overline{a}$ if $t>0$ and the opposite sense of $\overline{a}$ if $t<0$;
  - the magnitude $\|t\overline{a}\|=|t|\cdot\|\overline{a}\|$.

**Theorem 1.18** (see [6], p. 7). The multiplication of the free vectors with scalar has the following properties:

1. distributivity of scalar multiplication with respect to the vector addition:

$$t\left(\overline{a}+\overline{b}\right)=t\overline{a}+t\overline{b},\ (\forall)\,t\in\mathbb{R},\ (\forall)\,\overline{a},\overline{b}\in V_3;$$

2. distributivity of scalar multiplication with respect to the scalar addition:

$$(s+t)\,\overline{a}=s\overline{a}+t\overline{a},\ (\forall)\,s,t\in\mathbb{R},\ (\forall)\,\overline{a}\in V_3;$$

3. $s\,(t\overline{a}) = (st)\,\overline{a}$, $(\forall)\ s, t \in \mathbb{R}$, $(\forall)\ \overline{a} \in V_3$;

4. $1 \cdot \overline{a} = \overline{a}$, $(\forall)\,\overline{a} \in V_3$.

### *1.2.3 Vector Decomposition*

**Proposition 1.19 (the decomposition of a vector in a direction,** see [7], p. 8). Let be $\overline{a}, \overline{b} \in V_3 \backslash \{\overline{0}\}$. The vectors $\overline{a}$ and $\overline{b}$ are collinear if and only if $(\exists)\, t \in \mathbb{R}$ unique such that $\overline{b} = t\overline{a}$.

**Theorem 1.20** (see [7], p. 8). Let be $\overline{a}, \overline{b} \in V_3 \backslash \{\overline{0}\}$. The vectors $\overline{a}$ and $\overline{b}$ are collinear if and only if $(\exists)\ \alpha, \beta \in \mathbb{R}$ nonsimultaneous equals to zero (i.e. $\alpha^2 + \beta^2 \neq 0$) such that $\alpha\overline{a} + \beta\overline{b} = \overline{0}$.

The decomposition of a vector after two directions is the reverse operation to the addition of the two vectors.

**Proposition 1.21 (the decomposition of a vector after two noncollinear directions,** see [7], p. 9). Let be $\overline{a}, \overline{b}, \overline{c} \in V_3 \backslash \{\overline{0}\}$. If $\overline{a}, \overline{b}, \overline{c}$ are coplanar then $(\exists)\ \alpha, \beta \in \mathbb{R}$ uniquely determined such that $\overline{c} = \alpha\overline{a} + \beta\overline{b}$.

**Theorem 1.22** (see [7], p. 10). Let be $\overline{a}, \overline{b}, \overline{c} \in V_3 \backslash \{\overline{0}\}$. The vectors $\overline{a}, \overline{b}, \overline{c}$ are coplanar if and only if $(\exists)\ \alpha, \beta, \gamma \in \mathbb{R}$ nonsimultaneous equal to zero (namely $\alpha^2 + \beta^2 + \gamma^2 \neq 0$) such that $\alpha\overline{a} + \beta\overline{b} + \gamma\overline{c} = \overline{0}$.

**Proposition 1.23 (the decomposition of a vector after three noncoplanar directions,** see [7], p. 10). Let be $\overline{a}, \overline{b}, \overline{c}, \overline{d} \in V_3 \backslash \{\overline{0}\}$. If $\overline{a}, \overline{b}, \overline{c}$ are noncoplanar then $(\exists)\ \alpha, \beta, \gamma \in \mathbb{R}$ uniquely determined such that $\overline{d} = \alpha\overline{a} + \beta\overline{b} + \gamma\overline{c}$.

We suppose a point O in $E_3$ called **origin** and three non-coplanar versors $\overline{i}\,\overline{j}\,\overline{k}$ whose we attach the coordinate axes O$x$, O$y$, O$z$ that have the same sense as the sense of these versors (see Fig. 1.10). The ensemble $\left\{O, \overline{i}, \overline{j}, \overline{k}\right\}$ is called the **Cartesian reference** in $E_3$.

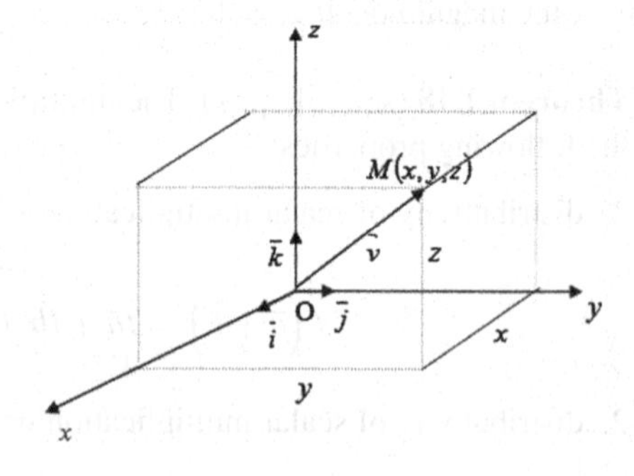


**Fig. 1.10** Representation of a cartesian reference in $E_3$

Whereas the versors $\overline{i}, \overline{j}, \overline{k}$ are non-coplanar, then under the Proposition 1.23, for any vector $\overline{v} \in V_3$ $(\exists)$ $r, s, t \in \mathbb{R}$ uniquely determined such that $\overline{v}$ is expressed as $\overline{v} = r\overline{i} + s\overline{j} + t\overline{k}$, called **analytical expression of the vector** $\overline{v}$. The numbers $(r, s, t)$ are called the **Euclidean coordinates** (the **components**) of $\overline{v}$ relative to the reference $\left\{O, \overline{i}, \overline{j}, \overline{k}\right\}$.

**Definition 1.24** (see [8], p. 449 and [10], p. 10) Let $M \in E_3$ be a fixed point. The vector $\overline{OM}$ is called the **position vector** of the point $M$. The coordinates of the position vector $\overline{OM}$ relative to the reference $\left\{O, \overline{i}, \overline{j}, \overline{k}\right\}$ are called the **coordinates of the point** $M$. If $\overline{OM} = x\overline{i} + y\overline{j} + z\overline{k}$ then one writes $M(x, y, z)$.

**Example 1.25** (see [9], p. 50 ). Find $\lambda \in \mathbb{R}$ such that the vectors

$$\begin{cases} \overline{v}_1 = 2\overline{i} + (\lambda + 2)\overline{j} + 3\overline{k} \\ \overline{v}_2 = \overline{i} + \lambda\overline{j} - \overline{k} \\ \overline{v}_3 = 4\overline{j} + 2\overline{k} \end{cases}$$

to be coplanar.

With this values of $\lambda$ decompose the vector $\overline{v}_1$ after the directions of the vectors $\overline{v}_2$ and $\overline{v}_3$.

**Solution**

Using the Theorem 1.22, $\overline{v}_1$ $\overline{v}_2$, $\overline{v}_3$ are coplanar if and only if $(\exists)$ $\alpha, \beta, \gamma \in \mathbb{R}$, $\alpha^2 + \beta^2 + \gamma^2 \neq 0$ such that $\alpha\overline{v}_1 + \beta\overline{v}_2 + \gamma\overline{v}_3 = \overline{0}$.

We obtain:

$$(2\alpha + \beta)\,\overline{i} + [(\lambda + 2)\,\alpha + \lambda\beta + 4\gamma]\,\overline{j} + (3\alpha - \beta + 2\gamma)\,\overline{k} = \overline{0},$$

i.e.

$$\begin{cases} 2\alpha + \beta = 0 \\ (\lambda + 2)\,\alpha + \lambda\beta + 4\gamma = 0 \\ 3\alpha - \beta + 2\gamma = 0; \end{cases}$$

the previous homogeneous system admits some non-trivial solutions $\Longleftrightarrow$

$$\begin{vmatrix} 2 & 1 & 0 \\ \lambda + 2 & \lambda & 4 \\ 3 & -1 & 2 \end{vmatrix} = 0 \Longleftrightarrow \lambda = -8.$$

If $\overline{v}_1$, $\overline{v}_2$, $\overline{v}_3$ are coplanar, from the Proposition 1.21 we have: $(\exists)\,\alpha', \beta' \in \mathbb{R}$ uniquely determined such that

$$\overline{v}_1 = \alpha'\overline{v}_2 + \beta'\overline{v}_3;$$

it results

$$2\bar{i}-6\bar{j}+3\bar{k}=\alpha'\bar{i}+\left(-8\alpha'+4\beta'\right)\bar{j}+\left(-\alpha'+2\beta'\right)\bar{k};$$

therefore

$$\begin{cases} \alpha'=2 \\ -8\alpha'+4\beta'=-6 \\ -\alpha'+2\beta'=3\Rightarrow 2+2\beta'-3\Rightarrow \beta'=\frac{5}{2}. \end{cases}$$

We achieve:

$$\bar{v}_1=2\bar{v}_2+\frac{5}{2}\bar{v}_3.$$

Using Sage we shall have:

```
sage: var('l','al','be','ga')
(l, al, be, ga)
sage: v1=vector([2,l+2,3])
sage: v2=vector([1,1,-1])
sage: v3=vector([0,4,2])
sage: u=al*v1+be*v2+ga*v3
sage: show(u)
```

$$(2\mathrm{al}+\mathrm{be},\ (l+2)\mathrm{al}+\mathrm{be}l+4\,\mathrm{ga},\ 3\,\mathrm{al}-\mathrm{be}+2\,\mathrm{ga})$$

```
sage: A=matrix([[2,1,0],[l+2,1,4],[3,-1,2]])
sage: la=solve(A.determinant()==0,l)
sage: show(la)
```

$$[l=(-8)]$$

```
sage: v1=v1.substitute(l=-8)
sage: v2=v2.substitute(l=-8)
sage: w=al*v2+be*v3
sage: solve([v1[0]==w[0],v1[1]==w[1],v1[2]==w[2]],al,be)
[[al == 2, be == (5/2)]]
```

We shall represent in Sage the three vectors $\bar{v}_1,\bar{v}_2,\bar{v}_3$:

```
sage: P=arrow3d((0,0,0),(1, -8, -1), 6 ,color='red')+text3d("v2",(1.3,-6.2,-1.3))
sage: Q=arrow3d((0,0,0),(0, 4, 2), 6, color='green')+text3d("v3",(0.5,1.8,2.7))
sage: R=arrow3d((0,0,0),(2, -6, 3), 4, color='blue')+text3d("v1",(2.5,-5.5,3.5))
sage: (P+Q+R).show(aspect_ratio=1)
```

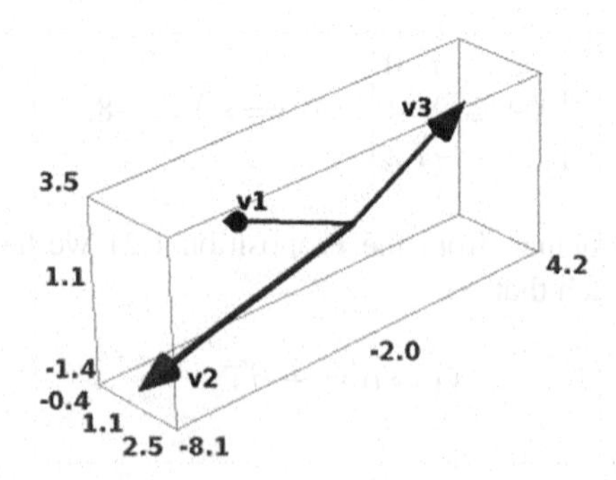


**Definition 1.26** (see [10], p. 12 ). If $A\,(x_1,y_1,z_1)$ , $B\,(x_2,y_2,z_2)$ are two given points from $E_3$ (see Fig. 1.11)

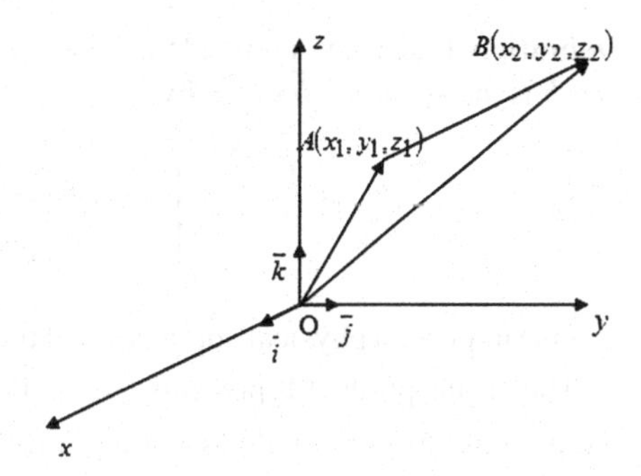


**Fig. 1.11** The distance between two points

then we have

$$\overline{AB} = \overline{OB} - \overline{OA} = (x_2 - x_1)\,\overline{i} + (y_2 - y_1)\overline{j} + (z_2 - z_1)\,\overline{k}, \tag{1.1}$$

and the **distance** from the points $A$ and $B$ denoted $d\,(A, B)$ is calculated using the formula:

$$d\,(A, B) = \left\|\overline{AB}\right\| = \sqrt{(x_2 - x_1)^2 + (y_2 - y_1)^2 + (z_2 - z_1)^2}. \tag{1.2}$$

**Definition 1.27** (see [1], p. 115). Let be $\overline{a}, \overline{b} \in V_3\ \{\overline{0}\}$, $O \in E_3$ and $\overrightarrow{OA} \in \overline{a}$, $\overrightarrow{OB} \in \overline{b}$. The angle $\varphi \in [0, \pi]$ determined by the oriented segments $\overrightarrow{OA}$ and $\overrightarrow{OB}$ is called the **angle between the free vectors** $\overline{a}$ and $\overline{b}$ (see Fig. 1.12).

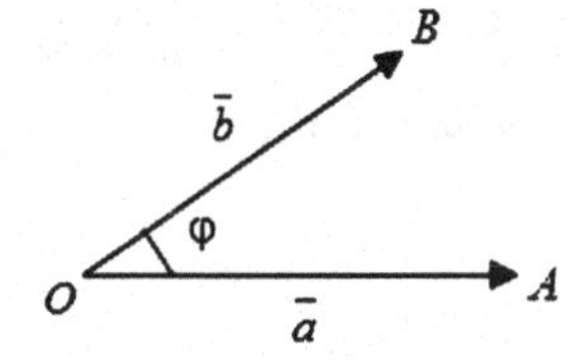


**Fig. 1.12** Representation of the angle between the respective free vectors

The free vectors $\overline{a}$ and $\overline{b}$ are called **orthogonal** if the angle between them is $\frac{\pi}{2}$.

## *1.2.4 Defining of the Products in the Set of the Free Vectors*

### 1.2.4.1 Scalar Product in $V_3$

We consider $\overline{a}, \overline{b} \in V_3 \backslash \{\overline{0}\}$. For $\overline{a} \neq \overline{0}, \overline{b} \neq \overline{0}$ one denotes by $\varphi \in [0, \pi]$ the angle between $\overline{a}$ and $\overline{b}$.

**Definition 1.28** (see [1], p. 116). **The scalar (dot) product of the free vectors** $\overline{a}$ and $\overline{b}$ is the scalar $\overline{a} \cdot \overline{b}$ given by

$$\overline{a} \cdot \overline{b} = \begin{cases} \|\overline{a}\| \left\|\overline{b}\right\| \cos\varphi, \overline{a} \neq \overline{0}, \overline{b} \neq \overline{0} \\ 0, \overline{a} = \overline{0}, \overline{b} = \overline{0}. \end{cases} \quad (1.3)$$

One important physical application of the scalar product is the calculation of work.

The scalar product represents (see [11], p. 126) the work done by the force $\overrightarrow{F}$ required to move a mobile to a straight line, with the director vector $\overrightarrow{d}$, that makes the angle $\varphi$ with the direction of the force $\overrightarrow{F}$.

**Proposition 1.29** (see [7], p. 15). The scalar product of the free vectors has the following properties:

1. commutativiy:
$$\overline{a} \cdot \overline{b} = \overline{b} \cdot \overline{a}, (\forall)\, \overline{a}, \overline{b} \in V_3;$$
2. $t\left(\overline{a} \cdot \overline{b}\right) = t\overline{a} \cdot \overline{b} = \overline{a} \cdot t\overline{b}, (\forall)\, \overline{a}, \overline{b} \in V_3, (\forall)\ t \in \mathbb{R};$
3. distributivity of scalar multiplication with respect to the vector addition:
$$\overline{a} \cdot \left(\overline{b} + \overline{c}\right) = \overline{a} \cdot \overline{b} + \overline{a} \cdot \overline{c}, (\forall)\, \overline{a}, \overline{b}, \overline{c} \in V_3;$$
$$\left(\overline{a} + \overline{b}\right) \cdot \overline{c} = \overline{a} \cdot \overline{b} + \overline{b} \cdot \overline{c}, (\forall)\, \overline{a}, \overline{b}, \overline{c} \in V_3;$$
4. $\begin{cases} \overline{a} \cdot \overline{a} > 0, (\forall)\, \overline{a} \in V_3 \setminus \left\{\overline{0}\right\}; \\ \overline{a} \cdot \overline{a} = 0 \Leftrightarrow \overline{a} = \overline{0}; \end{cases}$
5. $\overline{a} \cdot \overline{b} = 0 \Leftrightarrow \overline{a}$ and $\overline{b}$ are orthogonal, $(\forall)\ \overline{a}, \overline{b} \in V_3 \setminus \left\{\overline{0}\right\}$;
6. if $\overline{a}, \overline{b} \in V_3$,
$$\begin{cases} \overline{a} = a_1\overline{i} + a_2\overline{j} + a_3\overline{k} \\ \overline{b} = b_1\overline{i} + b_2\overline{j} + b_3\overline{k} \end{cases}$$

then we obtain the analytical expression of the scalar product:

$$\overline{a} \cdot \overline{b} = a_1 b_1 + a_2 b_2 + a_3 b_3. \quad (1.4)$$

Particularly,

$$\overline{a} \cdot \overline{a} = a_1^2 + a_2^2 + a_3^2 = \|\overline{a}\|^2. \quad (1.5)$$

7. the angle between the vectors $\overline{a}, \overline{b} \in V_3 \setminus \left\{\overline{0}\right\}$ is given by the formula:

$$\cos\varphi = \frac{\overline{a} \cdot \overline{b}}{\|\overline{a}\| \left\|\overline{b}\right\|} = \frac{a_1 b_1 + a_2 b_2 + a_3 b_3}{\sqrt{a_1^2 + a_2^2 + a_3^2} \cdot \sqrt{b_1^2 + b_2^2 + b_3^2}}, \varphi \in [0, \pi]; \quad (1.6)$$

one notices that the vectors and are orthogonal if and only if

$$a_1b_1 + a_2b_2 + a_3b_3 = 0.$$

**Example 1.30** (see [12], p. 161). Prove that the heights of a triangle are concurrent.
**Solution**

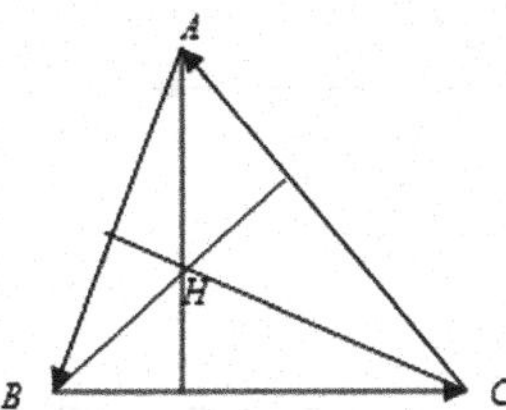


Let $H$ be the intersection of the heights from $A$ and $B$. We must show that $\overline{CH}$ is perpendicular on $\overline{AB}$, namely $\overline{CH} \cdot \overline{AB} = 0$.

Using the triangle rule we have:

$$\overline{AB} = \overline{AC} - \overline{BC}.$$

We shall obtain

$$\begin{aligned}\overline{CH} \cdot \overline{AB} = \overline{CH} \cdot \left(\overline{AC} - \overline{BC}\right) &= \overline{CH} \cdot \overline{AC} - \overline{CH} \cdot \overline{BC} \\ &= -\left(\overline{HB} + \overline{BC}\right) \cdot \overline{AC} + \left(\overline{HA} + \overline{AC}\right) \cdot \overline{BC},\end{aligned}$$

i.e.

$$\overline{CH} \cdot \overline{AB} = -\overline{HB} \cdot \overline{AC} - \overline{BC} \cdot \overline{AC} + \overline{HA} \cdot \overline{BC} + \overline{AC} \cdot \overline{BC} = -\overline{HB} \cdot \overline{AC} + \overline{HA} \cdot \overline{BC} = 0.$$

We have used the commutative property of the scalar product and the fact that $\overline{BH}$ is perpendicular on $\overline{AC}$ and $\overline{AH}$ is perpendicular on $\overline{BC}$.

We need the following Sage code to solve this problem:

```
sage: var("a1,a2,a3,b1,b2,b3,c1,c2,c3,h1,h2,h3")
(a1, a2, a3, b1, b2, b3, c1, c2, c3, h1, h2, h3)
sage: A=vector(SR,[a1,a2,a3]);B=vector(SR,[b1,b2,b3])
sage: C=vector(SR,[c1,c2,c3]);H=vector(SR,[h1,h2,h3])
sage: AC=C-A;BC=C-B;CH=H-C;AB=AC-BC;HB=B-H;HA=A-H
sage: E=CH.dot_product(AB).expand()
sage: u=-AC.dot_product(HB);v=HA.dot_product(BC)
sage: s=solve([u==0,v==0],h1,h2,h3)
sage: s1=s[0][0].right();s2=s[0][1].right();s3=s[0][2].right()
sage: E.subs(h1=s1,h2=s2,h3=s3).simplify_exp()
0
```

**Example 1.31** (see [12], p. 161). Prove that th diagonals in a rhombus are perpendicular.

**Solution**

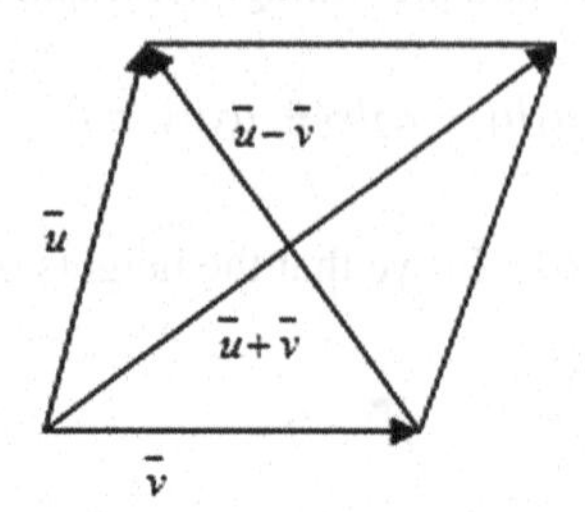


We shall deduce

$$(\overline{u}-\overline{v})\cdot(\overline{u}+\overline{v})=\overline{u}\cdot(\overline{u}+\overline{v})-\overline{v}\cdot(\overline{u}+\overline{v})=\overline{u}\cdot\overline{u}+\overline{u}\cdot\overline{v}-\overline{v}\cdot\overline{u}-\overline{v}\cdot\overline{v}=\|\overline{u}\|^2-\|\overline{v}\|^2.$$

As in the case of rhombus we have $\|\overline{u}\|=\|\overline{v}\|$ we shall obtain

$$(\overline{u}-\overline{v})\cdot(\overline{u}+\overline{v})=0,$$

i.e. the diagonals in a rhombus are perpendicular.

Solving in Sage we shall have:

```
sage: var("u1,u2,u3,v1,v2,v3")
(u1, u2, u3, v1, v2, v3)
sage: u=vector(SR,[u1,u2,u3]);v=vector(SR,[v1,v2,v3])
sage: ((u.norm()^2).simplify_exp()-(v.norm()^2).simplify_exp()==0).assume()
sage: E=(u-v).dot_product(u+v)
sage: j=assumptions();j[0].left()-E.expand()
0
```

#### 1.2.4.2 Cross Product in $V_3$

We consider $\overline{a},\overline{b}\in V_3\setminus\{\overline{0}\}$. For $\overline{a}\neq\overline{0}\,\overline{b}\neq\overline{0}$ one denotes by $\varphi\in[0,\pi]$ the angle between $\overline{a}$ and $\overline{b}$.

**Definition 1.32** (see [12], p. 175 and [1], p. 118) **. The cross product of the free vectors** $\overline{a}$ and $\overline{b}$ is the free vector $\overline{a}\times\overline{b}$ constructed as follows:

- The direction of $\overline{a}\times\overline{b}$ is orthogonal to the plane determined by vectors and $\overline{a}$ and $\overline{b}$;
- Its magnitude, equal to $\left\|\overline{a}\times\overline{b}\right\|$ is given by the formula

$$\left\|\overline{a}\times\overline{b}\right\|=\begin{cases}\|\overline{a}\|\left\|\overline{b}\right\|\sin\varphi, \overline{a},\overline{b}\text{ noncollinear}\\ 0, \overline{a},\overline{b}\text{ collinear.}\end{cases}\tag{1.7}$$

- Its sense is given by the *right-hand rule*: if the vector $\overline{a} \times \overline{b}$ is grasped in the right hand and the fingers curl around from $\overline{a}$ to $\overline{b}$ through the angle $\varphi$, the thumb points in the direction of $\overline{a} \times \overline{b}$ (Fig. 1.13).

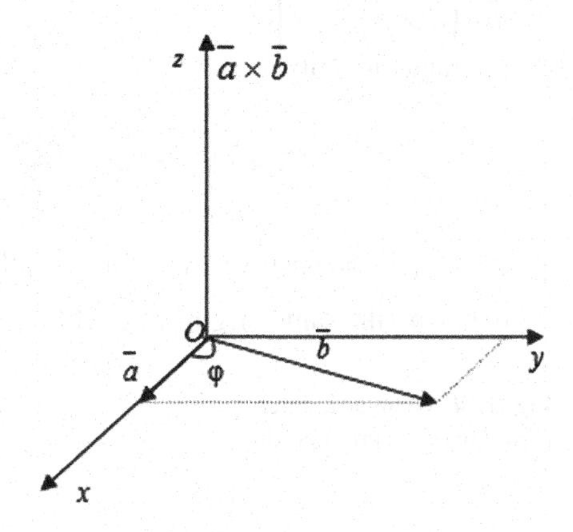


**Fig. 1.13** Graphical representation of cross product

**Proposition 1.33** (see [2], p. 17). The algebraic properties of the cross product of the free vectors are:

1. Anticommutativiy:
$$\overline{a} \times \overline{b} = -\left(\overline{b} \times \overline{a}\right), (\forall)\, \overline{a}, \overline{b} \in V_3;$$
2. $t\left(\frac{\overline{a}\times}{line b}\right) = t\overline{a} \times \overline{b} = \overline{a} \times t\overline{b},\ (\forall)\ \overline{a}, \overline{b} \in V_3,\ (\forall)\ t \in \mathbb{R};$
3. Distributivity with respect to the vector addition:
$$\overline{a} \times \left(\overline{b} + \overline{c}\right) = \overline{a} \times \overline{b} + \overline{a} \times \overline{c}, (\forall)\, \overline{a}, \overline{b}, \overline{c} \in V_3;$$
$$\left(\overline{a} + \overline{b}\right) \times \overline{c} = \overline{a} \times \overline{b} + \overline{b} \times \overline{c}, (\forall)\, \overline{a}, \overline{b}, \overline{c} \in V_3;$$
4. $\overline{a} \times \overline{a} = \overline{0},\ (\forall)\ \overline{a} \in V_3;$
5. $\overline{a} \times \overline{0} = \overline{0} \times \overline{a} = \overline{0},\ (\forall)\ \overline{a} \in V_3;$
6. if $\overline{a}, \overline{b} \in V_3$,
$$\begin{cases} \overline{a} = a_1\overline{i} + a_2\overline{j} + a_3\overline{k} \\ \overline{b} = b_1\overline{i} + b_2\overline{j} + b_3\overline{k} \end{cases}$$
then we obtain the analytical expression of the cross product:
$$\overline{a} \times \overline{b} = (a_2b_3 - a_3b_2)\,\overline{i} + (a_3b_1 - a_1b_3)\,\overline{j} + (a_1b_2 - a_2b_1)\,\overline{k} = \begin{vmatrix} \overline{i} & \overline{j} & \overline{k} \\ a_1 & a_2 & a_3 \\ b_1 & b_2 & b_3 \end{vmatrix} \quad (1.8)$$

**Proposition 1.34** (see [12], p. 170 and [2], p. 17).The geometric properties of the cross product of the free vectors are:

1. $\left.\begin{array}{l}\overline{a}\cdot\left(\overline{a}\times\overline{b}\right)=0\\ \overline{b}\cdot\left(\overline{a}\times\overline{b}\right)=0\end{array}\right\}$ namely $\overline{a}\times\overline{b}$ is orthogonal both on $\overline{a}$ and $\overline{b}$.
2. Lagrange identity:

$$\left\|\overline{a}\times\overline{b}\right\|^2=\|\overline{a}\|^2\left\|\overline{b}\right\|^2-\left(\overline{a}\cdot\overline{b}\right)^2,\ (\forall)\,\overline{a},\overline{b}\in V_3;$$

3. $\left\|\overline{a}\times\overline{b}\right\|$ is the positive area of the parallelogram determined by the vectors $\overline{a}$ and $\overline{b}$, having the same origin (Fig. 1.14).

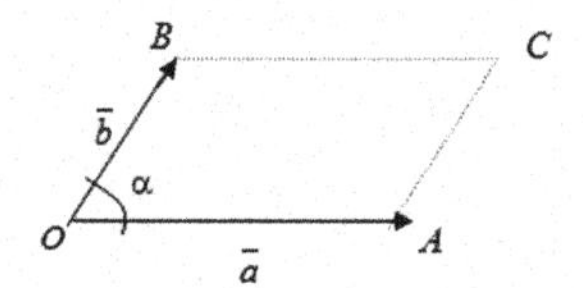


**Fig. 1.14** Geometric interpretation of the cross product

$$\mathrm{A}_{OBCA}=\left\|\overrightarrow{OA}\right\|\left\|\overrightarrow{OB}\right\|\sin\sphericalangle\left(\overrightarrow{OA},\overrightarrow{OB}\right)=\left\|\overrightarrow{OA}\times\overrightarrow{OB}\right\|.$$

But

$$\mathrm{A}_{OBCA}=2\mathrm{A}_{\Delta OAB}.$$

It results that

$$\mathrm{A}_{\Delta OAB}=\frac{\left\|\overrightarrow{OA}\times\overrightarrow{OB}\right\|}{2}.$$

**Example 1.35** (see [9], p. 52). Knowing two sides $\overline{AB}=3\overline{i}-4\overline{j}\,\overline{BC}=\overline{i}+5\overline{j}$ of a triangle, calculate the length of its height $\overline{CD}$.

**Solution**

$$\left\|\overline{AB}\right\|=\sqrt{9+16}=5$$

$$\overline{AB}\times\overline{BC}=\begin{vmatrix}\overline{i}&\overline{j}&\overline{k}\\3&-4&0\\1&5&0\end{vmatrix}=19\overline{k}$$

$$\mathrm{A}_{\Delta ABC}=\frac{\left\|\overline{AB}\times\overline{BC}\right\|}{2}=\frac{\left\|\overline{AB}\right\|\left\|\overline{CD}\right\|}{2}\Longrightarrow\left\|\overline{CD}\right\|=\frac{\left\|\overline{AB}\times\overline{BC}\right\|}{\left\|\overline{AB}\right\|}=\frac{19}{5}.$$

We shall give a solution using Sage, too:

```
sage: AB=vector(QQ,[3,-4,0])
sage: P=arrow3d((0,0,0),(3, -4, 0), 6 ,color='red')+text3d("AB",(2.3,-4.2,0.3))
sage: BC=vector(QQ,[1,5,0])
sage: Q=arrow3d((0,0,0),(1, 5, 0), 6 ,color='blue')+text3d("BC",(1.3,5.2,0.3))
sage: n=AB.norm()
sage: cp=AB.cross_product(BC)
sage: cp.norm()/n
19/5
sage: R=arrow3d((0,0,0),cp, 6 ,color='orange')+text3d("n",(0,0.3,19))
sage: (P+Q+R).show(aspect_ratio=1)
```

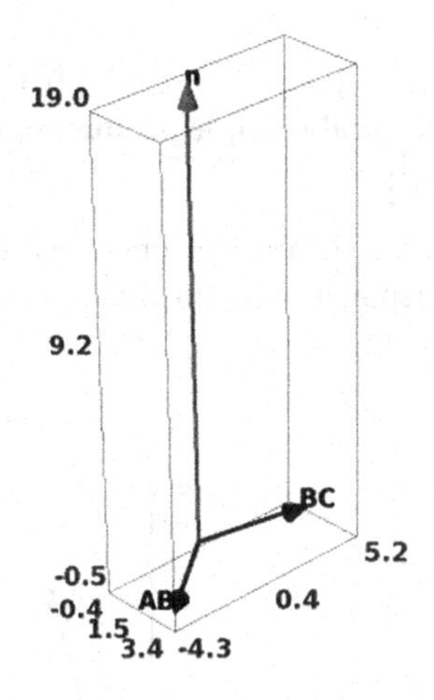


**Example 1.36** (see [5], p. 65). Let the vectors $\overline{a} = 3\overline{m} - \overline{n}\,\overline{b} = \overline{m} + 3\overline{n}$ be such that $\|\overline{m}\| = 3$ and $\|\overline{n}\| = 2$ and $\sphericalangle(\overline{m}, \overline{n}) = \frac{\pi}{2}$. Determine the area of the triangle formed by the vectors $\overline{a}$ and $\overline{b}$.

**Solution**

According to Proposition 1.34, the area of the triangle formed by the vectors $\overline{a}$ and $\overline{b}$ is:

$$A_\Delta = \frac{1}{2}\left\|\overline{a} \times \overline{b}\right\|.$$

We have

$$\begin{aligned}\overline{a} \times \overline{b} &= (3\overline{m} - \overline{n}) \times (\overline{m} + 3\overline{n}) = 3\overline{m} \times (\overline{m} + 3\overline{n}) - \overline{n} \times (\overline{m} + 3\overline{n}) \\ &= 3\underbrace{\overline{m} \times \overline{m}}_{=0} + 9\overline{m} \times \overline{n}\underbrace{-\overline{n} \times \overline{m}}_{=\overline{m}\times\overline{n}} - 3\underbrace{\overline{n} \times \overline{n}}_{=0} = 10\overline{m} \times \overline{n}.\end{aligned}$$

We shall obtain

$$A_\Delta = \frac{1}{2} \cdot 10 \cdot \|\overline{m} \times \overline{n}\| = 5\,\|\overline{m}\| \cdot \|\overline{n}\| \cdot \sin\frac{\pi}{2} = 30.$$

We can also achieve this result using Sage:

```
sage: m=3;n=2;u=pi/2
sage: 1/2*10*m*n*sin(u)
30
```

#### 1.2.4.3 Mixed Product in $V_3$

**Definition 1.37** (see [2], p. 19). Let $\overline{a}, \overline{b}, \overline{c} \in V_3 \setminus \{\overline{0}\}$ be three free vectors. The **mixed product** (also called the **scalar triple product** or **box product**) of these free vectors is the scalar $\overline{a} \cdot \left(\overline{b} \times \overline{c}\right)$.

**Remark 1.38** (see [1], p. 121). If the free vectors $\overline{a}, \overline{b}, \overline{c} \in V_3 \setminus \{\overline{0}\}$ are noncoplanar, then the volume of the parallelepiped determined by these three vectors (see Fig. 1.15) is given by $\left|\overline{a} \cdot \left(\overline{b} \times \overline{c}\right)\right|$.

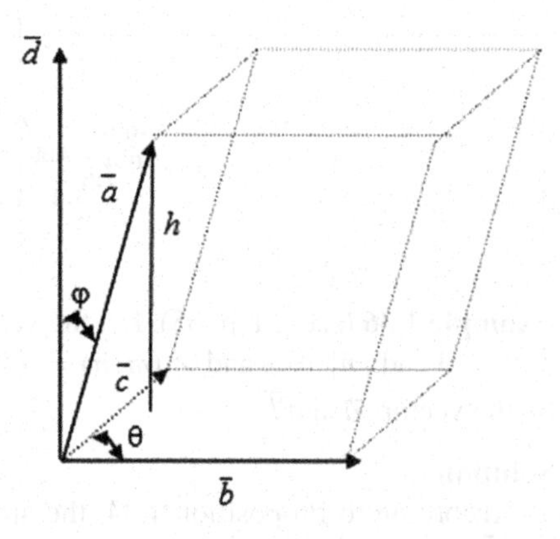


**Fig. 1.15** Geometric interpretation of the mixed product

If we denote by:

- $\theta$ the angle between the vectors $\overline{b}$ and $\overline{c}$,
- $\varphi$ the angle between the vectors $\overline{a}$ and $\overline{d} = \overline{b} \times \overline{c}$, then

$$\overline{a} \cdot \left(\overline{b} \times \overline{c}\right) = \overline{a} \cdot \overline{d} = \|\overline{a}\| \left\|\overline{d}\right\| \cdot \cos\varphi = \underbrace{\left\|\overline{b} \times \overline{c}\right\|}_{\text{A}} \underbrace{\|\overline{a}\| \cos\varphi}_{\pm 2h} = \pm \text{A}h = \pm \text{V}_{\text{paralelipiped}},$$

i.e.

$$\left|\overline{a} \cdot \left(\overline{b} \times \overline{c}\right)\right| = \text{V}_{\text{paralelipiped}}.$$

**Proposition 1.39** (see [2], p. 19 and [1], p. 121). The mixed product of the free vectors has the following properties:

1. $\overline{a} \cdot \left(\overline{b} \times \overline{c}\right) = \overline{c} \cdot \left(\overline{a} \times \overline{b}\right) = \overline{b} \cdot (\overline{c} \times \overline{a})$
2. $\overline{a} \cdot \left(\overline{b} \times \overline{c}\right) = -\overline{a} \cdot \left(\overline{c} \times \overline{b}\right)$
3. $t\overline{a} \cdot \left(\overline{b} \times \overline{c}\right) = \overline{a} \cdot \left(t\overline{b} \times \overline{c}\right) = \overline{a} \cdot \left(\overline{b} \times t\overline{c}\right)$, $(\forall)\ t \in \mathbb{R}$;
4. $\left(\overline{a} + \overline{b}\right) \cdot \left(\overline{c} \times \overline{d}\right) = \overline{a} \cdot \left(\overline{c} \times \overline{d}\right) + \overline{b} \cdot \left(\overline{c} \times \overline{d}\right)$
5. Lagrange identity:

$$\left(\overline{a} \times \overline{b}\right) \cdot \left(\overline{c} \times \overline{d}\right) = \begin{vmatrix} \overline{a} \cdot \overline{c} & \overline{a} \cdot \overline{d} \\ \overline{b} \cdot \overline{c} & \overline{b} \cdot \overline{d} \end{vmatrix}$$

6. $\overline{a} \cdot \left(\overline{b} \times \overline{c}\right) = 0$ if and only if:

   (a) at least one of the vectors $\overline{a}, \overline{b}, \overline{c}$ is null;
   (b) two of the three vectors are collinear;
   (c) the vectors $\overline{a}, \overline{b}, \overline{c}$ are coplanar.

7. if

$$\begin{cases} \overline{a} = a_1\overline{i} + a_2\overline{j} + a_3\overline{k} \\ \overline{b} = b_1\overline{i} + b_2\overline{j} + b_3\overline{k} \\ \overline{c} = c_1\overline{i} + c_2\overline{j} + c_3\overline{k} \end{cases}$$

   then one obtains the analytical expression of the mixed product:

$$\overline{a} \cdot \left(\overline{b} \times \overline{c}\right) = \begin{vmatrix} a_1 & a_2 & a_3 \\ b_1 & b_2 & b_3 \\ c_1 & c_2 & c_3 \end{vmatrix}. \tag{1.9}$$

**Example 1.40** (see [9], p. 52). Find $\lambda \in \mathbb{R}$ such that the volume of the parallelepiped determined by the vectors $\overline{a} = 2\overline{i} - 3\overline{j} + \overline{k}$, $\overline{b} = \overline{i} + \overline{j} - 2\overline{k}$, $\overline{c} = \lambda\overline{i} + 2\overline{j}$ be equal to 5.

**Solution**

The volume of the parallelepiped determined by these vectors is

$$\mathrm{V} = \pm \begin{vmatrix} 2 & -3 & 1 \\ 1 & 1 & -2 \\ \lambda & 2 & 0 \end{vmatrix} = \pm (10 + 5\lambda).$$

From the condition $V = 5$ we deduce $10 + 5\lambda = \pm 5$ i.e. $\lambda_1 = -1\ \lambda_2 = -3$.
The same solution can be obtained in Sage:

```
sage: la=var ('t')
sage: c=vector([t,2,0])
sage: a=vector([2,-3,1])
sage: b=vector([1,1,-2])
sage: V=matrix([a,b,c])
sage: t=solve(V.determinant()==5,la)
sage: t1=solve(V.determinant()==-5,la)
sage: print; t;t1
[t == -1]
[t == -3]
```

**Example 1.41** (see [13]). The following points $A(3,2,1), B(4,4,0), C(5,5,0), D(-1,5,-1)$ are given in space.

(a) Determine if $A, B, C, D$ are coplanar.
(b) If the points are not coplanar calculate the volume of the tetrahedron $ABCD$.
(c) In the same situation calculate the height of $A$ on the plane $(BCD)$.
(d) Prove that $\prec BAC \equiv \prec BAD$.

**Solution**

(a)+(b) The points $A, B, C, D$ are coplanar if the volume of the tetrahedron $ABCD$ is equal to 0.

We have

$$\overline{AB} = \bar{i} + 2\bar{j} - \bar{k}, \overline{AC} = 2\bar{i} + 3\bar{j} + 4\bar{k}, \overline{AD} = -4\bar{i} + 3\bar{j} - 2\bar{k}.$$

According to the Proposition 1.39, we shall obtain

$$\mathrm{V}_{ABCD} = \frac{\left|\overline{AB} \cdot \left(\overline{AC} \times \overline{AD}\right)\right|}{6} = \frac{1}{6} \cdot \left\| \begin{vmatrix} 1 & 2 & -1 \\ 2 & 3 & 4 \\ -4 & 3 & -2 \end{vmatrix} \right\| = 10;$$

therefore $A, B, C, D$ are non-coplanar, that form a tetrahedron.

(c) We know that

$$\mathrm{V}_{ABCD} = \frac{\mathrm{A}_{\Delta BCD} \cdot \mathrm{d}\left(A, (BCD)\right)}{6},$$

where

$$\mathrm{A}_{\Delta BCD} = \frac{1}{2} \left\| \overline{BC} \times \overline{BD} \right\|;$$

it results that

$$\mathrm{d}\left(A, (BCD)\right) = \frac{3\mathrm{V}_{ABCD}}{\mathrm{A}_{\Delta BCD}}.$$

As

$$\overline{BC} = \bar{i} + \bar{j} + 5\bar{k}, \overline{BD} = -5\bar{i} + \bar{j} - \bar{k}$$

we shall have

$$\overline{BC} \times \overline{BD} = \begin{vmatrix} \bar{i} & \bar{j} & \bar{k} \\ 1 & 1 & 5 \\ -5 & 1 & -1 \end{vmatrix} = -6\bar{i} - 24\bar{j} + 6\bar{k}$$

and

$$\left\| \overline{BC} \times \overline{BD} \right\| = \sqrt{36 + 576 + 36} = 18\sqrt{2};$$

therefore

$$\mathrm{A}_{\Delta BCD} = \frac{1}{2} \cdot 18\sqrt{2} = 9\sqrt{2}.$$

It will result

$$\mathrm{d}\left(A, (BCD)\right) = \frac{30}{9\sqrt{2}} = \frac{10}{3\sqrt{2}}.$$

(d) We have:

$$\left.\begin{array}{l} \cos \prec BAC = \cos \prec \left(\overline{AB}, \overline{AC}\right) = \frac{\overline{AB} \cdot \overline{AC}}{\left\|\overline{AB}\right\| \cdot \left\|\overline{AC}\right\|} = \frac{4}{\sqrt{6} \cdot \sqrt{29}} \\ \cos \prec BAD = \cos \prec \left(\overline{AB}, \overline{AD}\right) = \frac{\overline{AB} \cdot \overline{AD}}{\left\|\overline{AB}\right\| \cdot \left\|\overline{AD}\right\|} = \frac{4}{\sqrt{6} \cdot \sqrt{29}} \end{array}\right\} \Longrightarrow$$

$$\cos \prec BAC = \cos \prec BAD \Longrightarrow \prec BAC \equiv \prec BAD.$$

The solution of this problem in Sage is:

```
sage: A=vector([3,2,1]); B=vector([4,4,0]);
sage: C=vector([5,5,5]); D=vector([-1,5,-1]);
sage: AB=B-A
sage: AC=C-A
sage: AD=D-A
sage: u=AC.cross_product(AD)
sage: v=(AB.dot_product(u)).abs()/6; v
10
sage: BC=C-B
sage: BD=D-B
sage: p=BC.cross_product(BD)
sage: a=p.norm()/2;
sage: d=3*v/a; d
5/3*sqrt(2)
```

The following Sage code allows us to represent the tetrahedron $ABCD$:

```
sage: p1=polygon3d([(3,2,1), (4,4,0), (5,5,5),(-1,5,-1)])
sage: t1=text3d("A",(3.1,2,0.8))+text3d("B",(4.2,4,-0.2))
sage: t2=text3d("C",(5.1,5,5.2))+text3d("D",(-1.2,5,-0.8))
sage: p1+t1+t2
```

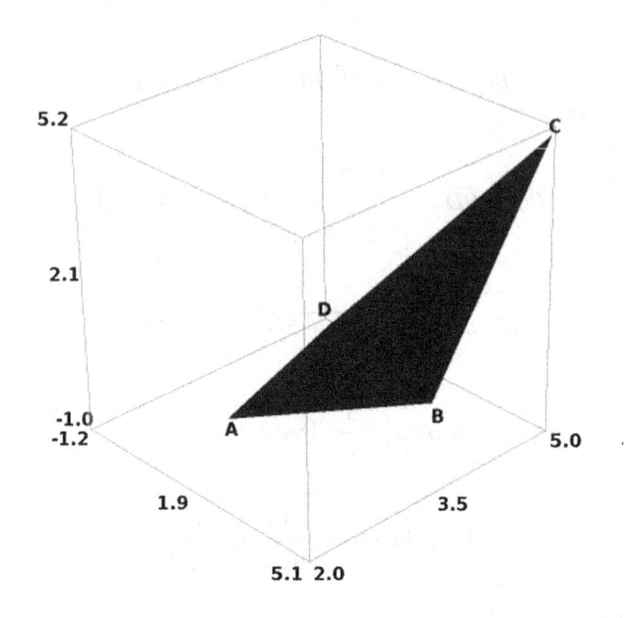


## 1.3 Vector Spaces

Linear algebra can be regarded as the theory of the vector spaces, as a vector space is a set of some objects or elements, that can be added together and multiplied by the numbers (the result remaining an element of the set), so that the ordinary rules of calculation to be valid.

An example of a vector space is the **geometric vector space** (the **free vector space**), which plays a central role in physics and technology and illustrates the importance of the vector spaces and linear algebra for all practical applications.

Let $K$ be a commutative field and $V$ be a non-empty set. The elements of $K$ are called scalars and we shall denote them by Greek letters, and the elements of $V$ are called vectors and we shall denote them by Latin letters, with bar above.

**Definition 1.42** (see [14], p. 1). The set $V$ is called a **vector space over the field** $K$ if the following are defined:

1. *an internal algebraic operation*, denoted additive "+", $+ : V \times V \to V$, called *addition*, in respect to which $V$ is a commutative group;
2. *an external algebraic operation*, denoted multiplicativ " • ", $\bullet: K \times V \to V$, called *multiplication by a scalar*, that satisfies the axioms:

   (a) $(\alpha+\beta)\,\overline{a} = \alpha\overline{a} + \beta\overline{a}$, $(\forall)\ \alpha,\ \beta \in K$ and $(\forall)\ \overline{a} \in V_3$
   (b) $\alpha\left(\overline{a}+\overline{b}\right) = \alpha\overline{a} + \alpha\overline{b}$, $(\forall)\ \alpha \in K$ and $(\forall)\ \overline{a},\ \overline{b} \in V_3$
   (c) $\alpha\,(\beta\overline{a}) = (\alpha\beta)\,\overline{a}$, $(\forall)\ \alpha,\ \beta \in K$ and $(\forall)\ \overline{a} \in V_3$
   (d) $1 \cdot \overline{a} = \overline{a}$, $(\forall)\ \overline{a} \in V_3$.

If $K$ is the field of the real numbers, $V$ is called the real vector space. In the case when $K$ is the field of the complex numbers, $V$ is called the complex vector space.

**Examples of vector spaces**

(1) **The real arithmetic vector space with n - dimensions** $(\mathbb{R}^n, +, \cdot)$

$$\mathbb{R}^n = \underbrace{\mathbb{R} \times \mathbb{R} \times \ldots \times \mathbb{R}}_{n\,\text{times}}$$

is the set of the ordered systems formed with n reale numbers, namely

$$\mathbb{R}^n = \left\{\overline{x} = \left(x^{(1)}, x^{(2)}, \ldots, x^{(n)}\right) | x^{(i)} \in \mathbb{R}, i = \overline{1, n}\right\}.$$

Let $\alpha \in \mathbb{R}$, $\overline{x}$, $\overline{y} \in \mathbb{R}^n$, $\overline{x} = \left(x^{(1)}, x^{(2)}, \ldots, x^{(n)}\right)$, $\overline{y} = \left(y^{(1)}, y^{(2)}, \ldots, y^{(n)}\right)$; then

$$+ : \mathbb{R}^n \times \mathbb{R}^n \to \mathbb{R}^n, \overline{x} + \overline{y} \stackrel{def}{=} \left(x^{(1)} + y^{(1)}, x^{(2)} + y^{(2)}, \ldots, x^{(n)} + y^{(n)}\right)$$

$$\cdot : \mathbb{R} \times \mathbb{R}^n \to \mathbb{R}^n, \alpha\overline{x} \stackrel{def}{=} \left(\alpha x^{(1)}, \alpha x^{(2)}, \ldots, \alpha x^{(n)}\right).$$

(2) **The vector space of the polynomials in the indeterminate $X$ with real coefficients, of degree** $\leq n$, i.e. $(\mathrm{R}_n[X], +, \cdot)$

$\mathrm{R}_n[X]$ means the set of the polynomials in the indeterminate $X$, with real coefficients, of degree $\leq n$.

Let $\alpha \in \mathbb{R}$, $P$, $Q \in \mathrm{R}_n[X]$,

$$P(X) = a_0 + a_1 X + \ldots + a_n X^n, Q(X) = b_0 + b_1 X + \ldots + b_n X^n;$$

then

$$+ : \mathrm{R}_n \times \mathrm{R}_n \to \mathrm{R}_n, P(X) + Q(X) \stackrel{def}{=} a_0 + b_0 + (a_1 + b_1) X + \ldots + (a_n + b_n) X^n \tag{1.10}$$

$$\cdot : \mathbb{R} \times \mathrm{R}_n \to \mathrm{R}_n, \alpha P(X) \stackrel{def}{=} \alpha a_0 + \alpha a_1 X + \ldots + \alpha a_n X^n. \tag{1.11}$$

(3) **The vector space $(\mathbf{M}_{m,n}(\mathbb{R}), +, \cdot)$ of the $m \times n$ matrices, with real coefficients.**

**Example 1.43** (see [13]). Show that the set of the matrices of real numbers with $m$ lines and $n$ columns forms a vector space on $\mathbb{R}$, toward the addition of the matrices and the scalar multiplication from $\mathbb{R}$.

**Solution**

*Stage I.* One proves that $(\mathbf{M}_{m,n}(\mathbb{R}), +, \cdot)$ is a commutative group (abelian group).

If $A, B \in \mathbf{M}_{m,n}(\mathbb{R})$ then $A + B \in \mathbf{M}_{m,n}(\mathbb{R})$, i.e. $\mathbf{M}_{m,n}(\mathbb{R})$ is a stable part in relation to the addition of the matrices (the addition is well defined). Since are easily to check the axioms on:

- Associativity:

$$(A+B)+C=A+(B+C)\,,(\forall)\,A,B,C\in \mathbf{M}_{m,n}(\mathbb{R});$$

- The existence of the neutral element:

$$A+\mathrm{O}=\mathrm{O}+A,\,(\forall)\,A\in \mathbf{M}_{m,n}(\mathbb{R}),\,\mathrm{O}=\begin{pmatrix}0\cdots 0\\ \vdots \quad \vdots\\ 0\cdots 0\end{pmatrix};$$

- The fact that any element is symmetrizable:

$$A+(-A)=(-A)+A=\mathrm{O},\;(\forall)\,A\in \mathbf{M}_{m,n}(\mathbb{R})$$

it results that $(\mathbf{M}_{m,n}(\mathbb{R}),+,\cdot)$ is a group.

Since the matrix addition is commutative, namely

$$A+B=B+A,\;(\forall)\,A,B\in \mathbf{M}_{m,n}(\mathbb{R})$$

it results that $(\mathbf{M}_{m,n}(\mathbb{R}),+,\cdot)$ is an abelian group.

*Stage II.* We check the axioms (a), (b), (c), which must satisfy by the scalar multiplication:

(a) $(\alpha+\beta)A=\alpha A+\beta A\;(\forall)\;\alpha,\;\beta\in\mathbb{R},\,(\forall)\;A,\;B\in \mathbf{M}_{m,n}(\mathbb{R})$
(b) $\alpha\,(A+B)=\alpha A+\alpha B,\;(\forall)\;\alpha\in\mathbb{R},\;(\forall)\,A,B\in \mathbf{M}_{m,n}(\mathbb{R})$
(c) $\alpha\,(\beta A)=(\alpha\beta)\,A\;(\forall)\;\alpha\in\mathbb{R},\,(\forall)\;A,\;B\in \mathbf{M}_{m,n}(\mathbb{R})$
(d) $1\cdot A=A\;(\forall)\;A\in \mathbf{M}_{m,n}(\mathbb{R})$
(4) **The vector space $(\mathcal{F},+,\cdot)$ of the functions defined on the set of real numbers with real values.**

If $\mathcal{F}=\{f|f:\mathbb{R}\to\mathbb{R}\}\,,f,\;g\in\mathcal{F},\;\alpha\in\mathbb{R}$ then

$$+:\mathcal{F}\times\mathcal{F}\to\mathcal{F},\,(f+g)\,(x)\overset{def}{=}f\,(x)+g\,(x)$$

$$\cdot:\mathbb{R}\times\mathcal{F}\to\mathcal{F},\,(\alpha f)\,(x)\overset{def}{=}\alpha f\,(x)\,.$$

(5) **The free vector space, denoted with $V_3$.**

**Theorem 1.44** (see [6], p. 9). If $V$ is a real vector space then the following statements occur:

(i) $0\cdot\overline{a}=\overline{0}\;(\forall)\;\overline{a}\in V$
(ii) $\alpha\cdot\overline{0}=\overline{0}\;(\forall)\;\alpha\in\mathbb{R}$
(iii) $(-1)\cdot\overline{a}=-\overline{a}\;(\forall)\;\overline{a}\in V$
(iv) if $(\forall)\;\alpha\in K,\;(\forall)\;\overline{a}\in V$ such that $\alpha\cdot\overline{a}=\overline{0}$ then $\alpha=0$ or $\overline{a}=\overline{0}$.

**Definition 1.45** (see [14], p. 6). A vector system $\{\overline{x}_1, \ldots, \overline{x}_m\}$ from the space vector $V$ over $K$ is **linearly dependent** if there are the scalars $\alpha^{(i)} \in K,\ \alpha^{(i)} \neq 0,\ (\forall)\, i = \overline{1, m}$, such that

$$\alpha^{(1)}\overline{x}_1 + \ldots + \alpha^{(m)}\overline{x}_m = \overline{0}.$$

If the previous relationship occurs only if

$$\alpha^{(1)} = \ldots = \alpha^{(m)} = 0$$

then the system is **liniarly independent**.

**Definition 1.46** (see [1], p. 6). Let $V$ be a vector space over the field $K$ and $S = \{\overline{x}_1, \ldots, \overline{x}_m\}$ be a system of vectors in $V$. We say that the vector $v \in V$ is a **linear combination** of elements from $S$ if

$$\overline{v} = \sum_{i=1}^{m} \alpha^{(i)}\overline{x}_i, \overline{x}_i \in S, \alpha^{(i)} \in K, i = \overline{1, m}.$$

**Definition 1.47** (see [14], p. 6). Let $V$ be a vector space over the field $K$. The finite system $S$ of vectors from $V$ is called **system of generators** for $V$ if any vector from $V$ is a linear combination of vectors from $S$.

**Definition 1.48** (see [6], p. 12). Let $V$ be a space vector over the field $K$. The finite system $B$ of vectors from $V$ is called **basis** of $V$ if:

(a) $B$ is linearly independent;
(b) $B$ is a system of generators for $V$.

**Example 1.49** (see [13]). $B = \left\{E_{ij}, i = \overline{1, m}, j = \overline{1, n}\right\}$ is a basis in $\mathbf{M}_{m,n}(\mathbb{R})$.

**Solution**

We consider

$$A \in \mathbf{M}_{m,n}(\mathbb{R}), A = \begin{pmatrix} a_{11} & \cdots & a_{1n} \\ \vdots & & \vdots \\ a_{m1} & \cdots & a_{mn} \end{pmatrix}.$$

We can write

$$A = \begin{pmatrix} a_{11} & 0 & \cdots & 0 & 0 \\ 0 & 0 & \cdots & 0 & 0 \\ \cdots & \cdots & \cdots & \cdots & \cdots \\ 0 & 0 & \cdots & 0 & 0 \end{pmatrix} + \begin{pmatrix} 0 & a_{12} & \cdots & 0 & 0 \\ 0 & 0 & \cdots & 0 & 0 \\ \cdots & \cdots & \cdots & \cdots & \cdots \\ 0 & 0 & \cdots & 0 & 0 \end{pmatrix} + \cdots +$$

$$\begin{pmatrix} 0 & 0 & \cdots & 0 & a_{1n} \\ 0 & 0 & \cdots & 0 & 0 \\ \cdots & \cdots & \cdots & \cdots & \cdots \\ 0 & 0 & \cdots & 0 & 0 \end{pmatrix} + \cdots + \begin{pmatrix} 0 & 0 & \cdots & 0 & 0 \\ 0 & 0 & \cdots & 0 & 0 \\ \cdots & \cdots & \cdots & \cdots & \cdots \\ 0 & 0 & \cdots & 0 & a_{mn} \end{pmatrix}$$

$$= a_{11} \begin{pmatrix} 1 & 0 & \cdots & 0 & 0 \\ 0 & 0 & \cdots & 0 & 0 \\ \hdotsfor{5} \\ 0 & 0 & \cdots & 0 & 0 \end{pmatrix} + a_{12} \begin{pmatrix} 0 & 1 & \cdots & 0 & 0 \\ 0 & 0 & \cdots & 0 & 0 \\ \hdotsfor{5} \\ 0 & 0 & \cdots & 0 & 0 \end{pmatrix}$$

$$+ \cdots + a_{mn} \begin{pmatrix} 0 & 0 & \cdots & 0 & 0 \\ 0 & 0 & \cdots & 0 & 0 \\ \hdotsfor{5} \\ 0 & 0 & \cdots & 0 & 1 \end{pmatrix}.$$

We denote

$$E_{ij} = i \overset{\displaystyle j}{\begin{pmatrix} 0 & \cdots & 0 \\ \vdots & 1 & \vdots \\ 0 & \cdots & 0 \end{pmatrix}}.$$

Using this notation, we have

$$A = a_{11}E_{11} + a_{12}E_{12} + \cdots + a_{mn}E_{mn},$$

namely

$$B = \left\{E_{ij}, i = \overline{1,m}, j = \overline{1,n}\right\}$$

is a system of generators.

We consider the linear null combination of the matrices from $B$:

$$\sum_{i=1}^{m}\sum_{j=1}^{n}\alpha_{ij}E_{ij} = \mathrm{O} \Longleftrightarrow \begin{pmatrix} a_{11} & \cdots & a_{1n} \\ \vdots & & \vdots \\ a_{m1} & \cdots & a_{mn} \end{pmatrix} = \mathrm{O} \Longleftrightarrow a_{ij} = 0, (\forall)\, i = \overline{1,m}, j = \overline{1,n};$$

therefore

$$B = \left\{E_{ij}, i = \overline{1,m}, j = \overline{1,n}\right\}$$

is linearly independent.

It results that

$$B = \left\{E_{ij}, i = \overline{1,m}, j = \overline{1,n}\right\}$$

is a basis in $\mathbf{M}_{m,n}(\mathbb{R})$.

A solution in Sage will be given to prove that, in the case of $\mathbf{M}_{2,3}(\mathbb{R})$:

```
sage: var("a00 a01 a02 a10 a11 a12")
(a00, a01, a02, a10, a11, a12)
sage: def MatrixCallable(M):
...     N = Matrix(SR,M)
...     def m(*a,**k):
...         return Matrix([[e(*a,**k) for e in row] for row in N])
...     return m
sage: M=[[a00,a01,a02],[a10,a11,a12]]
sage: N=MatrixCallable(M)
sage: E1=N(a00=1,a01=0,a02=0,a10=0,a11=0,a12=0)
sage: E2=N(a00=0,a01=1,a02=0,a10=0,a11=0,a12=0)
sage: E3=N(a00=0,a01=0,a02=1,a10=0,a11=0,a12=0)
sage: E4=N(a00=0,a01=0,a02=0,a10=1,a11=0,a12=0)
sage: E5=N(a00=0,a01=0,a02=0,a10=0,a11=1,a12=0)
sage: E6=N(a00=0,a01=0,a02=0,a10=0,a11=0,a12=1)
sage: Z=a00*E1+a01*E2+a02*E3+a10*E4+a11*E5+a12*E6
sage: Z==A
True
sage: solve([Z[0][0],Z[0][1],Z[0][2],Z[1][0],Z[1][1],Z[1][2]],a00,a01,a02,a10,a11,a12)
[[a00 == 0, a01 == 0, a02 == 0, a10 == 0, a11 == 0, a12 == 0]]
```

**Definition 1.50** (see [6], p. 13). Let $V$ be a vector space over $K$.

(a) The number $n \in \mathbb{N}^*$ of the vectors from a basis of a vector space $V$ is called the **finite dimension** of $V$ (over $K$ ) and one denotes by:

$\dim V$ or $\dim_K V$.

(b) In the case when $V$ is not finite generated is said to be of **infinite dimension** and one denotes $\dim_K V = \infty$.

**Remark 1.51** (see [6], p. 13). $\dim \mathbf{M}_{m,n}(\mathbb{R}) = m \cdot n$.
**Remark 1.52** (see [1], p. 10). Two arbitrary bases in $V$ have the same number of vectors.

**Example 1.53** (see [13]). Let

$$\mathrm{R}_n[X] = \left\{a_0 + a_1X + \ldots + a_nX^n | \alpha_i \in \mathbb{R}, i = \overline{0,n}, n \in \mathbb{N}^*\right\}$$

be the set of polynomials in the indeterminate $X$, with real coefficients, of degree $\leq n$.

(i) Show that the set $\mathrm{R}_n[X]$ is a real vector space, denoted by $(\mathrm{R}_n[X], +, \cdot)$ with the usual operations of addition of polynomials and of multiplication of the polynomials with scalars from $\mathbb{R}$.
(ii) Find the dimension of $\mathrm{R}_n[X]$.

**Solution**

(i) Let be $\alpha \in \mathbb{R}$, $P, Q \in \mathrm{R}_n[X]$, $P(X) = a_0 + a_1X + \ldots + a_nX^n$, $Q(X) = b_0 + b_1X + \ldots + b_nX^n$.

*Stage I.* It proves that $(\mathrm{R}_n[X], +, \cdot)$ is an abelian group, where "+" and "·" are defined in (1.10) and respectively (1.11).

As

$$\deg\text{ree}\,(P+Q) \le \max\left(\deg\text{ree}\,(P), \deg\text{ree}\,(Q)\right) \le n$$

it results

$$P + Q \in \mathrm{R}_n[X].$$

We showed that for $P, Q \in \mathrm{R}_n[X]$ we have $P + Q \in \mathrm{R}_n[X]$, namely $\mathrm{R}_n[X]$ is a stable parts in relation to the addition of the polynomials (the addition is well defined).

As the axioms on associativity, the existence of neutral element and the fact that any element is simetrizable are easily checked it results that $(\mathrm{R}_n[X], +, \cdot)$ is a group.

As the addition of the polynomials is commutative it results that $(\mathrm{R}_n[X], +, \cdot)$ is an abelian group.

*Stage II.* We check the axioms (a), (b), (c), (d) from the Definition 1.42, which must satisfy the scalar multiplication.

(a) $(\alpha + \beta)P = \alpha P + \beta P$, $(\forall)\ \alpha, \beta \in \mathbb{R}$ and $(\forall)\ P \in \mathrm{R}_n[X]$
(b) $\alpha(P + Q) = \alpha P + \alpha Q$, $(\forall)\ \alpha \in \mathbb{R}$ and $(\forall)\ P, Q \in \mathrm{R}_n[X]$
(c) $\alpha(\beta P) = (\alpha\beta)P$, $(\forall)\ \alpha, \beta \in \mathbb{R}$ and $(\forall)\ P \in \mathrm{R}_n[X]$
(d) $1 \cdot P = P$, $(\forall)\ P \in \mathrm{R}_n[X]$.

(ii) We denote

$$\begin{cases} P_0(X) = 1 \\ P_1(X) = X \\ \vdots \\ P_n(X) = X^n. \end{cases}$$

Let be the null linear combination:

$$\alpha^{(0)}P_0(X) + \alpha^{(1)}P_1(X) + \ldots + \alpha^{(n)}P_n(X) = O_{R_n[X]}.$$

As

$$\alpha^{(0)} + \alpha^{(1)}X + \ldots + \alpha^{(n)}X^n = O_{R_n[X]}$$

it results that $\alpha^{(0)} = \ldots = \alpha^{(n)} = 0$; so the family of polynomials $B = \{P_0, P_1, \ldots, P_n\}$ is linearly independent.

Note that $B = \{P_0, P_1, \ldots, P_n\}$ generates $\mathrm{R}_n[X]$ since $(\forall)\ P \in \mathrm{R}_n[X]$ :

$$P(X) = a_0 + a_1 X + \ldots + a_n X^n = \alpha^{(0)} P_0(X) + \alpha^{(1)} P_1(X) + \ldots + \alpha^{(n)} P_n(X);$$

therefore $B = \{P_0, P_1, \ldots, P_n\}$ is a system of generators.

As $B = \{P_0, P_1, \ldots, P_n\}$ is linearly independent and a a system of generators for $\mathrm{R}_n[X]$, from the Definition 1.48 it results that $B$ is a base for $\mathrm{R}_n[X]$.

It results that $\dim \mathrm{R}_n[X] = n + 1$.

We can prove that in Sage, too:

```
sage: R.<x>= RR['x'];var("a0,a1,a2,a3,b0,b1,b2,b3,al,be")
(a0, a1, a2, a3, b0, b1, b2, b3, al, be)
sage: P(x)=a0+a1*x+a2*x^2+a3*x^3;Q(x)=b0+b1*x+b2*x^2+b3*x^3
sage: (P(x)+Q(x)).simplify_exp()
(a3 + b3)*x^3 + (a2 + b2)*x^2 + (a1 + b1)*x + a0 + b0
sage: ((al+be)*P(x)).simplify_exp()-(al*P(x)+be*P(x)).simplify_exp()
0
sage: (al*(P(x)+Q(x))).simplify_exp()-(al*P(x)+al*Q(x)).simplify_exp()
0
sage: (al*(be*P(x))).simplify_exp()-((al*be)*P(x)).simplify_exp()
0
sage: 1*P(x)-P(x)
0
sage: z1=P(x).coefficient(x,0);z2=P(x).coeff(x);
sage: z3=P(x).coeff(x^2);z4=P(x).coeff(x^3)
sage: solve([z1,z2,z3,z4],a0,a1,a2,a3)
[[a0 == 0, a1 == 0, a2 == 0, a3 == 0]]
```

**Example 1.54** (see [13]). Prove that the set

$$V = \{f : \mathbb{R} \to \mathbb{R} | f(x) = \alpha \sin(x + \beta), \alpha, \beta \in \mathbb{R}\}$$

equipped with the normal operations of addition of functions and of multiplication of functions with real numbers is a vector space. Determine the dimension of $V$.

**Solution**

Let be $f, g \in V$, $f(x) = \alpha \sin(x + \beta)$, $g(x) = \alpha' \sin(x + \beta')$. We shall check if $(V, +)$ is a commutative group.

We show only that $f + g \in V$, namely $(\exists)\ \alpha_0, \beta_0 \in \mathbb{R}$ such that

$$(f + g)(x) = \alpha_0 \sin(x + \beta_0), (\forall) x \in \mathbb{R}.$$

It must be the case that

$$\alpha \sin\left(x+\beta\right)+\alpha' \sin\left(x+\beta'\right)=\alpha_0 \sin\left(x+\beta_0\right),\ (\forall)\, x\in\mathbb{R},$$

i.e.

$$\begin{aligned}&\alpha \sin x\cos\beta+\alpha\sin\beta\cos x+\alpha'\sin x\cos\beta'+\alpha'\sin\beta'\cos x\\&=\alpha_0\left(\sin x\cos\beta_0+\sin\beta_0\cos x\right)\Leftrightarrow\\&\left(\alpha\cos\beta+\alpha'\cos\beta'\right)\sin x+\left(\alpha\sin\beta+\alpha'\sin\beta'\right)\cos x\\&=\alpha_0\cos\beta_0\sin x+\alpha_0\sin\beta_0\cos x,\ (\forall)\, x\in\mathbb{R}.\end{aligned}$$

We denote:

$$\begin{cases}A=\alpha\cos\beta+\alpha'\cos\beta'\\B=\alpha\sin\beta+\alpha'\sin\beta'.\end{cases}$$

We deduce

$$\begin{cases}A=\alpha_0\cos\beta_0\\B=\alpha_0\sin\beta_0.\end{cases}$$

It results that

$$\frac{\sin\beta_0}{\cos\beta_0}=\frac{B}{A}\Rightarrow \operatorname{tg}\beta_0=\frac{B}{A}\Rightarrow\beta_0=\operatorname{arctg}\frac{B}{A}+k\pi,\ k\in\mathbb{Z},$$

where tg $x$ means the tangent function.

Although an infinite number of values are obtained for $\beta_0$, the function $\alpha_0\sin\left(x+\beta_0\right)$ has the same value, so the sum of two functions from $V$ is well defined. Immediately, we can determine

$$\alpha_0=\frac{A}{\cos\beta_0}.$$

Therefore $(\exists)\ \alpha_0,\beta_0\in\mathbb{R}$ such that

$$(f+g)(x)=\alpha_0\sin\left(x+\beta_0\right),\ (\forall)\, x\in\mathbb{R}.$$

(ii) Let be $f\in V\Rightarrow(\exists)\ \alpha,\beta\in\mathbb{R}$ such that

$$f(x)=\alpha\sin\left(x+\beta\right),\ (\forall)\, x\in\mathbb{R}.$$

We have

$$f(x)=\alpha\sin x\cos\beta+\alpha\sin\beta\cos x=\left(\alpha\cos\beta\right)\sin x+\left(\alpha\sin\beta\right)\cos x.$$

It is known that

$$\cos x = \sin\left(\frac{\pi}{2} - x\right) = -\sin\left(x - \frac{\pi}{2}\right).$$

We denote

$$f_1(x) = \sin x, f_2(x) = -\sin\left(x - \frac{\pi}{2}\right).$$

Hence

$$f = \alpha f_1 \cos\beta + \alpha f_2 \sin\beta.$$

We prove that $f_1, f_2 \in V$.
As

$$\begin{cases} f_1(x) = \sin x = 1 \cdot \sin(x+0) \Rightarrow f_1 \in V, \text{ for } \alpha = 1, \beta = 0; \\ f_2(x) = -\sin\left(x - \frac{\pi}{2}\right) = (-1)\sin\left(x - \frac{\pi}{2}\right) \Rightarrow f_1 \in V, \text{ for } \alpha = -1, \beta = -\frac{\pi}{2}. \end{cases}$$

Thereby, $B = \{f_1, f_2\}$ is a system of generators for $V$.
We consider the null linear combination

$$\alpha^{(1)} f_1 + \alpha^{(2)} f_2 = 0_V \Leftrightarrow \alpha^{(1)} f_1(x) + \alpha^{(2)} f_2(x) = 0, (\forall) x \in \mathbb{R};$$

it results that

$$\alpha^{(1)} \sin x + \alpha^{(2)} \cos x = 0, (\forall) x \in \mathbb{R}.$$

If $x = 0$ from the previous relation we get $\alpha^{(2)} = 0$, while for $x = \frac{\pi}{2}$ we obtain $\alpha^{(1)} = 0$.

It results that $B = \{f_1, f_2\}$ is linear independently. Therefore $B = \{f_1, f_2\}$ is a basis for $V \Rightarrow \dim V = 2$.

We can check the previous results in Sage, too:

```
sage: x,al,be,ap,bp,a0,b0,a,b,al1,al2=var('x,al,be,ap,bp,a0,b0,a,b,al1,al2')
sage: f=al*(sin(x+be));g=ap*sin(x+bp);h=a0*sin(x+b0)
sage: A=(f+g).simplify_trig().coeff(sin(x));e1=h.simplify_trig().coeff(sin(x))
sage: B=(f+g).simplify_trig().coeff(cos(x));e2=h.simplify_trig().coeff(cos(x))
sage: A;B;e1;e2
al*cos(be) + ap*cos(bp)
al*sin(be) + ap*sin(bp)
a0*cos(b0)
a0*sin(b0)
sage: b0=arctan(B/A); b0
arctan((al*sin(be) + ap*sin(bp))/(al*cos(be) + ap*cos(bp)))
sage: solve(f1==A,a0)
```

```
[a0 == (al*cos(be) + ap*cos(bp))/cos(b0)]
sage: ff=a*(sin(x+b))
sage: g1=ff.simplify_trig().coeff(sin(x));g2=ff.simplify_trig().coeff(cos(x))
sage: f1=sin(x);f2=-sin(x-pi/2)
sage: ff.simplify_trig()-(g1*f1+g2*f2).simplify()
0
sage: f1-ff.subs(a=1,b=0)
0
sage: f2-ff.subs(a=-1,b=-pi/2)
0
sage: F=(al1*f1+al2*f2).simplify()
sage: solve(F.subs(x=0),al2);solve(F.subs(x=pi/2),al1)
[al2 == 0]
[al1 == 0]
```

**Example 1.55** (see [13]). Let $V$ be a real vector space.

(i) Show that $\mathbb{C}V = V \times V = \{(\overline{u}_1, \overline{u}_2)\}$ is complex vector space (the complexification of the real vector space $V$ ) in relation to the operations:

$+ : \mathbb{C}V \times \mathbb{C}V \to \mathbb{C}V$ defined by $(\overline{u}_1, \overline{u}_2) + (\overline{v}_1, \overline{v}_2) = (\overline{u}_1 + \overline{v}_1, \overline{u}_2 + \overline{v}_2)$, $\cdot : \mathbb{C} \times \mathbb{C}V \to \mathbb{C}V$ defined by $(\alpha + i\beta) \cdot (\overline{v}_1, \overline{v}_2) = (\alpha\overline{v}_1 - \beta\overline{v}_2, \alpha\overline{v}_2 + \beta\overline{v}_1)$, $(\forall)\ \alpha,\ \beta \in \mathbb{R}$.

(ii) If $\dim_{\mathbb{R}} V = n < \infty$ determine $\dim_{\mathbb{C}} \mathbb{C}V$.

**Solution**

(i) Check the axioms from the definition of the vector space.
(ii) Let be $B = \{\overline{a}_1, \overline{a}_2, \ldots, \overline{a}_n\}$ a basis for $V$. We intend to show that $B' = \{i\overline{a}_1, i\overline{a}_2, \ldots, i\overline{a}_n\}$ a basis for $\mathbb{C}V$.

Let be $(\overline{u}_1, \overline{u}_2) \in \mathbb{C}V,\ \overline{u}_1, \overline{u}_2 \in V$.
As $B$ is a basis for $V$ it results that

- $(\forall)\ \overline{u}_1 \in V,\ (\exists)\ x^{(i)} \in K,\ i = \overline{1,n}$ unique, such that

$$\overline{u}_1 = x^{(1)}\overline{a}_1 + \ldots + x^{(n)}\overline{a}_n;$$

- $(\forall)\ \overline{u}_2 \in V,\ (\exists)\ y^{(i)} \in K,\ i = \overline{1,n}$ unique, such that

$$\overline{u}_2 = y^{(1)}\overline{a}_1 + \ldots + y^{(n)}\overline{a}_n.$$

We compute

$$\begin{aligned}(\overline{u}_1, \overline{u}_2) &= \left(x^{(1)}\overline{a}_1 + \ldots + x^{(n)}\overline{a}_n, y^{(1)}\overline{a}_1 + \ldots + y^{(n)}\overline{a}_n\right)\\ &= \left(x^{(1)}\overline{a}_1, y^{(1)}\overline{a}_1\right) + \ldots + \left(x^{(n)}\overline{a}_n, y^{(n)}\overline{a}_n\right)\\ &= \left(y^{(1)} - ix^{(1)}\right) \cdot \left(\overline{0}, \overline{a}_1\right) + \ldots + \left(y^{(n)} - ix^{(n)}\right) \cdot \left(\overline{0}, \overline{a}_n\right).\end{aligned}$$

It results that $B' = \left\{\left(\overline{0}, \overline{a}_1\right), \ldots, \left(\overline{0}, \overline{a}_n\right)\right\}$ is a system of generators for $\mathbb{C}\, V$.
Let be a null combination of elements from $B'$ :

$$\left(\alpha^{(1)} + i\beta^{(1)}\right) \cdot \left(\overline{0}, \overline{a}_1\right) + \ldots + \left(\alpha^{(n)} + i\beta^{(n)}\right) \cdot \left(\overline{0}, \overline{a}_n\right) = \left(\overline{0}, \overline{0}\right). \tag{1.12}$$

From (1.12) we have

$$\left(-\beta^{(1)}\overline{a}_1, \alpha^{(1)}\overline{a}_1\right) + \ldots + \left(-\beta^{(n)}\overline{a}_n, \alpha^{(n)}\overline{a}_n\right) = \left(\overline{0}, \overline{0}\right) \Leftrightarrow$$

$$\left(-\beta^{(1)}\overline{a}_1 - \ldots - \beta^{(n)}\overline{a}_n, \alpha^{(1)}\overline{a}_1 + \ldots + \alpha^{(n)}\overline{a}_n\right) = \left(\overline{0}, \overline{0}\right),$$

namely

$$-\beta^{(1)}\overline{a}_1 - \ldots - \beta^{(n)}\overline{a}_n = \overline{0} \tag{1.13}$$

and

$$\alpha^{(1)}\overline{a}_1 + \ldots + \alpha^{(n)}\overline{a}_n = \overline{0}. \tag{1.14}$$

As $\overline{a}_1, \ldots, \overline{a}_n$ are linearly independent from (1.13) and (1.14) it results

$$\begin{cases} \beta^{(1)} = \ldots = \beta^{(n)} = 0 \\ \alpha^{(1)} = \ldots = \alpha^{(n)} = 0. \end{cases} \tag{1.15}$$

From (1.12) and (1.15) we deduce that $B' = \left\{\left(\overline{0}, \overline{a}_1\right), \ldots, \left(\overline{0}, \overline{a}_n\right)\right\}$ is linearly independent.

Thus, $B'$ is a basis and $\dim_{\mathbb{C}} \mathbb{C}V = n$.

Checking this problem in Sage, for $n = 3$ we have:

```
sage: var("u1,u2,v1,v2,al,be,x1,x2,x3,y1,y2,y3,a1,a2,a3,al1,al2,al3,be1,be2,be3")
(u1, u2, v1, v2, al, be, x1, x2, x3, y1, y2, y3, a1, a2, a3, al1, al2, al3, be1, be2, be3)
sage: i=CDF(I);uu=u1+i*u2;vv=v1+i*v2
sage: s=vector(((uu+vv).real().simplify_exp(),(uu+vv).imag().simplify_exp()));s
(u1 + v1, u2 + v2)
sage: pp=((al+be*i)*vv).simplify_exp()
sage: p=vector((pp.real().simplify_exp(),pp.imag().simplify_exp()));p
(al*v1 - be*v2, al*v2 + be*v1)
```

```
sage: p1=pp.subs(al=y1,be=-x1,v1=0,v2=a1)
sage: p2=pp.subs(al=y2,be=-x2,v1=0,v2=a2)
sage: p3=pp.subs(al=y3,be=-x3,v1=0,v2=a3)
sage: pv1=vector((p1.real().simplify_exp(),p1.imag().simplify_exp()));pv1
(a1*x1, a1*y1)
sage: pv2=vector((p2.real().simplify_exp(),p2.imag().simplify_exp()));pv2
(a2*x2, a2*y2)
sage: pv3=vector((p3.real().simplify_exp(),p3.imag().simplify_exp()));pv3
(a3*x3, a3*y3)
sage: pv1+pv2+pv3
(a1*x1 + a2*x2 + a3*x3, a1*y1 + a2*y2 + a3*y3)
sage: p4=pp.subs(al=al1,be=be1,v1=0,v2=a1)
sage: p5=pp.subs(al=al2,be=be2,v1=0,v2=a2)
sage: p6=pp.subs(al=al3,be=be3,v1=0,v2=a3)
sage: pn=vector(((p4+p5+p6).real().simplify_exp(),(p4+p5+p6).imag().simplify_exp()));pn
(-a1*be1 - a2*be2 - a3*be3, a1*al1 + a2*al2 + a3*al3)
sage: pn[0].subs(be1=0,be2=0,be3=0);pn[1].subs(al1=0,al2=0,al3=0)
0
0
```

**Proposition 1.56** (see [1], p. 12). Let $V$ be a space vector of $n$ dimension and $B = \{\overline{a}_1, \ldots, \overline{a}_n\} \subset V$. Then:

(a) if $B$ is liniarly independent it results that $B$ is a basis,
(b) if $B$ is system of generators it results that $B$ is a basis.

**Theorem 1.57** (see [14], p. 7). Let $V$ be a space vector over $K$ and $B = \{\overline{a}_1, \ldots, \overline{a}_n\} \subset V$. Then $B$ is a basis of $V$ if and only if any vectors from $V$ can be written in an unique way as a linear combination of the vectors from $B$.

**Definition 1.58** (see [14], p. 7). The unique scalars $x^{(i)} \in K$, $i = \overline{1, m}$ that appear as coefficients in the writing of the vector $\overline{x} \in V$ as a linear combination of the vectors from the basis $B$ are called **the coordinates of the vector $\overline{x}$ relative to the basis $B$.**

We shall denote by $\overline{x}_B$ the column matrix formed with the coordinates of the vector $\overline{x}$ relative to the basis $B$; therefore

$$\overline{x}_B = \begin{pmatrix} x^{(1)} \\ x^{(2)} \\ \vdots \\ x^{(m)} \end{pmatrix}.$$

**Remark 1.59** (see [6], p. 14). The writing of a vector in a basis is unique.

**Example 1.60** (see [9], p. 31). In $\mathbb{R}^4$ the following vectors are given:

$$\overline{v}_1 = (1, 1, 2, 1), \overline{v}_2 = (1, -1, 0, 1), \overline{v}_3 = (0, 0, -1, 1), \overline{v}_4 = (1, 2, 2, 0).$$

(i) Show that these vectors form a basis.
(ii) Find the coordinates of the vector $\overline{v} = (1, 1, 1, 1)$ relative to this basis.

**Solution**

(i) $\dim \mathbb{R}^4 = 4 \Rightarrow$ it suffices to show that the vectors are linearly independent. Let be the null linear combination

$$\alpha^{(1)}\overline{v}_1 + \alpha^{(2)}\overline{v}_2 + \alpha^{(3)}\overline{v}_3 + \alpha^{(4)}\overline{v}_4 = \overline{0}.$$

It results that

$$\left(\alpha^{(1)}, \alpha^{(1)}, 2\alpha^{(1)}, \alpha^{(1)}\right) + \left(\alpha^{(2)}, -\alpha^{(2)}, 0, \alpha^{(2)}\right) + \left(0, 0, -\alpha^{(3)}, \alpha^{(3)}\right) + \left(\alpha^{(4)}, 2\alpha^{(4)}, 2\alpha^{(4)}, 0\right) = (0, 0, 0, 0).$$

One obtains the system:

$$\begin{cases} \alpha^{(1)} + \alpha^{(2)} + \alpha^{(4)} = 0 \\ \alpha^{(1)} - \alpha^{(2)} + 2\alpha^{(4)} = 0 \\ 2\alpha^{(1)} - \alpha^{(3)} + 2\alpha^{(4)} = 0 \\ \alpha^{(1)} + \alpha^{(2)} + \alpha^{(3)} = 0. \end{cases}$$

As

$$d = \begin{vmatrix} 1 & 1 & 0 & 1 \\ 1 & -1 & 0 & 2 \\ 2 & 0 & -1 & 2 \\ 1 & 1 & 1 & 0 \end{vmatrix} = -4 \neq 0$$

it results that the system has a unique solution

$$\alpha^{(1)} = \alpha^{(2)} = \alpha^{(3)} = \alpha^{(4)} = 0.$$

It follows that the vectors are linearly independent.
(ii) Let be

$$\overline{v} = \alpha^{(1)}\overline{v}_1 + \alpha^{(2)}\overline{v}_2 + \alpha^{(3)}\overline{v}_3 + \alpha^{(4)}\overline{v}_4.$$

We have

$$(1, 1, 1, 1) = \left(\alpha^{(1)}, \alpha^{(1)}, 2\alpha^{(1)}, \alpha^{(1)}\right) + \left(\alpha^{(2)}, -\alpha^{(2)}, 0, \alpha^{(2)}\right) + \left(0, 0, -\alpha^{(3)}, \alpha^{(3)}\right) + \left(\alpha^{(4)}, 2\alpha^{(4)}, 2\alpha^{(4)}, 0\right).$$

One obtains the system:

$$\begin{cases} \alpha^{(1)} + \alpha^{(2)} + \alpha^{(4)} = 1 \\ \alpha^{(1)} - \alpha^{(2)} + 2\alpha^{(4)} = 1 \\ 2\alpha^{(1)} - \alpha^{(3)} + 2\alpha^{(4)} = 1 \\ \alpha^{(1)} + \alpha^{(2)} + \alpha^{(3)} = 1 \end{cases}$$

whose solution is

$$\alpha^{(1)} = \frac{1}{4}, \alpha^{(2)} = \frac{1}{4}, \alpha^{(3)} = \frac{1}{2}, \alpha^{(4)} = \frac{1}{2}.$$

Solving this problem with Sage, we achieve:

```
sage: V=VectorSpace(RR,4)
sage: v1=V([1,1,2,1]);v2=V([1,-1,0,1]);
sage: v3=V([0,0,-1,1]);v4=V([1,2,2,0]);
sage: V.linear_dependence([v1,v2,v3,v4])==[]
True
sage: W=V.subspace_with_basis([v1,v2,v3,v4])
sage: v=V([1,1,1,1])
sage: W.coordinate_vector(v).n(digits=3)
(0.250, 0.250, 0.500, 0.500)
```

Hence

$$\overline{v}_B = \begin{pmatrix} 1/4 \\ 1/4 \\ 1/2 \\ 1/2 \end{pmatrix}.$$

Let $B_1 = \{\overline{e}_1, \ldots, \overline{e}_n\}$, $B_1 = \left\{\overline{f}_1, \ldots, \overline{f}_n\right\}$ be two bases of $V\,\overline{x} \in V$ and

$$\overline{x} = x^{(1)}\overline{a}_1 + \ldots + x^{(n)}\overline{a}_n \tag{1.16}$$

$$\overline{x} = y^{(1)}\overline{f}_1 + \ldots + y^{(n)}\overline{f}_n. \tag{1.17}$$

We suppose that the vectors from $B_2$ can be written as a linear combination of the vectors from the basis $B_1$:

$$\overline{f}_1 = \alpha_1^{(1)}\overline{e}_1 + \ldots + \alpha_1^{(n)}\overline{e}_n \tag{1.18}$$

$$\vdots$$

$$\overline{f}_n = \alpha_n^{(1)}\overline{e}_1 + \ldots + \alpha_n^{(n)}\overline{e}_n.$$

Writing the column coefficients of these linear combinations we get (see [14], p. 8) the matrix

$$M_{(B_1,B_2)} = \begin{pmatrix} \alpha_1^{(1)} & \dots & \alpha_n^{(1)} \\ \alpha_1^{(2)} & \dots & \alpha_n^{(2)} \\ \dots & \dots & \dots \\ \alpha_1^{(n)} & \dots & \alpha_n^{(n)} \end{pmatrix}, \qquad (1.19)$$

which represents the **transition matrix from the basis $B_1$ to the basis $B_2$.**
**Remark 1.61** (see [6], p. 18). The matrix $M_{(B_1,B_2)}$ is always a nonsingular matrix due to the linear independence of the basis vectors.

**Example 1.62** (see [13]). Find for the space of polynomials by at most four degree, the transition matrix from the basis $B_1 = \{1, X, X^2, X^3, X^4\}$ to the basis $B_2 = \{1, (X+1), (X+1)^2, (X+1)^3, (X+1)^4\}$.

**Solution**

The vectors from $B_2$ can be written as a linear combination of the vectors from $B_1$ thus:

$$\begin{aligned} 1 &= 1 \\ X+1 &= 1+X \\ (X+1)^2 &= 1+2X+X^2 \\ (X+1)^3 &= 1+3X+3X^2+X^3 \\ (X+1)^4 &= 1+4X+6X^2+4X^3+X^4. \end{aligned}$$

It results that

$$M_{(B_1,B_2)} = \begin{pmatrix} 1 & 1 & 1 & 1 & 1 \\ 0 & 1 & 2 & 3 & 4 \\ 0 & 0 & 1 & 3 & 6 \\ 0 & 0 & 0 & 1 & 4 \\ 0 & 0 & 0 & 0 & 1 \end{pmatrix}.$$

The same matrix can be achieved using Sage:

```
sage: R.<x> = RR['x'];K=R^5
sage: M=K.span([[1,x,x^2,x^3,x^4]])
sage: U=M([1,x+1, (x+1)^2, (x+1)^3, (x+1)^4]);
sage: v0=vector([U[0][0],U[0][1],U[0][2],U[0][3],U[0][4]]);
sage: v1=vector([U[1][0],U[1][1],U[1][2],U[1][3],U[1][4]]);
sage: v2=vector([U[2][0],U[2][1],U[2][2],U[2][3],U[2][4]]);
sage: v3=vector([U[3][0],U[3][1],U[3][2],U[3][3],U[3][4]]);
sage: v4=vector([U[4][0],U[4][1],U[4][2],U[4][3],U[4][4]]);
sage: A = column_matrix([v0,v1, v2, v3,v4]).n(digits=3)
sage: A
[ 1.00  1.00  1.00  1.00  1.00]
[0.000  1.00  2.00  3.00  4.00]
[0.000 0.000  1.00  3.00  6.00]
[0.000 0.000 0.000  1.00  4.00]
[0.000 0.000 0.000 0.000  1.00]
```

Substituting (1.18) into (1.17) we obtain

$$\begin{aligned}\overline{x} &= y^{(1)}\left(\alpha_1^{(1)}\overline{e}_1+\ldots+\alpha_1^{(n)}\overline{e}_n\right)+\ldots+y^{(n)}\left(\alpha_n^{(1)}\overline{e}_1+\ldots+\alpha_n^{(n)}\overline{e}_n\right)\\ &= \left(y^{(1)}\alpha_1^{(1)}+\ldots+y^{(n)}\alpha_n^{(1)}\right)\overline{e}_1+\ldots+\\ &\left(y^{(1)}\alpha_1^{(n)}+\ldots+y^{(n)}\alpha_n^{(n)}\right)\overline{e}_n.\end{aligned} \tag{1.20}$$

Due to the uniqueness of writing a vector into a basis, from (1.16) and (1.20) it results (see [6], p. 18)

$$\begin{cases} x^{(1)} = y^{(1)}\alpha_1^{(1)}+\ldots+y^{(n)}\alpha_n^{(1)}\\ \quad\vdots\\ x^{(n)} = y^{(1)}\alpha_1^{(n)}+\ldots+y^{(n)}\alpha_n^{(n)}.\end{cases} \tag{1.21}$$

The formulas (1.21) are (see [6], p. 18) the **formulas of changing a vector coordinates when the basis of a vector space one changes.**

The relations (1.21) can be written in the matrix form as:

$$\overline{x}_{B_1} = \mathrm{M}_{(B_1,B_2)}\cdot\overline{x}_{B_2} \tag{1.22}$$

or

$$\overline{x}_{B_2} = \mathrm{M}_{(B_1,B_2)}^{-1}\cdot\overline{x}_{B_1}. \tag{1.23}$$

**Example 1.63** (see [13]). In the arithmetic vector space $\mathbb{R}^3$ the following vectors are considered:

$$\begin{aligned}&\overline{a}_1=(2,-1,2)\,,\overline{a}_2=(1,-1,2)\,,\overline{a}_3=(0,3,2)\,,\\ &\overline{b}_1=(0,1,-1)\,,\overline{b}_2=(2,1,1)\,,\overline{b}_3=(-1,2,1)\,,\\ &\overline{x}=(-1,2,3)\,.\end{aligned}$$

(a) Prove that $B_1=\{\overline{a}_1,\overline{a}_2,\overline{a}_3\}$ is a basis of $\mathbb{R}^3$.
(b) Determine the coordinates of $\overline{x}$ relative to the basis $B_1$.
(c) Prove that $B_2=\left\{\overline{b}_1,\overline{b}_2,\overline{b}_3\right\}$ is a new basis of $\mathbb{R}^3$ and write the transition matrix from the basis $B_1$ to the basis $B_2$.
(d) Write the formulas of changing a vector coordinates when one passes from the basis $B_1$ to the basis $B_2$.

**Solution**

(a) $\dim\mathbb{R}^3=3\Rightarrow$ it is enough to prove that the vectors are linearly independent. Consider the null linear combination

$$\alpha^{(1)}\overline{a}_1 + \alpha^{(2)}\overline{a}_2 + \alpha^{(3)}\overline{a}_3 = \overline{0}.$$

The, it results that

$$\left(2\alpha^{(1)}, -\alpha^{(1)}, 2\alpha^{(1)}\right) + \left(\alpha^{(2)}, -\alpha^{(2)}, 2\alpha^{(2)}\right) + \left(0, 3\alpha^{(3)}, 2\alpha^{(3)}\right) = (0,0,0).$$

One obtains the system:

$$\begin{cases} 2\alpha^{(1)} + \alpha^{(2)} = 0 \\ -\alpha^{(1)} - \alpha^{(2)} + 3\alpha^{(3)} = 0 \\ 2\alpha^{(1)} + 2\alpha^{(2)} + 2\alpha^{(3)} = 0. \end{cases}$$

As

$$d = \begin{vmatrix} 2 & 1 & 0 \\ -1 & -1 & 3 \\ 2 & 2 & 2 \end{vmatrix} = -8 \neq 0$$

it results that the system has a unique solution

$$\alpha^{(1)} = \alpha^{(2)} = \alpha^{(3)} = 0.$$

It follows that the vectors are linearly independent.

(b) Let

$$\overline{x} = \alpha^{(1)}\overline{a}_1 + \alpha^{(2)}\overline{a}_2 + \alpha^{(3)}\overline{a}_3.$$

We have

$$(-1,2,3) = \left(2\alpha^{(1)}, -\alpha^{(1)}, 2\alpha^{(1)}\right) + \left(\alpha^{(2)}, -\alpha^{(2)}, 2\alpha^{(2)}\right) + \left(0, 3\alpha^{(3)}, 2\alpha^{(3)}\right).$$

We obtain the system:

$$\begin{cases} 2\alpha^{(1)} + \alpha^{(2)} = -1 \\ -\alpha^{(1)} - \alpha^{(2)} + 3\alpha^{(3)} = 2 \\ 2\alpha^{(1)} + 2\alpha^{(2)} + 2\alpha^{(3)} = 3 \end{cases}$$

whose solution is

$$\alpha^{(1)} = -\frac{13}{8}, \alpha^{(2)} = \frac{9}{4}, \alpha^{(3)} = \frac{7}{8}.$$

Hence

$$\overline{x}_B = \begin{pmatrix} -13/8 \\ 9/4 \\ 7/8 \end{pmatrix}.$$

(c) One proceeds similarly that $B_2$ is also a basis.

We know that

$$M_{(B_1,B_2)} = \begin{pmatrix} \alpha_1^{(1)} & \alpha_2^{(1)} & \alpha_3^{(1)} \\ \alpha_1^{(2)} & \alpha_2^{(2)} & \alpha_3^{(2)} \\ \alpha_1^{(3)} & \alpha_2^{(3)} & \alpha_3^{(3)} \end{pmatrix}.$$

We can write

$$\overline{b}_1 = \alpha_1^{(1)}\overline{a}_1 + \alpha_1^{(2)}\overline{a}_2 + \alpha_1^{(3)}\overline{a}_3 \tag{1.24}$$

$$\overline{b}_2 = \alpha_2^{(1)}\overline{a}_1 + \alpha_2^{(2)}\overline{a}_2 + \alpha_2^{(3)}\overline{a}_3 \tag{1.25}$$

$$\overline{b}_3 = \alpha_3^{(1)}\overline{a}_1 + \alpha_3^{(2)}\overline{a}_2 + \alpha_3^{(3)}\overline{a}_3. \tag{1.26}$$

From (1.24) we deduce

$$(0, 1, -1) = \alpha_1^{(1)}\,(2, -1, 2) + \alpha_1^{(2)}\,(1, -1, 2) + \alpha_1^{(3)}\,(0, 3, 2)\,. \tag{1.27}$$

From (1.27) it results the system

$$\begin{cases} 2\alpha_1^{(1)} + \alpha_1^{(2)} = 0 \\ -\alpha_1^{(1)} - \alpha_1^{(2)} + 3\alpha_1^{(3)} = 1 \\ 2\alpha_1^{(1)} + 2\alpha_1^{(2)} + 2\alpha_1^{(3)} = -1 \end{cases}$$

which has the solution

$$\alpha_1^{(1)} = 0.625, \alpha_1^{(2)} = -1.25, \alpha_1^{(3)} = 0.125.$$

From (1.25) we deduce

$$(2, 1, 1) = \alpha_2^{(1)}\,(2, -1, 2) + \alpha_2^{(2)}\,(1, -1, 2) + \alpha_2^{(3)}\,(0, 3, 2)\,. \tag{1.28}$$

From (1.28) it results the system

$$\begin{cases} 2\alpha_2^{(1)} + \alpha_2^{(2)} = 2 \\ -\alpha_2^{(1)} - \alpha_2^{(2)} + 3\alpha_2^{(3)} = 1 \\ 2\alpha_2^{(1)} + 2\alpha_2^{(2)} + 2\alpha_2^{(3)} = 1 \end{cases}$$

which has the solution

$$\alpha_2^{(1)} = 1.875, \alpha_2^{(2)} = -1.25, \alpha_2^{(3)} = 0.125.$$

From (1.26) we deduce

$$(-1, 2, 1) = \alpha_3^{(1)}\,(2, -1, 2) + \alpha_3^{(2)}\,(1, -1, 2) + \alpha_3^{(3)}\,(0, 3, 2)\,. \tag{1.29}$$

From (1.29) it results the system

$$\begin{cases} 2\alpha_3^{(1)} + \alpha_3^{(2)} = -1 \\ -\alpha_3^{(1)} - \alpha_3^{(2)} + 3\alpha_3^{(3)} = 2 \\ 2\alpha_3^{(1)} + 2\alpha_3^{(2)} + 2\alpha_3^{(3)} = 1 \end{cases}$$

which has the solution

$$\alpha_3^{(1)} = -0.875, \alpha_3^{(2)} = 0.75, \alpha_3^{(3)} = 0.625.$$

One obtains

$$M_{(B_1,B_2)} = \begin{pmatrix} 0.625 & 1.875 & -0.875 \\ -1.25 & -1.75 & 0.75 \\ 0.125 & 0.375 & 0.625 \end{pmatrix}.$$

(d) Let

$$\overline{v}_{B_1} = x^{(1)}\overline{a}_1 + x^{(2)}\overline{a}_2 + x^{(3)}\overline{a}_3$$

$$\overline{v}_{B_2} = y^{(1)}\overline{b}_1 + y^{(2)}\overline{b}_2 + y^{(3)}\overline{b}_3.$$

We shall have

$$\begin{pmatrix} x^{(1)} \\ x^{(2)} \\ x^{(3)} \end{pmatrix} = M_{(B_1,B_2)} \cdot \begin{pmatrix} y^{(1)} \\ y^{(2)} \\ y^{(3)} \end{pmatrix} \Rightarrow \begin{cases} x^{(1)} = 0.625y^{(1)} + 1.875y^{(2)} - 0.875y^{(3)} \\ x^{(2)} = -1.25y^{(1)} - 1.75y^{(2)} + 0.75y^{(3)} \\ x^{(3)} = 0.125y^{(1)} + 0.375y^{(2)} + 0.625y^{(3)}. \end{cases}$$

We shall give a solution in Sage, too:

```
sage: V=VectorSpace(RR,3)
sage: a1=V([2,-1,2]);a2=V([1,-1,2]);a3=V([0,3,2]);x=V([-1,2,3]);
sage: V.linear_dependence([a1, a2, a3])==[]
True
sage: W = V.subspace_with_basis([a1, a2,a3])
sage: W.coordinate_vector(x)
(-1.62500000000000, 2.25000000000000, 0.875000000000000)
sage: b1=V([0,1,-1]);b2=V([2,1,1]);b3=V([-1,2,1])
sage: V.linear_dependence([b1, b2, b3])==[]
True
sage: v1=W.coordinate_vector(b1);v2=W.coordinate_vector(b2);v3=W.coordinate_vector(b3);
sage: A = column_matrix([v1, v2, v3]).n(digits=3)
sage: A
[ 0.625    1.88 -0.875]
[ -1.25   -1.75  0.750]
[ 0.125  0.375  0.625]
sage: x1, x2, x3,y1,y2,y3 = var('x1, x2, x3,y1,y2,y3')
sage: yy = column_matrix([y1, y2, y3]);
sage: xx=A*yy
sage: xx
[ 0.625*y1 + 1.88*y2 - 0.875*y3]
[ -1.25*y1 - 1.75*y2 + 0.750*y3]
[0.125*y1 + 0.375*y2 + 0.625*y3]
```

**Example 1.64.** In the vector space $M_{2,2}(\mathbb{R})$ the following matrices are considered:

$$C_1 = \begin{pmatrix} 1 & 0 \\ 0 & 0 \end{pmatrix}, C_2 = \begin{pmatrix} 1 & -1 \\ 0 & 0 \end{pmatrix}, C_3 = \begin{pmatrix} 1 & -1 \\ 1 & 0 \end{pmatrix}, C_4 = \begin{pmatrix} 1 & -1 \\ 1 & -1 \end{pmatrix},$$
$$A_1 = \begin{pmatrix} 0 & 0 \\ 0 & 1 \end{pmatrix}, A_2 = \begin{pmatrix} 0 & 0 \\ -1 & 1 \end{pmatrix}, A_3 = \begin{pmatrix} 0 & 1 \\ -1 & 1 \end{pmatrix}, A_4 = \begin{pmatrix} -1 & -1 \\ -1 & 1 \end{pmatrix}.$$

One requires:

(a) Prove that both $B_1 = \{C_1, C_2, C_3, C_4\}$ and $B_2 = \{A_1, A_2, A_3, A_4\}$ are bases for $M_{2,2}(\mathbb{R})$.
(b) Find the transition matrix from the basis $B_1$ to the basis $B_2$.
(c) Write the formulas of changing a vector coordinates when one passes from the basis $B_1$ to the basis $B_2$.

**Solution**

(a) Let

$$\alpha^{(1)}A_1 + \alpha^{(2)}A_2 + \alpha^{(3)}A_3 = O_2.$$

We have

$$\alpha^{(1)}\begin{pmatrix} 0 & 0 \\ 0 & 1 \end{pmatrix} + \alpha^{(2)}\begin{pmatrix} 0 & 0 \\ -1 & 1 \end{pmatrix} + \alpha^{(3)}\begin{pmatrix} 0 & 1 \\ -1 & 1 \end{pmatrix} + \alpha^{(4)}\begin{pmatrix} -1 & -1 \\ -1 & 1 \end{pmatrix} = \begin{pmatrix} 0 & 0 \\ 0 & 0 \end{pmatrix} \Leftrightarrow$$

$$\begin{pmatrix} 0 & 0 \\ 0 & \alpha^{(1)} \end{pmatrix} + \begin{pmatrix} 0 & 0 \\ -\alpha^{(2)} & \alpha^{(2)} \end{pmatrix} + \begin{pmatrix} 0 & \alpha^{(3)} \\ -\alpha^{(3)} & \alpha^{(3)} \end{pmatrix} + \begin{pmatrix} -\alpha^{(4)} & -\alpha^{(4)} \\ -\alpha^{(4)} & \alpha^{(4)} \end{pmatrix} = \begin{pmatrix} 0 & 0 \\ 0 & 0 \end{pmatrix}.$$

It results that $-\alpha^{(4)} = 0$; therefore $\alpha^{(4)} = 0$.
We obtain the system:

$$\begin{cases} \alpha^{(3)} - \alpha^{(4)} = 0 \Rightarrow \alpha^{(3)} = \alpha^{(4)} \\ -\alpha^{(2)} - \alpha^{(3)} - \alpha^{(4)} = 0 \Rightarrow \alpha^{(2)} = 0 \\ \alpha^{(1)} + \alpha^{(2)} + \alpha^{(3)} + \alpha^{(4)} = 0 \Rightarrow \alpha^{(1)} = 0. \end{cases}$$

Hence: $\alpha^{(1)} = \alpha^{(2)} = \alpha^{(3)} = \alpha^{(4)} = 0 \Rightarrow B_2$ linearly independent.

$$\left.\begin{array}{l} \dim M_{2,2}(\mathbb{R}) = 4 \\ B_2 \text{has 4 elements} \end{array}\right\} \overset{\text{Proposition 1.56}}{\Rightarrow} B_2 \text{ basis.}$$

Similarly, one proves that $B_1$ is a basis, too.

(b) Using (1.19) we have

$$M_{(B_1,B_2)} = \begin{pmatrix} \alpha_1^{(1)} & \alpha_2^{(1)} & \alpha_3^{(1)} & \alpha_4^{(1)} \\ \alpha_1^{(2)} & \alpha_2^{(2)} & \alpha_3^{(2)} & \alpha_4^{(2)} \\ \alpha_1^{(3)} & \alpha_2^{(3)} & \alpha_3^{(3)} & \alpha_4^{(3)} \\ \alpha_1^{(4)} & \alpha_2^{(4)} & \alpha_3^{(4)} & \alpha_4^{(4)} \end{pmatrix}.$$

On the basis of the formulas from (1.18) we obtain:

$$A_1 = \alpha_1^{(1)} C_1 + \alpha_1^{(2)} C_2 + \alpha_1^{(3)} C_3 + \alpha_1^{(4)} C_4 \tag{1.30}$$

$$A_2 = \alpha_2^{(1)} C_1 + \alpha_2^{(2)} C_2 + \alpha_2^{(3)} C_3 + \alpha_2^{(4)} C_4 \tag{1.31}$$

$$A_3 = \alpha_3^{(1)} C_1 + \alpha_3^{(2)} C_2 + \alpha_3^{(3)} C_3 + \alpha_3^{(4)} C_4 \tag{1.32}$$

$$A_4 = \alpha_4^{(1)} C_1 + \alpha_4^{(2)} C_2 + \alpha_4^{(3)} C_3 + \alpha_4^{(4)} C_4. \tag{1.33}$$

From (1.30) we deduce

$$\begin{pmatrix} 0 & 0 \\ 0 & 1 \end{pmatrix} = \alpha_1^{(1)} \begin{pmatrix} 1 & 0 \\ 0 & 0 \end{pmatrix} + \alpha_1^{(2)} \begin{pmatrix} 1 & -1 \\ 0 & 0 \end{pmatrix} + \alpha_1^{(3)} \begin{pmatrix} 1 & -1 \\ 1 & 0 \end{pmatrix} + \alpha_1^{(4)} \begin{pmatrix} 1 & -1 \\ 1 & -1 \end{pmatrix}. \tag{1.34}$$

From (1.34) it results the system:

$$\begin{cases} \alpha_1^{(1)} + \alpha_1^{(2)} + \alpha_1^{(3)} + \alpha_1^{(4)} = 0 \Rightarrow \alpha_1^{(1)} = 0 \\ -\alpha_1^{(2)} - \alpha_1^{(3)} - \alpha_1^{(4)} = 0 \Rightarrow \alpha_1^{(2)} = 0 \\ \alpha_1^{(3)} + \alpha_1^{(4)} = 0 \Rightarrow \alpha_1^{(3)} = 1 \\ -\alpha_1^{(4)} = 0 \Rightarrow \alpha_1^{(4)} = -1. \end{cases}$$

Finally, one obtains

$$M_{(B_1,B_2)} = \begin{pmatrix} 0 & 0 & 1 & -2 \\ 0 & 1 & 0 & 2 \\ 1 & 0 & 0 & 0 \\ -1 & -1 & -1 & -1 \end{pmatrix}.$$

(c) Let

$$\overline{x}_{B_1} = x^{(1)} C_1 + x^{(2)} C_2 + x^{(3)} C_3 + x^{(4)} C_4$$

$$\overline{x}_{B_2} = y^{(1)} A_1 + y^{(2)} A_2 + y^{(3)} A_3 + y^{(4)} A_4.$$

Substituting these relations into (1.22) we shall have

$$\begin{pmatrix} x^{(1)} \\ x^{(2)} \\ x^{(3)} \\ x^{(4)} \end{pmatrix} = M_{(B_1,B_2)} \cdot \begin{pmatrix} y^{(1)} \\ y^{(2)} \\ y^{(3)} \\ y^{(4)} \end{pmatrix} \Rightarrow \begin{cases} x^{(1)} = y^{(3)} - 2y^{(4)} \\ x^{(2)} = y^{(2)} + 2y^{(4)} \\ x^{(3)} = y^{(1)} \\ x^{(4)} = -y^{(1)} - y^{(2)} - y^{(3)} - y^{(4)}. \end{cases}$$

The solution using Sage is:

```
sage: M = MatrixSpace(RR,2)
sage: C1=M([[1,0],[0,0]]);C2=M([[1,-1],[0,0]]);C3=M([[1,-1],[1,0]]);C4=M([[1,-1],[1,-1]]);
sage: A1=M([[0,0],[0,1]]);A2=M([[0,0],[-1,1]]);A3=M([[0,1],[-1,1]]);A4=M([[-1,-1],[-1,1]]);
sage: var('al1','al2','al3','al4')
(al1, al2, al3, al4)
sage: O=matrix(2)
sage: U=al1*A1+al2*A2+al3*A3+al4*A4; V=al1*C1+al2*C2+al3*C3+al4*C4;
sage: solve([U[0][0]==O[0][0],U[0][1]==O[0][1],U[1][0]==O[1][0],U[1][1]==O[1][1]],al1,al2,al3,al4)
[[al1 == 0, al2 == 0, al3 == 0, al4 == 0]]
sage: solve([V[0][0]==A1[0][0],V[0][1]==A1[0][1],V[1][0]==A1[1][0],V[1][1]==A1[1][1]],al1,al2,al3,al4)
[[al1 == 0, al2 == 0, al3 == 1, al4 == -1]]
sage: solve([V[0][0]==A2[0][0],V[0][1]==A2[0][1],V[1][0]==A2[1][0],V[1][1]==A2[1][1]],al1,al2,al3,al4)
[[al1 == 0, al2 == 1, al3 == 0, al4 == -1]]
sage: solve([V[0][0]==A3[0][0],V[0][1]==A3[0][1],V[1][0]==A3[1][0],V[1][1]==A3[1][1]],al1,al2,al3,al4)
[[al1 == 1, al2 == 0, al3 == 0, al4 == -1]]
sage: solve([V[0][0]==A4[0][0],V[0][1]==A4[0][1],V[1][0]==A4[1][0],V[1][1]==A4[1][1]],al1,al2,al3,al4)
[[al1 == -2, al2 == 2, al3 == 0, al4 == -1]]
sage: v0=vector([0,0,1,-1]);v1=vector([0,1,0,-1]);
sage: v2=vector([1,0,0,-1]);v3=vector([-2,2,0,-1]);
sage: A = column_matrix([v0,v1, v2, v3])
sage: A
[ 0  0  1 -2]
[ 0  1  0  2]
[ 1  0  0  0]
[-1 -1 -1 -1]
sage: var('y1','y2','y3','y4')
(y1, y2, y3, y4)
sage: y = column_matrix([y1, y2, y3,y4]);
sage: xx=A*y
sage: xx
[          y3 - 2*y4]
[          y2 + 2*y4]
[                 y1]
[-y1 - y2 - y3 - y4]
```

## 1.4 Vector Subspaces

**Definition 1.65** (see [14], p. 8). Let $V$ be a vector space over the $n$ -dimensional $K$ field. The nonempty subset $W$ of $V$ it's called a **vector subspace** of $V$ if the following conditions are satisfied:

(1) $\overline{x}+\overline{y}\in W\,(\forall)\,\overline{x},\overline{y}\in W$
(2) $\alpha\overline{x}\in W\,(\forall)\,\alpha\in K,\ (\forall)\,\overline{x}\in W.$

**Proposition 1.66** (see [14], p. 8). Let $V$ be a vector space over the $n$-dimensional $K$ field. The nonempty subset $W$ of $V$ it's called a vector subspace of $V$ if and only if

$$\alpha\overline{x}+\beta\overline{y}\in W,\ (\forall)\alpha,\beta\in K,\ (\forall)\overline{x},\overline{y}\in W.$$

**Examples of vector subspaces**

(1) $W_1=\left\{\left(x^{(1)},x^{(2)},\ldots,x^{(n)}\right)\in\mathbb{R}^n|x^{(1)}+x^{(2)}+\ldots+x^{(n)}=0\right\}$ is a subspace vector of $\mathbb{R}^n$.
(2) $\mathbf{M}_n^s(\mathbb{R})=\left\{A\in\mathbf{M}_n(\mathbb{R})|A^t=A\right\}$ the *vector subspace of the symmetric matrices* is a vector subspace of $\mathbf{M}_n(\mathbb{R})$.

We shall prove that the subset $\mathbf{M}_n^s(\mathbb{R})$ is a vector subspace of $\mathbf{M}_n(\mathbb{R})$ and we shall determine its dimension, highlighting a basis of it.

Let $\alpha, \beta \in \mathbb{R}$ and $A, B \in \mathbf{M}_n^s(\mathbb{R})$. It results that $A^t = A$ and $B^t = B$.
We shall have

$$(\alpha A + \beta B)^t = \alpha A^t + \beta B^t = \alpha A + \beta B \Rightarrow \alpha A + \beta B \in \mathbf{M}_n^s(\mathbb{R}),$$

namely $\mathbf{M}_n^s(\mathbb{R})$ is a vector subspace of $\mathbf{M}_n(\mathbb{R})$.

$$A \in \mathbf{M}_n^s(\mathbb{R}) \Rightarrow A = \begin{pmatrix} a_{11} & a_{12} & \dots & a_{1n} \\ a_{12} & a_{22} & \dots & a_{2n} \\ \vdots & \vdots & \vdots & \vdots \\ a_{1n} & a_{2n} & \dots & a_{nn} \end{pmatrix}.$$

We can write

$$A = a_{11} \underbrace{\begin{pmatrix} 1 & 0 & \dots & 0 & 0 \\ 0 & 0 & \dots & 0 & 0 \\ \dots & \dots & \dots & \dots & \dots \\ 0 & 0 & \dots & 0 & 0 \end{pmatrix}}_{=F_1} + a_{12} \underbrace{\begin{pmatrix} 0 & 1 & \dots & 0 & 0 \\ 1 & 0 & \dots & 0 & 0 \\ \dots & \dots & \dots & \dots & \dots \\ 0 & 0 & \dots & 0 & 0 \end{pmatrix}}_{=F_2}$$

$$+ \dots + a_{1n} \underbrace{\begin{pmatrix} 0 & 0 & \dots & 0 & 1 \\ 0 & 0 & \dots & 0 & 0 \\ \dots & \dots & \dots & \dots & \dots \\ 1 & 0 & \dots & 0 & 0 \end{pmatrix}}_{=F_n} + a_{22} \underbrace{\begin{pmatrix} 0 & 0 & \dots & 0 & 0 \\ 0 & 1 & \dots & 0 & 0 \\ \dots & \dots & \dots & \dots & \dots \\ 0 & 0 & \dots & 0 & 0 \end{pmatrix}}_{=F_{n+1}}$$

$$+ \dots + a_{nn} \underbrace{\begin{pmatrix} 0 & 0 & \dots & 0 & 0 \\ 0 & 0 & \dots & 0 & 0 \\ \dots & \dots & \dots & \dots & \dots \\ 0 & 0 & \dots & 0 & 1 \end{pmatrix}}_{=F_{n(n+1)/2}}.$$

Using this notation, we have

$$A = a_{11}F_1 + a_{12}F_2 + \dots + a_{1n}F_n + a_{22}F_{n+1} + \dots + a_{nn}F_{n(n+1)/2}$$

namely B $= \left\{F_i, i = \overline{1, n(n+1)/2}\right\}$ is a system of generators.
We consider the linear combination of the null matrices from $B$

$$\begin{pmatrix} \alpha_{11} \dots \alpha_{1n} \\ \vdots \quad\quad \vdots \\ \alpha_{1n} \dots \alpha_{nn} \end{pmatrix} = \mathrm{O} \Leftrightarrow \alpha_{ij} = \alpha_{ji} = 0, (\forall)\, i = \overline{1, n}, j = \overline{1, n};$$

therefore B $= \left\{F_i, i = \overline{1, n(n+1)/2}\right\}$ is linearly independent.
It results that B is a basis in $\mathbf{M}_n^s(\mathbb{R})$; hence

$$\dim\ \mathbf{M}_n^s(\mathbb{R}) = \frac{n(n+1)}{2}.$$

We can check that in Sage, too:

```
sage: var("a00 a01 a02 a10 a11 a12 a20 a21 a22 b00 b01 b02 b11 b12 b22 al be")
(a00, a01, a02, a10, a11, a12, a20, a21, a22, b00, b01, b02, b11, b12, b22, al, be)
sage: M=matrix(SR,[[a00,a01,a02],[a10,a11,a12],[a20,a21,a22]])
sage: M1=M.transpose()
sage: s=solve([M1[1][0]==M[1][0],M1[2][0]==M[2][0],M1[2][1]==M[2][1]],a10,a20,a21)
sage: A=M1.subs(a10=s[0][0].right(),a20=s[0][1].right(),a21=s[0][2].right())
sage: B=A.subs(a00=b00,a01=b01,a02=b02,a11=b11,a12=b12,a22=b22)
sage: (al*A+be*B).transpose()==(al*A).transpose()+(be*B).transpose()
True
sage: F1=A.subs(a00=1,a01=0,a02=0,a11=0,a12=0,a22=0)
sage: F2=A.subs(a00=0,a01=1,a02=0,a11=0,a12=0,a22=0)
sage: F3=A.subs(a00=0,a01=0,a02=1,a11=0,a12=0,a22=0)
sage: F4=A.subs(a00=0,a01=0,a02=0,a11=1,a12=0,a22=0)
sage: F5=A.subs(a00=0,a01=0,a02=0,a11=0,a12=1,a22=0)
sage: F6=A.subs(a00=0,a01=0,a02=0,a11=0,a12=0,a22=1)
sage: Z=a00*F1+a01*F2+a02*F3+a11*F4+a12*F5+a22*F6
sage: Z==A
True
sage: solve([Z[0][0],Z[0][1],Z[0][2],Z[1][1],Z[1][2],Z[2][2]],a00,a01,a02,a11,a12,a22)
[[a00 == 0, a01 == 0, a02 == 0, a11 == 0, a12 == 0, a22 == 0]]
```

(3) $\mathbf{M}_n^a(\mathbb{R}) = \left\{A \in \mathbf{M}_n(\mathbb{R}) | A^t = -A\right\}$ the *vector subspace of the antisymmetric matrices* is a vector subspace of $\mathbf{M}_n(\mathbb{R})$.

We shall prove that the subset $\mathbf{M}_n^a(\mathbb{R})$ is a vector subspace of $\mathbf{M}_n(\mathbb{R})$ and we shall determine its dimension, highlighting a basis of its.

Let $\alpha, \beta \in \mathbb{R}$ and $A, B \in \mathbf{M}_n^a(\mathbb{R})$. It results that $A^t = -A$ and $B^t = -B$.

We shall have

$$(\alpha A + \beta B)^t = \alpha A^t + \beta B^t = -\alpha A - \beta B \Rightarrow \alpha A + \beta B \in \mathbf{M}_n^a(\mathbb{R}),$$

i.e. $\mathbf{M}_n^a(\mathbb{R})$ is a vector subspace of $\mathbf{M}_n(\mathbb{R})$.

$$A \in \mathbf{M}_n^a(\mathbb{R}) \Rightarrow A = \begin{pmatrix} a_{11} & a_{12} & \dots & a_{1n} \\ -a_{12} & a_{22} & \dots & a_{2n} \\ \vdots & \vdots & \vdots & \vdots \\ -a_{1n} & -a_{2n} & \dots & a_{nn} \end{pmatrix}.$$

We can write

$$A = a_{12}\underbrace{\begin{pmatrix} 0 & 1 & \dots & 0 & 0 \\ -1 & 0 & \dots & 0 & 0 \\ \dots & \dots & \dots & \dots & \dots \\ 0 & 0 & \dots & 0 & 0 \end{pmatrix}}_{=G_1} + \dots + a_{1n}\underbrace{\begin{pmatrix} 0 & 0 & \dots & 0 & 1 \\ 0 & 0 & \dots & 0 & 0 \\ \dots & \dots & \dots & \dots & \dots \\ 1 & 0 & \dots & 0 & 0 \end{pmatrix}}_{=G_{n-1}}$$

$$+ \dots + a_{nn}\underbrace{\begin{pmatrix} 0 & 0 & \dots & 0 & 0 \\ 0 & 0 & \dots & 0 & 0 \\ \dots & \dots & \dots & \dots & \dots \\ 0 & 0 & \dots & 0 & 1 \end{pmatrix}}_{=G_{n(n-1)/2}}.$$

Using this notation, we have

$$A = a_{12}G_1 + \dots + a_{1n}G_{n-1} + \dots + a_{nn}G_{n(n-1)/2}$$

i.e. B= $\left\{G_i, i = \overline{1, n(n-1)/2}\right\}$ is a system of generators.

We consider the linear combination of the null matrices from B:

$$\begin{pmatrix} 0 & \alpha_{12} & \dots & \alpha_{1n} \\ -\alpha_{12} & 0 & \dots & \alpha_{2n} \\ \vdots & \vdots & \vdots & \vdots \\ -\alpha_{1n} & -\alpha_{2n} & \dots & 0 \end{pmatrix} = \mathrm{O} \Leftrightarrow \alpha_{ij} = 0, (\forall)\, i = \overline{1, n}, j = \overline{1, n}, i < j;$$

therefore B= $\left\{G_i, i = \overline{1, n(n-1)/2}\right\}$ is linearly independent.

It results that B is a basis in $\mathbf{M}_n^a(\mathbb{R})$; hence

$$\dim\ \mathbf{M}_n^a(\mathbb{R}) = \frac{n(n-1)}{2}.$$

Using Sage we shall have:

```
sage: var("a00 a01 a02 a10 a11 a12 a20 a21 a22 b00 b01 b02 b11 b12 b22 al be")
(a00, a01, a02, a10, a11, a12, a20, a21, a22, b00, b01, b02, b11, b12, b22, al, be)
sage: M=matrix(SR,[[a00,a01,a02],[a10,a11,a12],[a20,a21,a22]])
sage: M1=M.transpose()
sage: s=solve([M1[0][0]==-M[0][0],M1[1][0]==-M[1][0],M1[1][1]==-M[1][1],M1[2][0]==-M[2][0],
M1[2][1]==-M[2][1],M1[2][2]==-M[2][2]],a00,a10,a11,a20,a21,a22)
sage: s1=s[0][0].right();s2=s[0][1].right();s3=s[0][2].right();s4=s[0][3].right()
sage: s5=s[0][4].right();s6=s[0][5].right()
sage: A=M.subs(a00=s1,a10=s2,a11=s3,a20=s4,a21=s5,a22=s6);A
[ 0 a01 a02]
[-a01   0 a12]
[-a02 -a12   0]
```

```
sage: B=A.subs(a00=b00,a01=b01,a02=b02,a11=b11,a12=b12,a22=b22)
sage: (al*A+be*B).transpose()==(al*A).transpose()+(be*B).transpose()
True
sage: G1=A.subs(a01=1,a02=0,a11=0,a12=0,a22=0)
sage: G2=A.subs(a01=0,a02=1,a11=0,a12=0,a22=0)
sage: G3=A.subs(a01=0,a02=0,a11=1,a12=0,a22=0)
sage: G4=A.subs(a01=0,a02=0,a11=0,a12=1,a22=0)
sage: G5=A.subs(a01=0,a02=0,a11=0,a12=0,a22=1)
sage: Z=a01*G1+a02*G2+a11*G3+a12*G4+a22*G5
sage: Z==A
True
sage: solve([Z[0][1],Z[0][2],Z[1][2],Z[0][2]],a01,a02,a12)
[[a01 == 0, a02 == 0, a12 == 0]]
```

**Proposition 1.67** (see [6], p. 20). The set of the solutions corresponding to a linear and homogeneous system with $m$ equations and $n$ unknowns is a vector subspace of $\mathbb{R}^n$, having the dimension $n-r$ , $r$ being the rank of the associated matrix system $A$.

**Example 1.68** (see [13]). In the vector space $\mathbb{R}^5$ we consider

$$W = \left\{\overline{x} = (x_1, x_2, x_3, x_4, x_5) \in \mathbb{R}^5\right\}$$

which checks the system

$$\begin{cases} x_1 + x_2 + x_3 + x_4 + x_5 = 0 \\ x_2 - x_3 + x_4 + 2x_5 = 0 \\ x_1 + 2x_2 = 0. \end{cases}$$

Determine $\dim W$ and one of its basis.

**Solution**

Within the Proposition 1.67 we have $\dim W = n - r = 5 - 3 = 2$, the associated matrix of the system being

$$A = \begin{pmatrix} 1 & 1 & 1 & -1 & 1 \\ 0 & 1 & -1 & 1 & 2 \\ 1 & 2 & 0 & 0 & 0 \end{pmatrix}$$

and

$$\Delta = \left| \begin{pmatrix} 1 & 1 & 1 \\ 0 & 1 & 2 \\ 1 & 2 & 0 \end{pmatrix} \right| = -3 \neq 0.$$

It results the system

$$\begin{cases} x_1 + x_2 + x_5 = -x_3 + x_4 \\ x_2 + 2x_5 = x_3 - x_4 \\ x_1 + 2x_2 = 0 \Rightarrow x_1 = -2x_2. \end{cases}$$

We denote

$$x_3 = t, x_4 = u, t, u \in \mathbb{R};$$

we obtain

$$\begin{cases} -x_2 + x_5 = -t + u \\ x_2 + 2x_5 = t - u, \end{cases}$$

so we deduce that $x_5 = 0$.

Finally, we get the solution of the system

$$\begin{cases} x_1 = -2t + 2u \\ x_2 = t - u \\ x_3 = t \\ x_4 = u \\ x_5 = 0 \end{cases}, (\forall)\, t, u \in \mathbb{R}.$$

For

- $x_3 = 0$ and $x_4 = 1$ we obtain $\overline{a}_1 = (2, -1, 0, 1, 0)$;
- $x_3 = 1$ and $x_4 = 0$ we obtain $\overline{a}_2 = (-2, 1, 1, 0, 0)$

and $\mathrm{B}_1 = \{\overline{a}_1, \overline{a}_2\}$ is a basis of $W$.

We shall give the solution in Sage, too:

```
sage: var('x1','x2','x3','x4','x5')
(x1, x2, x3, x4, x5)
sage: solve([x1+x2+x3+x4+x5==0,x2-x3+x4+2*x5==0,x1+2*x2==0],x1,x2,x3,x4,x5)
[[x1 == r1 - 2*r2, x2 == -1/2*r1 + r2, x3 == r2, x4 == -3/2*r1, x5 == r1]]
sage: var ('r1','r2')
(r1, r2)
sage: x1 = r1 - 2*r2;x2= -1/2*r1 + r2; x3 = r2; x4 = -3/2*r1; x5 =r1
sage: xx1=x1.substitute(r1=0,r2=1);xx2=x2.substitute(r1=0,r2=1);
sage: xx3=x3.substitute(r1=0,r2=1);xx4=x4.substitute(r1=0,r2=1);xx5=x5.substitute(r1=0,r2=1);
sage: y1=x1.substitute(r1=1,r2=0);y2=x2.substitute(r1=1,r2=0);
sage: y3=x3.substitute(r1=1,r2=0);y4=x4.substitute(r1=1,r2=0);y5=x5.substitute(r1=1,r2=0);
sage: V = VectorSpace(RR,5)
sage: v1=V([y1,y2,y3,y4,y5]).n(digits=2);
sage: v2=V([xx1,xx2,xx3,xx4,xx5]).n(digits=2);
sage: S = V.subspace([V([xx1,xx2,xx3,xx4,xx5]),V([y1,y2,y3,y4,y5])])
sage: S.linear_dependence([v1,v2])==[]
True
```

**Proposition 1.69** (see [14], p. 8 and [6], p. 22). Let $V$ be a vector space over the field $K$ and $W_1$ $W_2$ be two of its vector subspaces. Then $W_1 \cap W_2$ is a vector subspace of $V$.

**Definition 1.70** (see [14], p. 8 and [6], p. 22). Let $V$ be a vector space over the field $K$ and $W_1$ $W_2$ be two of its vector subspaces. The **sum of two vector subspaces** $W_1$ and $W_2$ is defined as:

$$W_1 + W_2 = \{\overline{x}_1 + \overline{x}_2 | \overline{x}_1 \in W_1 \text{ and } \overline{x}_2 \in W_2\}.$$

**Proposition 1.71** (see [14], p. 8 and [6], p. 22). Let $V$ be a vector space over the field $K$ and $W_1$ $W_2$ be two of its vector subspaces. Then $W_1 + W_2$ is a vector subspace of $V$.

**Theorem 1.72 (Grassmann- the dimension formula,** see [14], p. 8 and [6], p. 22). Let $V$ be a finite $n$ dimensional vector space over the field $K$, and $W_1$ $W_2$ be two of its vector subspaces. Then there is the relation

$$\dim(W_1 + W_2) = \dim W_1 + \dim W_2 - \dim(W_1 \cap W_2).$$

**Definition 1.73** (see [14], p. 8 and [6], p. 23). Let $V$ be a $n$ dimensional finite vector space over the field $K$, and $W_1$ $W_2$ be two of its vector subspaces. The sum of the vector subspaces $W_1$ and $W_2$ is called the **direct sum,** denoted by $W_1 \oplus W_2$ if any vector from $W_1 + W_2$ can be uniquely written in the form:

$$\overline{x} = \overline{x}_1 + \overline{x}_2, \overline{x}_1 \in W_1, \overline{x}_2 \in W_2.$$

**Proposition 1.74** (see [14], p. 9 and [6], p. 23). Let $V$ be a $n$ dimensional finite vector space over the field $K$ and $W_1$ $W_2$ be two of its vector subspaces. Then $V = W_1 \oplus W_2$ if and only if the following conditions are satisfied:

(1) $W_1 \cap W_2 = \{\overline{0}\}$
(2) $\dim W_1 + \dim W_2 = \dim V$.

**Proposition 1.75** (see [6], p. 21). Let $V$ be a $n$ dimensional finite vector space over the field $K$. If $W$ is a vector space of $V$, then the dimension of $W$ is finite and $\dim W \leq \dim V$.

**Proposition 1.76** (see [6], p. 22). Let $V$ be a $n$ dimensional finite vector space over the field $K$. If $W$ is a vector space of $V$, then $\dim W = \dim V$ if and only if $W = V$.

**Example 1.77** (see [13]). In the vector space $R_3[X]$ we consider

$$W_1 = \{P \in R_3[X] | P(1) = 0\}, W_2 = \{P \in R_3[X] | \text{degree } P = 0\}.$$

(a) Show that $W_1$ and $W_2$ are two vector subspaces of $R_3[X]$.
(b) Determine the basis and dimension of each of these subspaces.
(c) Prove that $R_3[X] = W_1 \oplus W_2$.

**Solution**

(a) Let be $\alpha, \beta \in K$ and $P, Q \in W_1$. It results that $P(1) = 0$ and $Q(1) = 0$.

We shall have

$$(\alpha P + \beta Q)(1) = \alpha \underbrace{P(1)}_{=0} + \beta \underbrace{Q(1)}_{=0} = 0 \Rightarrow \alpha P + \beta Q \in W_1;$$

hence $W_1$ is a vector subspace of $\mathrm{R}_3[X]$.

Let be $\alpha, \beta \in K$ and $P, Q \in W_2$. It results that degree $P = 0$ and degree $Q = 0$. We shall have:

$$0 \leq \text{degree}(\alpha P + \beta Q) \leq \max\{\text{degree}(\alpha P), \text{degree}(\beta Q)\}$$
$$= \max\{\alpha \cdot \text{degree}(P), \beta \cdot \text{degree}(Q)\} = 0 \Rightarrow \text{degree}(\alpha P + \beta Q) = 0,$$

i.e. $\alpha P + \beta Q \in W_2$, therefore $W_2$ is a vector subspace of $\mathrm{R}_3[X]$.

(b) Let be $P \in W_1$. It results that $P(1) = 0$; we deduce

$$X - 1 | P \Rightarrow P = (X - 1)Q, \text{degree}(Q) \leq 2 \Rightarrow P = (X - 1)\left(a_0 + a_1 X + a_2 X^2\right)$$
$$\Rightarrow P = a_0 \underbrace{(X - 1)}_{=P_1} + a_1 \underbrace{X(X - 1)}_{=P_2} + a_2 \underbrace{X^2(X - 1)}_{=P_3} = a_0 P_1 + a_1 P_2 + a_2 P_3;$$

hence $\mathrm{B}_1 = \{P_1, P_2, P_3\}$ is a system of generators for $W_1$.

Let

$$\alpha_1 P_1 + \alpha_2 P_2 + \alpha_3 P_3 = 0 \Rightarrow \alpha_1 (X - 1) + \alpha_2 X (X - 1) + \alpha_3 X^2 (X - 1) = 0.$$

We obtain the system:

$$\begin{cases} -\alpha_1 = 0 \Rightarrow \alpha_1 = 0 \\ \alpha_1 - \alpha_2 = 0 \Rightarrow \alpha_2 = 0 \\ \alpha_2 - \alpha_3 = 0 \Rightarrow \alpha_3 = 0 \\ \alpha_3 = 0 \end{cases} \Rightarrow$$

$\mathrm{B}_1$ is a linearly independent system.

Therefore $\mathrm{B}_1$ is a basis of $W_1$ and dim $W_1 = 3$.

A solution in Sage will be given, too:

```
sage: R.<x>= RR['x'];var("a0,a1,a2,a3,b0,b1,b2,b3,al,be")
(a0, a1, a2, a3, b0, b1, b2, b3, al, be)
sage: P(x)=a0+a1*x+a2*x^2+a3*x^3;Q(x)=b0+b1*x+b2*x^2+b3*x^3
sage: A=(al*P+be*Q)(1).subs(a0+a1+a2+a3==0).subs(b0+b1+b2+b3==0)
sage: B=(al*P(1)+be*Q(1)).subs(a0+a1+a2+a3==0).subs(b0+b1+b2+b3==0)
sage: A-B
0
```

```
sage: Q1(x)=a0+a1*x+a2*x^2
sage: R(x)=expand((x-1)*Q1(x))
sage: expand(R(x)).collect(a0).collect(a1).collect(a2)
(x - 1)*a0 + (x^2 - x)*a1 + (x^3 - x^2)*a2
sage: r1=R(x).coeff(x^3);r2=R(x).coeff(x^2);r3=R(x).coeff(x)
sage: solve([r1,r2,r3,r4],a0,a1,a2)
[[a0 == 0, a1 == 0, a2 == 0]]
```

Let be $P \in W_2$. It results that degree $P = 0 \Rightarrow P =$ct$= t,\ t \in \mathbb{R}$; we deduce

$$P = t \cdot \underbrace{1}_{=P_1} = t \cdot P_1$$

therefore $B_2 = \{P_1\}$ is a system of generators for $W_2$.

Let

$$\alpha_0 P_1 = 0 \Rightarrow \alpha_0 = 0;$$

it results that $B_2$ is a linearly independent system.

Hence $B_2$ is a basis of $W_2$ and dim $W_2 = 1$.

(c) Within the Proposition 1.74

$$R_3[X] = W_1 \oplus W_2 \Leftrightarrow \begin{cases} W_1 \cap W_2 = \{P_0\} \\ \dim W_1 + \dim W_2 = \dim R_3[X], \end{cases}$$

where $Q_0 \in R_3[X]$ is the null polynomial.

Let $P \in W_1 \cap W_2$, namely:

- $P \in W_1$; it results $P(1) = 0$, hence

$$X - 1 | P \Rightarrow P = (X - 1)\, Q, \text{ degree } Q \leq 2$$

- $P \in W_2$; it results degree $P = 0$.

  We deduce

$$W_1 \cap W_2 = \{P_0\}.$$

  It is noticed that

$$\dim\ W_1 + \dim\ W_2 = 3 + 1 = 4 = \dim\ R_3\,[X].$$

## 1.5 Problems

1. Decompose the vector $\overline{v}_1 = \overline{i} - 3\overline{j} + 2\overline{k}$ after the directions of the vectors $\overline{a} = \overline{i} + \overline{j},\ \overline{b} = \overline{j} + \overline{k},\ \overline{c} = \overline{i} + \overline{j} + 3\overline{k}$.

**Solution**

We shall present the solution in Sage:

```
sage: v1=vector([1,-3,2]);a=vector([1,1,0]);b=vector([0,1,1]);c=vector([1,1,3])
sage: al1,al2,al3=var('al1,al2,al3')
sage: u=al1*a+al2*b+al3*c
sage: solve([u[0]==v1[0],u[1]==v1[1],u[2]==v1[2]],al1,al2,al3)
[[al1 == -1, al2 == -4, al3 == 2]]
```

2. Show that the points $A(3, -1, 1)$ $B(4, 1, 4)$ and $C(6, 0, 4)$ are the vertices of a right triangle.

**Solution**

Using Sage we shall have:

```
sage: A=vector([3,-1,1]);B=vector([4,1,4]);C=vector([6,0,4])
sage: AB=B-A
sage: AC=C-A
sage: BC=C-B
sage: AC.norm()^2==BC.norm()^2+AB.norm()^2
19 == 19
```

We need the following Sage code to represent the right triangle:

```
sage: P=polygon3d([(3,-1,1), (4,1,4), (6,0,4)])
sage: t1=text3d("A",(3.1,-1,1))
sage: t2=text3d("B",(4,1,4.1))
sage: t3=text3d("C",(6,0.1,4))
sage: show(P+t1+t2+t3)
```

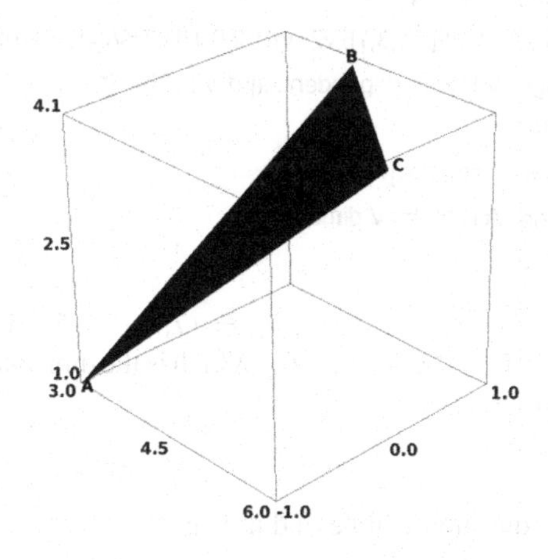

3. Calculate the height of the parallelepiped built on the vectors $\overline{v}_1, \overline{v}_2, \overline{v}_3$ taking as a basis the parallelogram built on vectors $\overline{v}_1$ and $\overline{v}_2$ knowing that

$$\overline{v}_1 = 2\overline{i} + \overline{j} - \overline{k}, \overline{v}_2 = 3\overline{i} + 2\overline{j} + \overline{k}, \overline{v}_3 = -\overline{j} + 2\overline{k}.$$

4. In the $n$ arithmetic vector space $\mathbb{R}^3$ consider the vectors:

$$\overline{a}_1 = (-1, -1, 2), \overline{a}_2 = (0, 1, -1), \overline{a}_3 = (2, 1, 1), \overline{a}_4 = (-1, -1, 7).$$

   (a) Prove that $S = \{\overline{a}_1, \overline{a}_2, \overline{a}_3, \overline{a}_4\}$ is a system of generators for $\mathbb{R}^3$.
   (b) Extract a subsystem $S'$ from $S$, that constitutes a basis of $\mathbb{R}^3$.

5. The following vectors are given: $\overline{OA} = 12\overline{i} - 4\overline{j} + 3\overline{k}$, $\overline{OB} = 3\overline{i} + 12\overline{j} - 4\overline{k}$, $\overline{OC} = 2\overline{i} + 3\overline{j} - 4\overline{k}$.

   (a) Find the lengths corresponding to the edges of $\Delta ABC$.
   (b) Prove that $\Delta AOC$ is a right triangle and $\Delta OAB$ is an isosceles triangle.
   (c) Calculate the scalar product $\overline{AB} \cdot \overline{BC}$.
   (d) Calculate the cross product $\overline{AB} \times \overline{BC}$.
   (e) Compute the area of $\Delta ABC$.
   (f) Find the length of the height $\overline{AA'}$ in $\Delta ABC$.

6. Investigate the linearly dependency of the following system of vectors: $\overline{v}_1 = (2, 1, 3, 1)$, $\overline{v}_2 = (1, 2, 0, 1)$, $\overline{v}_3 = (-1, 1, -3, 0)$ in $\mathbb{R}^4$. Does this system form a basis in $\mathbb{R}^4$?

**Solution**

The solution in Sage will be given:

```
sage: V=RR^4
sage: v1=V([2,1,3,1]);v2=V([1,2,0,1]);v3=V([-1,1,3,0])
sage: V linear_dependence([v1,v2,v3])==[]
True
sage: A=matrix([v1,v2,v3])
sage:: A rank()==V dimensioD()
False
```

7. Let be the vectors $\overline{v}_1 = (2, 4, 1, 3)$, $\overline{v}_2 = (7, 4, -9, 5)$, $\overline{v}_3 = (4, 8, -3, 7)$, $\overline{v}_4 = (5, 5, -5, 5)$, $\overline{v}_5 = (8, 4, -14, 6)$. Which is the dimension of that subspace generated by them?

**Solution**

The answer of this question will be find in Sage:

```
sage: V=VectorSpace(GF(5),4)
sage: v1=V([2,4,1,3]);v2=V([7,4,-9,5]);v3=V([4,8,-3,7]);
sage: v4=V([5,5,-5,5]);v5=V([8,4,-14,6])
sage: W=V subspace([v1,v2,v3,v4,v5])
sage: W dimension();W
3
Vector space of degree 4 and dimension 3 over Finite Field of size 5
Basis matrix:
[1 0 0 0]
[0 1 4 0]
[0 0 0 1]
```

8. Establish the transformation formulas of the coordinates when passing from the basis $B$ to the basis $B'$, if

$$B = \{\overline{u}_1 = (1, 2, -1, 0), \overline{u}_2 = (1, -1, 1, 1), \overline{u}_3 = (-1, 2, 1, 1), \overline{u}_4 = (-1, -1, 0, 1)\},$$
$$B' = \{\overline{v}_1 = (2, 1, 0, 1), \overline{v}_2 = (0, 1, 2, 2), \overline{v}_3 = (-2, 1, 1, 2), \overline{v}_4 = (1, 3, 1, 2)\}$$

are some bases in $\mathbb{R}^4$.

**Solution**

Using Sage we shall have:

```
sage: V=VectorSpace(RR,4)
sage: u1=V([1,2,-1,0]);u2=V([1,-1,1,1]);u3=V([-1,2,1,1]);u4=V([-1,-1,0,1]);
sage: W = V.subspace_with_basis([u1,u2,u3,u4])
sage: v1=V([2,1,0,1]);v2=V([0,1,2,2]);v3=V([-2,1,1,2]);v4=V([1,3,1,2]);
sage: w1=W.coordinate_vector(v1);w2=W.coordinate_vector(v2);
sage: w3=W.coordinate_vector(v3);w4=W.coordinate_vector(v4);
sage: A = column_matrix([w1,w2,w3,w4]).n(digits=3)
sage: A
[    1.00 1.11e-16 2.22e-16     1.00]
[    1.00     1.00 2.22e-16     1.00]
[5.55e-17     1.00     1.00     1.00]
[   0.000 2.22e-16     1.00    0.000]
sage: x1,x2,x3,x4,y1,y2,y3,y4 = var('x1,x2,x3,x4,y1,y2,y3,y4')
sage: yy = column_matrix([y1, y2, y3,y4]);
sage: xx=A*yy
sage: xx
[y1 + (1.11e-16)*y2 + (2.22e-16)*y3 + y4]
[              y1 + y2 + (2.22e-16)*y3 + y4]
[              (5.55e-17)*y1 + y2 + y3 + y4]
[                     (2.22e-16)*y2 + y3]
```

9. In the arithmetic vector space $\mathbb{C}^3$ the following vectors are considered:

$$\overline{a}_1 = (1, 0, 1), \overline{a}_2 = (i, 1, 0), \overline{a}_3 = (i, 2, i + 1), \overline{x} = (1, -1, i)$$
$$\overline{b}_1 = (1, 1 + i, 1 + 2i), \overline{b}_2 = (0, -1, i), \overline{b}_3 = (i + 1, i - 1, i).$$

(a) Prove that $B_1 = \{\overline{a}_1, \overline{a}_2, \overline{a}_3\}$ is a basis of $\mathbb{C}^3$.
(b) Determine the coordinates of $\overline{x}$ relative to the basis $B_1$.
(c) Prove that $B_2 = \{\overline{b}_1, \overline{b}_2, \overline{b}_3\}$ is a new basis of $\mathbb{C}^3$ and write the transition matrix from the basis $B_1$ to the basis $B_2$.
(d) Write the formulas of changing a vector coordinates when one passes from the basis $B_1$ to the basis $B_2$.

10. In the vector space $\mathbb{R}^4$ let $W = \{\overline{x} = (x_1, x_2, x_3, x_4) \in \mathbb{R}^4\}$, verifying the system

$$\begin{cases} x_1 - x_2 + 2x_3 + 3x_4 = 0 \\ x_3 - 4x_4 = 0. \end{cases}$$

Determine $\dim W$ and one of its bases.

## References

1. V. Balan, *Algebră liniară, geometrie analitică, ed* (Fair Partners, Bucureşti, 1999)
2. P. Matei, *Algebră liniară. Gometrie analitică şi diferenţială, ed* (Agir, Bucureşti, 2002)
3. I. Iatan, *Advances Lectures on Linear Algebra with Applications* (Lambert Academic Publishing, Saarbrücken, 2011)
4. C. Udrişte, *Algebră liniară, geometrie analitică* (Geometry Balkan Press, Bucureşti, 2005)
5. Gh Atanasiu, Gh Munteanu, M. Postolache, *Algebr ă liniară, geometrie analitică şi diferenţială, ecuaţii diferenţiale, ed* (ALL, Bucureşti, 1998)
6. I. Vladimirescu, M. Popescu, *Algebră liniară şi geometrie n- dimensională, ed* (Radical, Craiova, 1996)
7. P. Matei, *Algebră liniară şi geometrie analitică. Culegere de probleme, ed.* (MatrixRom, Bucureşti, 2007)
8. V. Postelnicu, S. Coatu, *Mică enciclopedie matematică, ed* (Tehnică, Bucureşti, 1980)
9. S. Chiriţă, *Probleme de matematici superioare, ed* (Didactică şi Pedagogică, Bucureşti, 1989)
10. T. Didenco, *Geometrie analitică şi diferenţială* (Academia Militară, Bucureşti, 1977)
11. L. Constantinescu, C. Petrişor, *Geometrie şi trigonometrie, ed* (Didactică şi pedagogică, Bucureşti, 1975)
12. W.K. Nicholson, *Linear Algebra and with Applications* (PWS Publishing Company, Boston, 1995)
13. I. Vladimirescu, M. Popescu, M. Sterpu, *Algebră liniară şi geometrie analitică* (Universitatea din Craiova, Note de curs şi aplica ţii, 1993)
14. I. Vladimirescu, M. Popescu, *Algebră liniară şi geometrie analitică, ed* (Universitaria, Craiova, 1993)

# Chapter 2
# Plane and Straight Line in $E_3$

## 2.1 Equations of a Straight Line in $E_3$

**Definition 2.1** (see [2], p. 448). A **vector** is a translation of the three-dimensional space; therefore it must be studied the basics of the three-dimensional Euclidean geometry: the points, the straight lines and the planes.

Let $R = \{O; \overline{i}, \overline{j}, \overline{k}\}$ be a Cartesian reference. For $M \in E_3$, the coordinates of the point $M$ are the coordinates of the position vector $\overline{OM}$. If

$$\overline{OM} = x\overline{i} + y\overline{j} + z\overline{k}$$

then $M(x, y, z)$.

### *2.1.1 Straight Line Determined by a Point and a Nonzero Vector*

A straight line from $E_3$ can be determined by (see [6]):

(1) a point and a nonzero free vector;
(2) two distinct points;
(3) the intersection of the two planes.

Let be $M_0 \in E_3$, $M_0(x_0, y_0, z_0)$ and $\overline{v} \in V_3 \backslash \{\overline{0}\}$, $\overline{v} = a\overline{i} + b\overline{j} + c\overline{k}$. We intend to find the equation of the straight line determined by the point $M_0$ and by the nonzero vector $\overline{v}$, denoted by $d = (M_0, \overline{v})$ and represented in Fig. 2.1.

G. A. Anastassiou and I. F. Iatan, *Intelligent Routines II*,
Intelligent Systems Reference Library 58, DOI: 10.1007/978-3-319-01967-3_2,

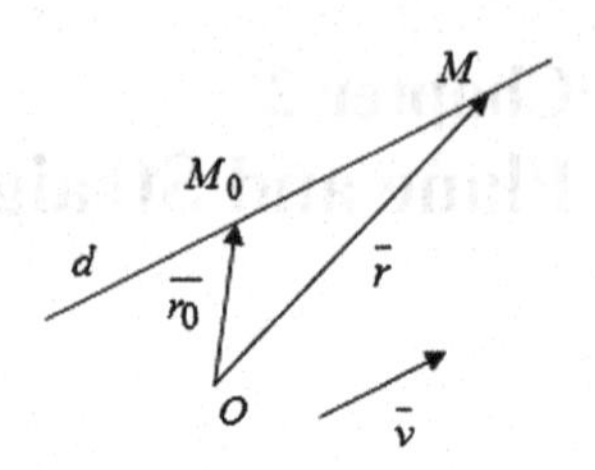


**Fig. 2.1** The straight line determined by a point and a nonzero vector

We denote

$$\overline{r}_0 = \overline{OM}_0 = x_0\overline{i} + y_0\overline{j} + z_0\overline{k}.$$

Let be $M \in E_3$, $M(x, y, z)$ and we denote

$$\overline{r} = \overline{OM} = x\overline{i} + y\overline{j} + z\overline{k}.$$

The point $M \in d \Leftrightarrow \overline{M_0M}$ and $\overline{v}$ are collinear$\Leftrightarrow$

$$\overline{M_0M} \times \overline{v} = \overline{0}. \tag{2.1}$$

However

$$\overline{M_0M} = \overline{r} - \overline{r}_0. \tag{2.2}$$

From (2.1) and (2.2) it results (see [6]) the *vector equation* of the straight line $d = (M_0, \overline{v})$ :

$$(\overline{r} - \overline{r}_0) \times \overline{v} = \overline{0};$$

$\overline{v}$ is called the *direction vector* of the straight line.

If $\overline{M_0M}$ and $\overline{v}$ are collinear then $(\exists)\, t \in \mathbb{R}$ unique, such that

$$\overline{M_0M} = t\overline{v}.$$

Taking account the relation (2.2) it results that $\overline{r} - \overline{r}_0 = t\overline{v}$; we obtain (see [6]) the *vector parametric equation* of the straight line $d$:

$$\overline{r} - \overline{r}_0 = t\overline{v},\; t \in \mathbb{R}. \tag{2.3}$$

The Eq. (2.3) can be written as

$$x\overline{i} + y\overline{j} + z\overline{k} = x_0\overline{i} + y_0\overline{j} + z_0\overline{k} + ta\overline{i} + tb\overline{j} + tc\overline{k};$$

we deduce the *parametric equations* (see [4], p. 49) of the straight line $d$:

$$\begin{cases} x = x_0 + ta \\ y = y_0 + tb \\ z = z_0 + tc \end{cases}, \ t \in \mathbb{R}. \tag{2.4}$$

If in the relation (2.4) we eliminate the parameter $t$ we obtain (see [4], p. 49) the *Cartesian equations* of the straight line $d$:

$$\frac{x - x_0}{a} = \frac{y - y_0}{b} = \frac{z - z_0}{c}. \tag{2.5}$$

In the relation (2.5), there is the following convention: if one of the denominators is equal to 0, then we shall also cancel that numerator.

## 2.1.2 Straight Line Determined by Two Distinct Points

Let be:

- $M_1 \in E_3,\ M_1(x_1, y_1, z_1),\ \overline{r}_1 = \overline{OM}_1$ and
- $M_2 \in E_3,\ M_2(x_2, y_2, z_2),\ \overline{r}_2 = \overline{OM}_2$.

We want to determine the equation of straight line determined by the points $M_1$ and $M_2$, denoted by $d = (M_1, M_2)$ and represented in Fig. 2.2.

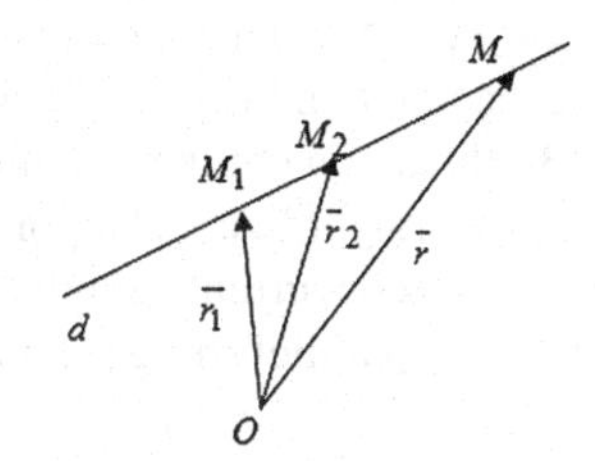


**Fig. 2.2** Straight line determined by two distinct points

Let be $M \in d$, $M(x, y, z)$ and we denote $\overline{r} = \overline{OM}$. We consider that the straight line is determined by $M_1$ and $\overline{M_1M_2}$.

As

$$\overline{M_1M} \times \overline{M_1M_2} = \overline{0}$$

it results that *vector equation* (see [6]) of the straight line $d$ is:

$$(\overline{r} - \overline{r}_1) \times (\overline{r}_2 - \overline{r}_1) = \overline{0}.$$

Since (within the Proposition 1.1 from the Chap. 1), $(\exists)\ t \in \mathbb{R}$ unique, such that

$$\overline{M_1M} = t\,\overline{M_1M_2}$$

it results that

$$\overline{r}-\overline{r}_1=t\left(\overline{r}_2-\overline{r}_1\right),\ (\forall)\ t\in\mathbb{R},$$

namely the *vector parametric equation* (see [5], p. 166) is of form:

$$\overline{r}=\overline{r}_1+t\left(\overline{r}_2-\overline{r}_1\right),\ (\forall)\ t\in\mathbb{R}; \tag{2.6}$$

the *parametric equations* (see [6]) are:

$$\begin{cases} x=x_1+t\left(x_2-x_1\right) \\ y=y_1+t\left(y_2-y_1\right),\ (\forall)\ t\in\mathbb{R} \\ z=z_1+t\left(z_2-z_1\right) \end{cases} \tag{2.7}$$

and the *Cartesian equations* (see [4], p. 49) will be:

$$\frac{x-x_1}{x_2-x_1}=\frac{y-y_1}{y_2-y_1}=\frac{z-z_1}{z_2-z_1}. \tag{2.8}$$

## 2.2 Plane in $E_3$

A plane can be determined (see [6]) in $E_3$ as follows:

(1) a point and a nonnull vector normal to plane,
(2) a point and two noncollinear vectors,
(3) three non collinear points,
(4) a straight line and a point that does not belong to the straight line,
(5) two concurrent straight lines,
(6) two parallel straight lines.

### 2.2.1 A a Point and a Non Zero Vector Normal to the Plane

Let be $M_0\in\pi$, $M_0\left(x_0,y_0,z_0\right)$ and the nonnull free vector $\overline{n}=a\overline{i}+b\overline{j}+c\overline{k}$ normal to the plane (see Fig. 2.3).

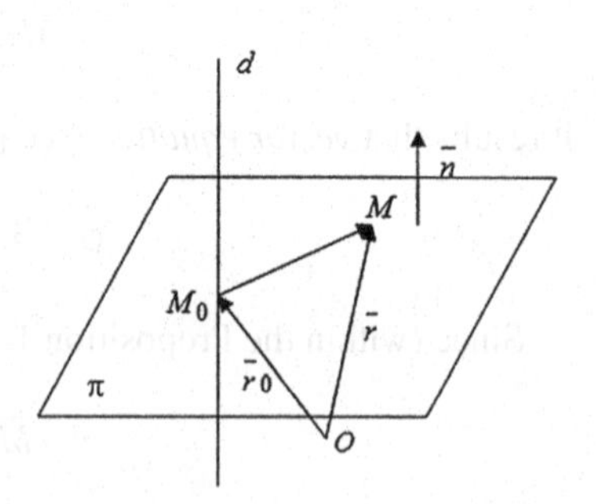


**Fig. 2.3** Plane determined by a point and a non zero vector normal to plane

**Definition 2.2** (see [6]). The straight line $d$ which passes through $M_0$ and has the direction of the vector $\overline{n}$ is called the **normal to the plane** through $M_0$; the vector $\overline{n}$ is the **normal vector** of the plane.

We propose to obtain the plane equation determined by the point $M_0$ and by the vector $\overline{n}$, denoted with $\pi = (M_0, \overline{n})$.

A point $M(x, y, z) \in \pi \Leftrightarrow \overline{M_0M}$ and $\overline{n}$ are orthogonal.

We denote

$$\overline{r} = \overline{OM} = x\overline{i} + y\overline{j} + z\overline{k}.$$

As $\overline{n}$ is perpendicular on $\overline{M_0M}$ it results

$$\overline{M_0M} \cdot \overline{n} = 0,$$

namely

$$(\overline{r} - \overline{r}_0) \cdot \overline{n} = 0;$$

from here we deduce the *normal equation* (see [6]) of the plane $\pi$:

$$\overline{r} \cdot \overline{n} - \overline{r}_0 \cdot \overline{n} = 0. \tag{2.9}$$

Writing the relation (2.9) as

$$ax + by + cz - ax_0 - by_0 - cz_0 = 0$$

we obtain the *Carthesian equation* (see [6]) of the plane $\pi$:

$$a(x - x_0) + b(y - y_0) + c(z - z_0) = 0. \tag{2.10}$$

If we denote

$$ax_0 - by_0 - cz_0 = -d$$

then from the equation (2.10) one deduces the *general Carthesian equation* (see [6]) of the plane $\pi$:

$$ax + by + cz + d = 0. \tag{2.11}$$

### 2.2.2 Plane Determined by a Point and Two Noncollinear Vectors

Let $\overline{u}, \overline{v}$ be two noncollinear vectors, namely $\overline{u} \times \overline{v} \neq \overline{0}$, of the form

$$\overline{u} = l_1\overline{i} + m_1\overline{j} + n_1\overline{k}, \overline{v} = l_2\overline{i} + m_2\overline{j} + n_2\overline{k}$$

and let be $M_0 \in \pi$, $M_0(x_0, y_0, z_0)$ (see Fig. 2.4).

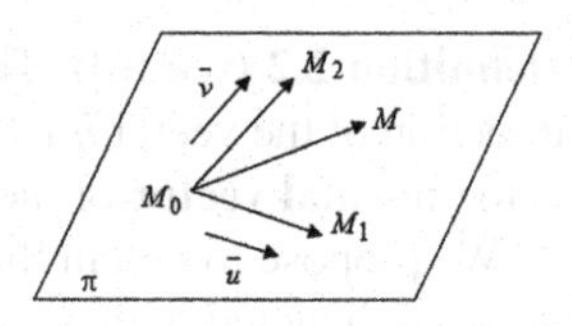


**Fig. 2.4** Plane determined by a point and two non collinear vectors

We want to find the equation of the plane determined by the point $M_0$ and by the free vectors $\overline{u}$ and $\overline{v}$, denoted by $\pi = (M_0, \overline{u}, \overline{v})$.

Let

- $\overrightarrow{M_0M_1}$ be a representative for the free vector $\overline{u}$,
- $\overrightarrow{M_0M_2}$ be a representative for the free vector $\overline{v}$.

A point $M(x, y, z) \in \pi \Leftrightarrow \overline{M_0M}, \overline{M_0M_1}, \overline{M_0M_2}$ are coplanar. The coplanarity of these vectors can be expressed as:

(a) using the Proposition 1.21 from the Chap. 1, $(\exists)\ t_1, t_2 \in \mathbb{R}$ uniquely determined, such that

$$\overline{M_0M} = t_1\overline{M_0M_1} + t_2\overline{M_0M_2}; \tag{2.12}$$

(b) using the Proposition 1.34 from the Chap. 1, $\overline{M_0M}$ is perpendicular on $\overline{u} \times \overline{v}$, namely

$$\overline{M_0M} \cdot (\overline{u} \times \overline{v}) = 0. \tag{2.13}$$

Writing the relation (2.12) on the form:

$$\overline{r} - \overline{r}_0 = t_1\overline{u} + t_2\overline{v}$$

we deduce the *vector parametric equation* (see [6]) of the plane $\pi$:

$$\overline{r} = \overline{r}_0 + t_1\overline{u} + t_2\overline{v},\ (\forall)\ t_1, t_2 \in \mathbb{R} \tag{2.14}$$

and then the *parametric equations* of the plane $\pi$:

$$\begin{cases} x = x_0 + t_1 l_1 + t_2 l_2 \\ y = y_0 + t_1 m_1 + t_2 m_2, \ (\forall)\ t_1, t_2 \in \mathbb{R}. \\ z = z_0 + t_1 n_1 + t_2 n_2 \end{cases} \tag{2.15}$$

From the relation (2.13) we obtain the *vector equation* (see [6]) of the plane $\pi$:

$$(\overline{r} - \overline{r}_0) \cdot (\overline{u} \times \overline{v}) = 0. \tag{2.16}$$

As

$$\overline{M_0M} = (x - x_0)\,\overline{i} + (y - y_0)\,\overline{j} + (z - z_0)\,\overline{k} \tag{2.17}$$

we have

$$\overline{M_0M} \cdot (\overline{u} \times \overline{v}) = \begin{vmatrix} x - x_0 & y - y_0 & z - z_0 \\ l_1 & m_1 & n_1 \\ l_2 & m_2 & n_2 \end{vmatrix};$$

so from the relation (2.13) we deduce the *Carthesian equations* (see [6]) of the plane $\pi$:

$$\begin{vmatrix} x - x_0 & y - y_0 & z - z_0 \\ l_1 & m_1 & n_1 \\ l_2 & m_2 & n_2 \end{vmatrix} = 0. \tag{2.18}$$

### 2.2.3 Plane Determined by Three Noncollinear Points

Let $M_0, M_1, M_2 \in E_3$ be three noncollinear points, $M_0\,(x_0, y_0, z_0)$, $M_1\,(x_1, y_1, z_1)$, $M_2\,(x_2, y_2, z_2)$. It results that $\overline{M_0M_1}, \overline{M_0M_2}$ are noncollinear. We propose to obtain the equation of the plane determined by these points, which is represented in Fig. 2.5 and it is denoted by $\pi = (M_0, M_1, M_2)$.

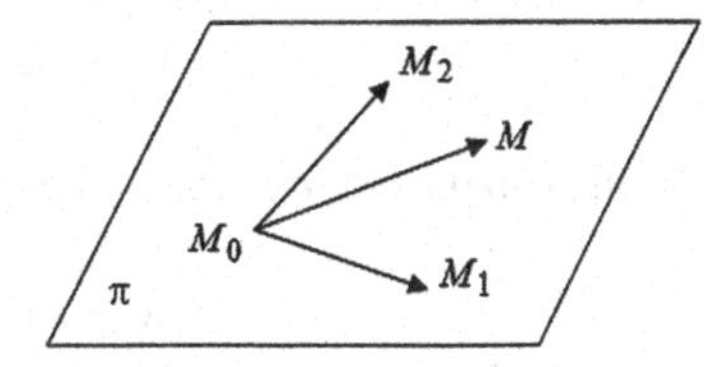


**Fig. 2.5** Plane determined by three noncollinear points

We note that $\pi = (M_0, M_1, M_2)$ coincides with $\pi_1 = \left(M_0, \overline{M_0M_1}, \overline{M_0M_2}\right)$, namely we are in the case presented in the previous paragraph. We have

$$\begin{aligned} \overline{M_0M_1} &= (x_1 - x_0)\,\overline{i} + (y_1 - y_0)\,\overline{j} + (z_1 - z_0)\,\overline{k}, \\ \overline{M_0M_2} &= (x_2 - x_0)\,\overline{i} + (y_2 - y_0)\,\overline{j} + (z_2 - z_0)\,\overline{k}. \end{aligned}$$

A point $M\,(x, y, z) \in \pi \Leftrightarrow \overline{M_0M}, \overline{M_0M_1}, \overline{M_0M_2}$ are coplanar, namely

$$\overline{M_0M} \cdot \left(\overline{M_0M_1} \times \overline{M_0M_2}\right) = 0.$$

Using $\overline{M_0M}$ from (2.17) we obtain the following *Cartesian equation* (see [6]) of the plane $\pi$:

$$\begin{vmatrix} x-x_0 & y-y_0 & z-z_0 \\ x_1-x_0 & y_1-y_0 & z_1-z_0 \\ x_2-x_0 & y_2-y_0 & z_2-z_0 \end{vmatrix} = 0. \tag{2.19}$$

### 2.2.4 Plane Determined by a Straight Line and a Point that Doesn't Belong to the Straight Line

Let $d \subset E_3$ and a point $M_0 \notin d$ (see Fig. 2.6).

We want to obtain the equation of the plane determined by the straight line $d$ and by the point $M_0$, denoted by $\pi = (M_0, d)$.

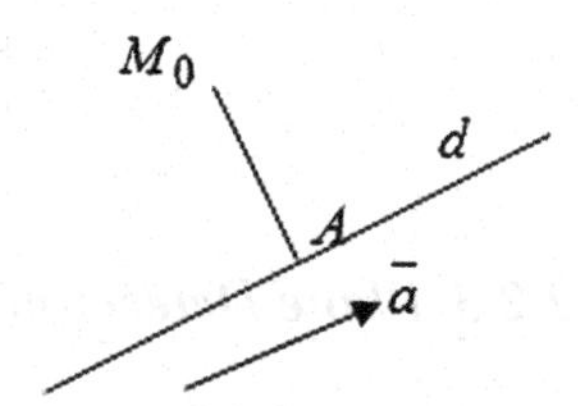


**Fig. 2.6** Plane determined by a straight line and a point that doesn't belong to straight line

Let be $A \in d$, hence we have $d = (A, \overline{a})$. We note that $\pi = (M_0, d)$ coincides with $\pi_1 = \left(M_0, \overline{a}, \overline{M_0 A}\right)$. If $\overline{r}_0$ is the position vector of the point $M_0$ (denoted with $M_0(\overline{r}_0)$), $A(\overline{r}_A)$ and $M(x, y, z) \in \pi$ then the *vector equation* (see [6]) of the plane $\pi$ is:

$$(\overline{r} - \overline{r}_0) \cdot [\overline{a} \times (\overline{r}_A - \overline{r}_0)] = 0 \tag{2.20}$$

and the *Carthesian equation* (see [6]) of the plane $\pi$:

$$\begin{vmatrix} x-x_0 & y-y_0 & z-z_0 \\ a_1 & a_2 & a_3 \\ x_A-x_0 & y_A-y_0 & z_A-z_0 \end{vmatrix} = 0, \tag{2.21}$$

where $\overline{a} = a_1\overline{i} + a_2\overline{j} + a_3\overline{k}$ and $A(x_A, y_A, z_A)$.

### 2.2.5 Plane Determined by Two Concurrent Straight Lines

Let $d_1 \cap d_2 = \{P\}$, see Fig. 2.7; the straight line

- $d_1$ is the straight line which passes through $P$ and has the direction vector $\overline{a}_1$, $d_1 = (P, \overline{a}_1)$,
- $d_2$ is the straight line which passes through $P$ and has the direction vector $\overline{a}_2$, $d_2 = (P, \overline{a}_2)$.

We want to find the equation of the plane determined by the straight lines $d_1$ and $d_2$.

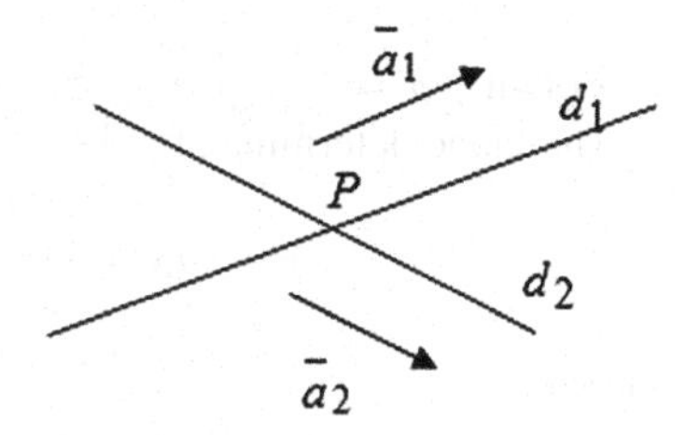


**Fig. 2.7** Plane determined by two concurrent straight lines

Noting that $\pi = (d_1, d_2)$ coincides with $\pi = (P, \overline{a}_1, \overline{a}_2)$, i.e. with the plane which passes through $P$ and has the direction vectors $\overline{a}_1$ and $\overline{a}_2$. If $M(x, y, z) \in \pi$ we deduce that the *vector equation* (see [6]) of the plane is:

$$(\overline{r} - \overline{r}_P) \cdot (\overline{a}_1 \times \overline{a}_2) = 0; \tag{2.22}$$

the *Cartesian equation* of the plane $\pi$ will be:

$$\begin{vmatrix} x - x_P & y - y_P & z - z_P \\ l_1 & m_1 & n_1 \\ l_2 & m_2 & n_2 \end{vmatrix} = 0, \tag{2.23}$$

where: $\overline{a}_1 = l_1\overline{i} + m_1\overline{j} + n_1\overline{k}$, $\overline{a}_2 = l_2\overline{i} + m_2\overline{j} + n_2\overline{k}$, $P(x_P, y_P, z_P)$.

**Example 2.3** Check if the following straight lines are concurrent:

$$\begin{cases} d_1 : \frac{x-1}{2} = \frac{y-7}{1} = \frac{z-5}{4} \\ d_2 : \frac{x-6}{3} = \frac{y+1}{-2} = \frac{z}{1} \end{cases}$$

and then write the equation of the plane which they determine.

**Solution**

We note that the direction vectors of the two straight lines are: $\overline{a}_1 = 2\overline{i} + \overline{j} + 4\overline{k}$ and respectively $\overline{a}_2 = 3\overline{i} - 2\overline{j} + \overline{k}$. As

$$\overline{a}_1 \times \overline{a}_2 = \begin{vmatrix} \overline{i} & \overline{j} & \overline{k} \\ 2 & 1 & 4 \\ 3 & -2 & 1 \end{vmatrix} = 9\overline{i} + 10\overline{j} - 7\overline{k} \neq \overline{0}$$

it results that the vectors $\overline{a}_1$ and $\overline{a}_2$ are noncollinear, namely $d_1 \cap d_2 \neq \emptyset$.

Let $P = d_1 \cap d_2$. Since:

- $P \in d_1$ we obtain $x_P - 1 = 2y_P - 14$;
- $P \in d_2$ we obtain $-2x_P + 12 = 3y_P + 3$.

Solving the system

$$\begin{cases} x_P - 1 &= 2y_P - 14 \\ -2x_P + 12 &= 3y_P + 3 \end{cases}$$

we obtain: $x_P = -3$, $y_P = 5$, $z_P = -3$.

The plane determined by the straight lines $d_1$ and $d_2$ will have the equation

$$\pi : 9(x+3) + 10(y-5) - 7(z+3) = 0,$$

namely

$$\pi : 9x + 10y - 7z - 44 = 0.$$

We shall present the solution of this problem in Sage, too:

```
sage: a1=vector([2,1,4]);a2=vector([3,-2,1])
sage: var("x y z")
(x, y, z)
sage: p=implicit_plot((x-1)/2 ==y-7==(z-5)/4, (x,-10,10), (y,-10,10))
sage: q=implicit_plot((x-6)/3 ==(y+1)/(-2)==z, (x,-10,10), (y,-10,10))
sage: (p+q).show(aspect_ratio=0.5)
```

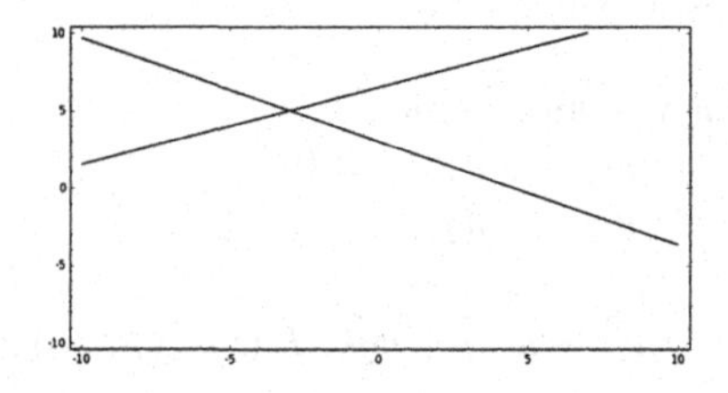

```
sage: a1.cross_product(a2)
(9, 10, -7)
sage: eqn=[x-1==2*y-14, -2*x+12==3*y+3]
sage: s = solve(eqn, x,y); s
[[x == -3, y == 5]]
sage: M=matrix(SR,3,[x+3, y-5, z+3, 2,1,4,3,-2,1])
sage: M
[x + 3 y - 5 z + 3]
[    2     1     4]
[    3    -2     1]
sage: M.det() ==0
9*x + 10*y - 7*z - 44 == 0
```

## 2.2.6 Plane Determined by Two Parallel Straight Lines

Let $d_1, d_2 \in E_3$, $d_1 || d_2$, see Fig. 2.8; the straight line:

- $d_1$ is the straight line which passes through $A_1$ and has the direction vector $\overline{a}$, $d_1 = (A_1, \overline{a})$,

- $d_2$ is the straight line which passes through $A_2$ and has the direction vector $\overline{a}$, $d_2 = (A_2, \overline{a})$.

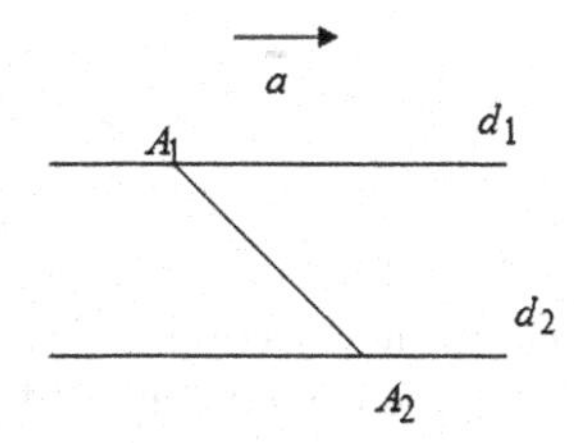


**Fig. 2.8** Plane determined by two parallel straight lines

The plane determined by $d_1$ and $d_2$ is the plane determined by $A_1$ and the two non collinear vectors $\overline{a}$ and $\overline{A_1A_2}$.

If $M(x, y, z) \in \pi$ then the *vector equation* (see [6]) of the plane $\pi$ is:

$$\left(\overline{r} - \overline{r}_{A_1}\right) \cdot \left(\overline{a} \times \overline{A_1A_2}\right) = 0. \tag{2.24}$$

The *Cartesian equation* of the plane $\pi$:

$$\begin{vmatrix} x - x_{A_1} & y - y_{A_1} & z - z_{A_1} \\ a_1 & a_2 & a_3 \\ x_{A_2} - x_{A_1} & y_{A_2} - y_{A_1} & z_{A_2} - z_{A_1} \end{vmatrix} = 0, \tag{2.25}$$

where: $\overline{a} = a_1\overline{i} + a_2\overline{j} + a_3\overline{k}$, $A_1\left(x_{A_1}, y_{A_1}, z_{A_1}\right)$, $A_2\left(x_{A_2}, y_{A_2}, z_{A_2}\right)$.

### 2.2.7 The Straight Line Determined by the Intersection of Two Planes

We consider $\pi_1, \pi_2 \in E_3$ (see Fig. 2.9) having the equations:

$$\begin{cases} \pi_1 : a_1x + b_1y + c_1z + d_1 = 0 \\ \pi_2 : a_2x + b_2y + c_2z + d_2 = 0. \end{cases}$$

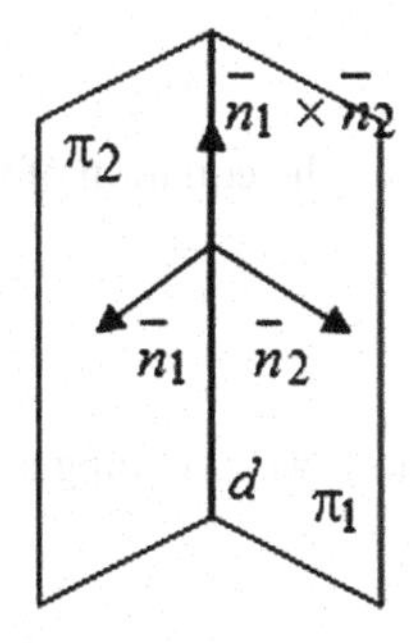


**Fig. 2.9** Straight line determined by the intersection of two planes

The intersection of the planes $\pi_1$ and $\pi_2$ is the set of solutions of the system of equations determined by the equations of $\pi_1$ and $\pi_2$.

We denote

$$A = \begin{pmatrix} a_1 & b_1 & c_1 \\ a_2 & b_2 & c_2 \end{pmatrix}.$$

If rank $(A) = 2$ it results a compatible system which is undetermined simple and the intersection of the two planes is a straight line. If rank $(A) \neq$ rank $(\overline{A})$, $\overline{A}$ being the extend matrix of the system it results an incompatible system, therefore $\pi_1 \cap \pi_2 = \emptyset$, namely $\pi_1 || \pi_2$.

Let

- $\overline{n}_1$be the normal to $\pi_1$, $\overline{n}_1 = a_1\overline{i} + b_1\overline{j} + c_1\overline{k}$,
- $\overline{n}_2$ be the normal to $\pi_2$, $\overline{n}_2 = a_2\overline{i} + b_2\overline{j} + c_2\overline{k}$.

We have

$$\left.\begin{array}{l} d \subset \pi_1 \Rightarrow \overline{n}_1 \perp d \\ d \subset \pi_2 \Rightarrow \overline{n}_2 \perp d \end{array}\right\} \Rightarrow \text{The straight line } d \text{ has the direction vector } \overline{n}_1 \times \overline{n}_2.$$

We denote

$$\overline{u} = \overline{n}_1 \times \overline{n}_2, \overline{u} = l\overline{i} + m\overline{j} + n\overline{k}.$$

We have

$$\overline{n}_1 \times \overline{n}_2 = \begin{vmatrix} \overline{i} & \overline{j} & \overline{k} \\ a_1 & b_1 & c_1 \\ a_2 & b_2 & c_2 \end{vmatrix} = (b_1c_2 - b_2c_1)\overline{i} + (a_2c_1 - a_1c_2)\overline{j} + (a_1b_2 - a_2b_1)\overline{k}.$$

We deduce

$$\begin{cases} l = \begin{vmatrix} b_1 & c_1 \\ b_2 & c_2 \end{vmatrix} \\ m = \begin{vmatrix} c_1 & a_1 \\ c_2 & a_2 \end{vmatrix} \\ n = \begin{vmatrix} a_1 & b_1 \\ a_2 & b_2 \end{vmatrix}. \end{cases}$$

The equation of the straight line is (see [6]):

$$\frac{x - x_0}{l} = \frac{y - y_0}{m} = \frac{z - z_0}{n}, \tag{2.26}$$

$(x_0, y_0, z_0)$ being a solution of the system.

**Example 2.4** Write the equation of a plane which:

(a) passes through the point $M(-2, 3, 4)$ and is parallel with the vectors $\overline{v}_1 = \overline{i} - 2\overline{j} + \overline{k}$ and $\overline{v}_2 = 3\overline{i} + 2\overline{j} + 4\overline{k}$;
(b) passes through the point $M(1, -1, 1)$ and is perpendicular on the planes $\pi_1 : x - y + z - 1 = 0$ and $\pi_2 : 2x + y + z + 1 = 0$.

**Solution**

(a) The vector equation of the plane is

$$\pi : (\overline{r} - \overline{r}_M) \cdot (\overline{v}_1 \times \overline{v}_2) = 0.$$

Where as

- $\overline{v}_1 \times \overline{v}_2 = \begin{vmatrix} \overline{i} & \overline{j} & \overline{k} \\ 1 & -2 & 1 \\ 3 & 2 & 4 \end{vmatrix} = -10\overline{i} - \overline{j} + 8\overline{k},$
- $\overline{r} - \overline{r}_M = (x + 2)\,\overline{i} + (y - 3)\,\overline{j} + (z - 4)\,\overline{k}$

we obtain

$$\pi : -10\,(x + 2) - (y - 3) + 8\,(z - 4) = 0$$

or

$$\pi : -10x - y + 8z - 49 = 0.$$

This equation can be determined with Sage, too:

```
sage: v1=vector([1,-2,1]);v2=vector([3,2,4])
sage: vp= v1.cross_product(v2)
sage: var("x y z")
(x, y, z)
sage: r=vector([x,y,z]);M=vector([-2,3,4])
sage: (r-M).dot_product(vp)==0
-10*x - y + 8*z - 49 == 0
```

and it can also be plotted:

```
sage: var("x y z")
(x, y, z)
sage: pl=plot3d((49+10*x+y)/8,(x,-10,10),(y,0,8),rgbcolor="lightblue")
sage: A=point3d((-2,3,4),color='red',size=20)
sage: v1=arrow3d((3,3,3),(4,5,4),6,color='red')+text3d("v1",(1.2,2,0.4))
sage: v2=arrow3d((0,0,0),(3,2,4),6,color='red')+text3d("v2",(3.3,2.7,2.5))
sage: A+pl+v1+v2
```

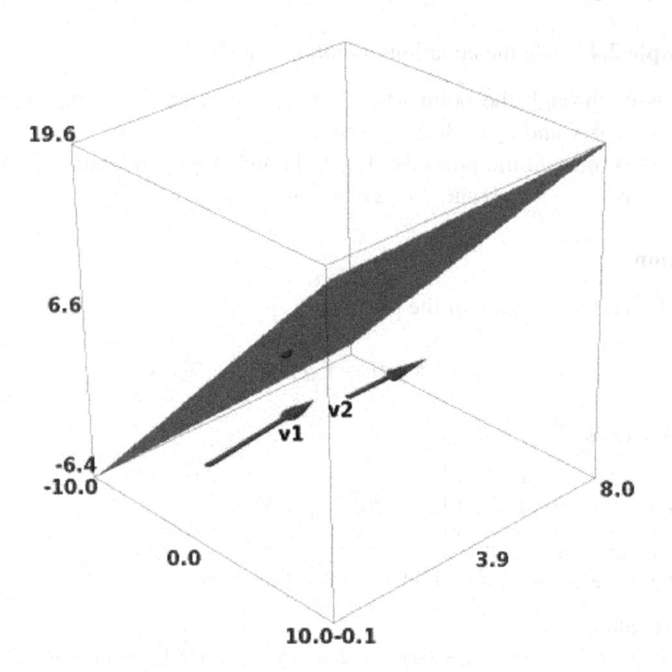


(b) The normal of the required plane is $\overline{n} = \overline{n}_1 \times \overline{n}_2$, where $\overline{n}_1 = \overline{i} - \overline{j} + \overline{k}$ and $\overline{n}_2 = 2\overline{i} + \overline{j} + \overline{k}$; therefore $\overline{n} = -2\overline{i} + \overline{j} + 3\overline{k}$.

The Cartesian equation of the plane will be

$$\pi : (-2)(x-1) + (y+1) + 3(z-1) = 0 \Leftrightarrow \pi : -2x + y + 3z = 0.$$

We can also find this equation in Sage:

```
sage: n1=vector([1,-1,1]);n2=vector([2,1,1]);
sage: n=n1.cross_product(n2)
sage: var("x y z")
(x, y, z)
sage: M1=vector([1,-1,1]);r=vector([x,y,z]);
sage: (r-M1).dot_product(n)==0
-2*x + y + 3*z == 0
```

A graphical solution in Sage is:

```
sage: var("x y z")
(x, y, z)
sage: M=point3d((1,-1,1),color='red',size=20)
sage: p1=plot3d(1-x+y,(x,-1,1),(y,-1,1),rgbcolor="lightblue")
sage: p2=plot3d(1-2*x-y,(x,-1,1),(y,-1,1),rgbcolor="lightblue")
sage: p=plot3d((2*x-y)/3,(x,-1.5,1.5),(y,-2,2),rgbcolor="blue")
sage: M+p1+p2+p
```

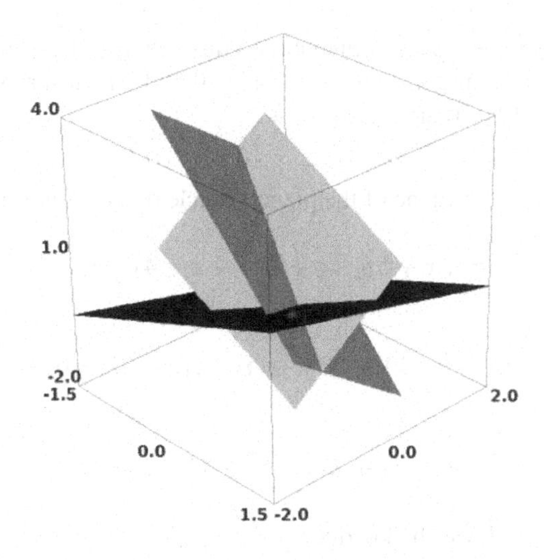


## 2.3 Plane Fascicle

Let be $d = \pi_1 \cap \pi_2$

$$\begin{cases} \pi_1 : a_1x + b_1y + c_1z + d_1 = 0 \\ \pi_2 : a_2x + b_2y + c_2z + d_2 = 0. \end{cases}$$

**Definition 2.5** (see [1], p. 62 and [2], p. 681). The set of the planes which contain the straight line $d$ is called a **plane fascicle** of axis $d$ (see Fig. 2.10). The straight line $d$ is called the **fascicle axis** and $\pi_1, \pi_2$ are called the **base planes** of the fascicle.

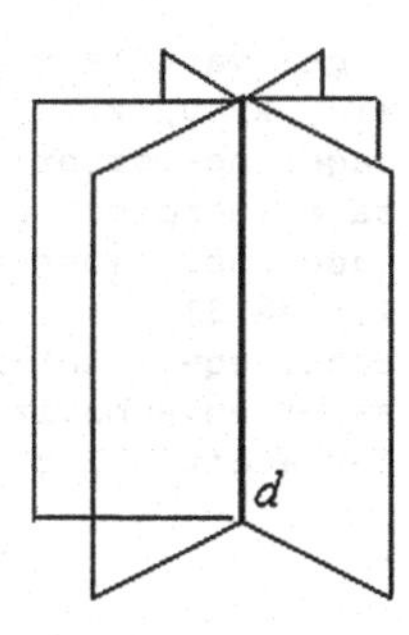


**Fig. 2.10** Plane fascicle

An arbitrary plane of the fascile has the equation of the form:

$$\pi : a_1x + b_1y + c_1z + d_1 + \lambda (a_2x + b_2y + c_2z + d_2) = 0, \ \lambda \in \mathbb{R}^*.$$

**Example 2.6** Determine a plane which passes through the intersection of the planes $\pi_1 : x + y + 5z = 0$ and $\pi_2 : x - z + 4 = 0$ and which forms with the plane $\pi : x - 4y - 8z + 12 = 0$ an angle $\varphi = \frac{\pi}{2}$.

**Solution**

Let be $d = d_1 \cap d_2$. A plane of the plane fascicle of axis $d$ has the equation

$$\pi' : x + 5y + z + \lambda\,(x - z + 4) = 0,$$

namely

$$\pi' : (1 + \lambda)\,x + 5y + (1 - \lambda)\,z + 4\lambda = 0.$$

It results that

$$\overline{n'} = (1 + \lambda)\,\overline{i} + 5\overline{j} + (1 - \lambda)\,\overline{k}.$$

As $\overline{n} = \overline{i} - 4\overline{j} - 8\overline{k}$ we shall deduce

$$\cos \sphericalangle \left(\pi, \pi'\right) = \cos\frac{\pi}{2} = \frac{\overline{n}\cdot\overline{n'}}{\|\overline{n}\|\cdot\left\|\overline{n'}\right\|} = \frac{-27 + 9\lambda}{9\sqrt{27 + 2\lambda^2}} \Leftrightarrow \lambda = 3.$$

The equation of the required plane is

$$\pi' : (1 + 3)\,x + 5y + (1 + 3)\,z + 4\cdot 3 = 0$$

or

$$\pi' : 4x + 5y - 2z + 12 = 0.$$

The solution in Sage of this problem is:

```
sage: var("la x y z");n=vector([1,-4,-8])
(la, x, y, z)
sage: np=vector([1+la,5,1-la]);nnp=expand(sqrt((1+la)^2+25+(1-la)^2))
sage: u=np.dot_product(n)/(n.norm()*nnp)
sage: solve(u==cos(pi/2),la)
[la == 3]
sage: eq=(1+la)*x+5*y+(1-la)*z+4*la
sage: eq.substitute(la=3)==0
4*x + 5*y - 2*z + 12 == 0
```

## 2.4 Distances in $E_3$

### 2.4.1 Distance from a Point to a Straight Line

Let be $d = (A, \overline{a})$ with $A\,(x_A, y_A, z_A)\,,\, \overline{a} = a_1\overline{i} + a_2\overline{j} + a_3\overline{k}$ and $M \in E_3$.

Let $\overrightarrow{AA'}$ be a representative for $\overline{a}$. The equation of the straight line is:

$$\frac{x - x_A}{a_1} = \frac{y - y_A}{a_2} = \frac{z - z_A}{a_3}.$$

We build the parallelogram $AA'PM$ (see Fig. 2.11).

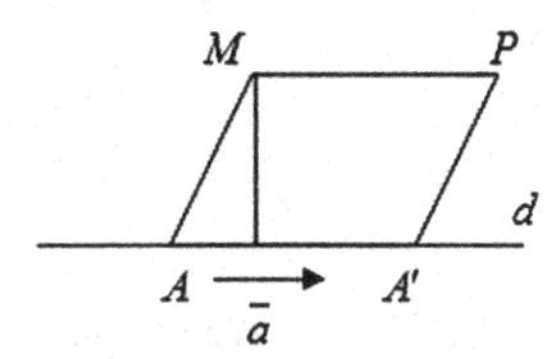


**Fig. 2.11** The distance from a point to a straight line

We know that

$$\mathrm{A}_{AA'PM} = \left\| \overline{AA'} \times \overline{MA} \right\|. \tag{2.27}$$

However

$$\mathrm{A}_{AA'PM} = \left\| \overline{AA'} \right\| \cdot \rho\left(M, d\right). \tag{2.28}$$

From (2.27) and (2.28) it results that the *distance formula from a point to a straight line* is [6]

$$\rho\left(M, d\right) = \frac{\left\| \overline{AA'} \times \overline{MA} \right\|}{\left\| \overline{AA'} \right\|} = \frac{\left\| \overline{a} \times \overline{MA} \right\|}{\left\| \overline{a} \right\|}. \tag{2.29}$$

## *2.4.2 Distance from a Point to a Plane*

We consider the plane $\pi : ax + by + cz + d = 0$ and the point $M_0\left(x_0, y_0, z_0\right)$, $M_0 \notin \pi$. Let $M_1$ be the projection of $M_0$ on the plane $\pi$, $M_1\left(x_1, y_1, z_1\right)$, see Fig. 2.12.

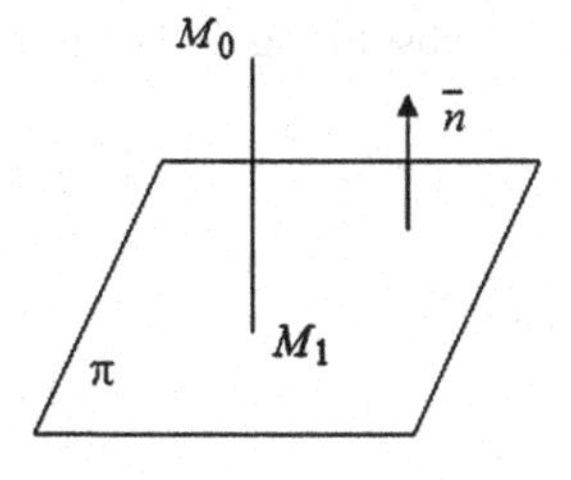


**Fig. 2.12** The distance from a point to a plane

The distance from the point $M_0$ to the plane $\pi$ is

$$\rho\left(M_0,\pi\right)=\left\|\overline{M_0M_1}\right\|.$$

Let $d = (M_0, \overline{n})$ be the normal line to the plane which passes through $M_0$, $\overline{n} = a\overline{i} + b\overline{j} + c\overline{k}$ . The equation of this straight line is

$$d : \frac{x - x_0}{a} = \frac{y - y_0}{b} = \frac{z - z_0}{c} = t$$

or

$$d : \begin{cases} x = x_0 + ta \\ y = y_0 + tb, \ t \in \mathbb{R}. \\ z = z_0 + tc \end{cases}$$

As $M_1 \in d$ we deduce

$$d : \frac{x_1 - x_0}{a} = \frac{y_1 - y_0}{b} = \frac{z_1 - z_0}{c} = t \Rightarrow$$

$$\begin{cases} x_1 = x_0 + ta \\ y_1 = y_0 + tb, \ t \in \mathbb{R}. \\ z_1 = z_0 + tc \end{cases} \tag{2.30}$$

Since $M_1 \in \pi \Rightarrow ax_1 + by_1 + cz_1 + d = 0 \Rightarrow$

$$ax_1 + by_1 + cz_1 = -d \tag{2.31}$$

Multiplying the first equation of (2.31) with $a$, the second with $b$ and the third with $c$ we have:

$$\begin{cases} ax_1 = ax_0 + ta^2 \\ by_1 = by_0 + tb^2, \ t \in \mathbb{R}. \\ cz_1 = cz_0 + tc^2 \end{cases} \tag{2.32}$$

Adding the three equations of (2.32) it results

$$ax_1 + by_1 + cz_1 = ax_0 + by_0 + cz_0 + t\left(a^2 + b^2 + c^2\right). \tag{2.33}$$

Substituting (2.31) into (2.33) we deduce

$$ax_0 + by_0 + cz_0 + d = -t\left(a^2 + b^2 + c^2\right),$$

namely

$$t = -\frac{ax_0 + by_0 + cz_0 + d}{a^2 + b^2 + c^2}. \tag{2.34}$$

We have

$$\overline{M_0M_1} = (x_1 - x_0)\,\overline{i} + (y_1 - y_0)\,\overline{j} + (z_1 - z_0)\,\overline{k};$$

therefore

$$\left\|\overline{M_0M_1}\right\| = \sqrt{(x_1 - x_0)^2 + (y_1 - y_0)^2 + (z_1 - z_0)^2} \overset{(2.30)}{=} \sqrt{t^2\left(a^2 + b^2 + c^2\right)} \Rightarrow$$

$$\left\|\overline{M_0M_1}\right\| = |t|\,\sqrt{a^2 + b^2 + c^2}. \tag{2.35}$$

Substituting (2.34) into (2.35) we can deduce [6] the *distance formula from a point to a plane*:

$$\rho\left(M_0, \pi\right) = \frac{|ax_0 + by_0 + cz_0 + d|}{\sqrt{a^2 + b^2 + c^2}}. \tag{2.36}$$

**Example 2.7** (see [6]). One gives:

- the plane $\pi : x + y - z + 2 = 0$,
- the straight line

$$d : \begin{cases} x - y - 1 = 0 \\ x + 2y + z - 4 = 0 \end{cases}$$

- the point $A = (1, 1, 2)$.

(a) Compute the distance from the point $A$ to the plane $\pi$.
(b) Find the distance from the point $A$ to the straight line $d$.

**Solution**

(a) Using the formul (2.36) we achieve:

$$\rho\left(A, \pi\right) = \frac{2}{\sqrt{3}} = \frac{2\sqrt{3}}{3}.$$

(b) The distance from the point $A$ to the straight line $d$ (Fig. 2.13) is computed using the formula

$$\rho\left(A, d\right) = \frac{\left\|\overline{MA'} \times \overline{AM}\right\|}{\left\|\overline{MA'}\right\|} = \frac{\left\|\overline{a} \times \overline{AM}\right\|}{\left\|\overline{a}\right\|}.$$

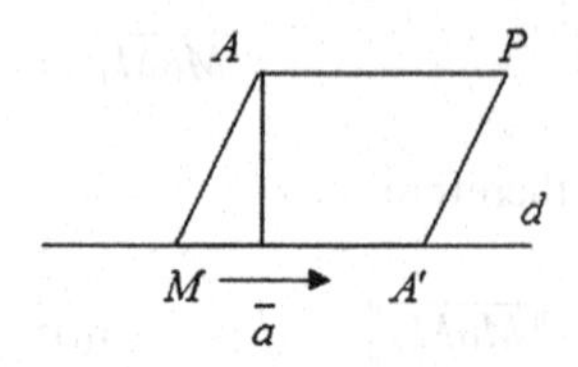


**Fig. 2.13** The distance from a point to a straight line

The direction vector of the straight line $d$ is

$$\overline{a} = \begin{vmatrix} \overline{i} & \overline{j} & \overline{k} \\ 1 & -1 & 1 \\ 1 & 2 & 1 \end{vmatrix} = -\overline{i} - \overline{j} + 3\overline{k}.$$

We obtain: $\|\overline{a}\| = \sqrt{11}$.
From $M \in d$ we have

$$\begin{cases} x_M - y_M - 1 = 0 \\ x_M + 2y_M + z_M - 4 = 0 \end{cases} \Leftrightarrow \begin{cases} x_M - y_M - 1 = 0 \\ x_M + 2y_M = 4 - z_M. \end{cases}$$

Denoting $z_M = u \in \mathbb{R}$ we deduce

$$3y_M = 3 - u \Leftrightarrow y_M = 1 - \frac{1}{3}u$$

and

$$x_M = y_M + 1 = 2 - \frac{1}{3}u.$$

We can suppose that $u = 0$; we obtain: $\begin{cases} x_M = 2 \\ y_M = 1 \\ z_M = 0 \end{cases} \Longrightarrow M(2, 1, 0)$. We have:

$\overline{AM} = \overline{i} - \overline{2k}$ and

$$\overline{a} \times \overline{AM} = \begin{vmatrix} \overline{i} & \overline{j} & \overline{k} \\ -1 & -1 & 3 \\ 1 & 0 & -2 \end{vmatrix} = 2\overline{i} + \overline{j} + \overline{k};$$

therefore $\|\overline{a} \times \overline{AM}\| = \sqrt{6}$. We shall obtain

$$\rho(A, d) = \frac{\sqrt{6}}{\sqrt{11}} = 0.739.$$

Solving this problem with Sage, we shall have:

```
sage: var("x y z");
(x, y, z)
sage: plane=plot3d(x+y+2, (x, -10, 0), (y, -10, 10));
sage: A = point3d((1, 1, 2), color = 'red', size = 10)
sage: p1=implicit_plot3d(x-y-1==0, (x, -10, 10), (y, -10, 10),(z, -10, 0),rgbcolor="orange");
sage: p2=plot3d(4-x-2*y, (x, -10, 10), (y, -10, 10),rgbcolor="orange");
sage: p1+p2+plane+A
```

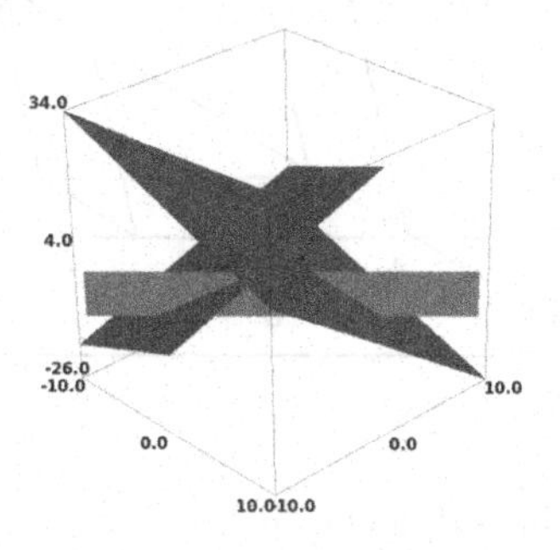


```
sage: a=1;b=1;c=-1;d=2;
sage: f=a*x+b*y+c*z+d
sage: (f.subs(x=1,y=1,z=2)).abs()/sqrt(a^2+b^2+c^2)
2/3*sqrt(3)
sage: a1=vector([1,-1,0]);a2=vector([1,2,1]);
sage: aa=a1.cross_product(a2); aa
(-1, -1, 3)
sage: na=aa.norm(); na
sqrt(11)
sage: eqn=[x-y-1==0,x+2*y+z-4==0];
sage: s=solve(eqn,x,y,z);s
[[x == -1/3*r3 + 2, y == -1/3*r3 + 1, z == r3]]
sage: s[0][0].subs(r2=0);s[0][1].subs(r2=0);s[0][2].subs(r2=0);
x == -1/3*r3 + 2
y == -1/3*r3 + 1
z == r3
sage: A=vector([1,1,2]); M=vector([2,1,0])
sage: AM=M-A; AM
(1, 0, -2)
sage: cp=aa.cross_product(AM); cp
(2, 1, 1)
sage: ncp=cp.norm(); ncp
sqrt(6)
sage: rho=ncp/na; rho.n(digits=3)
0.739
```

## 2.4.3 Distance Between Two Straight Lines

Let $d_1, d_2$ be two noncoplanar straight lines (see Fig. 2.14).

Fig. 2.14 the distance between two straight lines

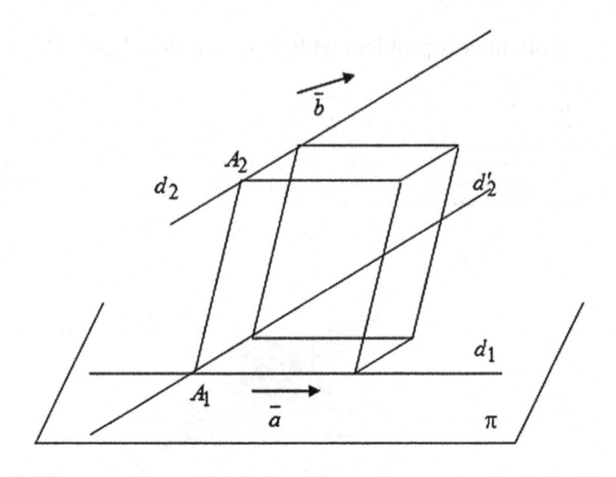


The distance between the straight lines $d_1$ and $d_2$ is

$$\rho\left(d_1, d_2\right) = \rho\left(A_1, A_2\right) = \rho\left(A_2, \pi\right),$$

where:

- $\pi$ is the plane which passes through $d_1$ and it is parallel with $d_2$,
- $\rho\left(A_2, \pi\right)$ is the height which corresponds to the vertex $A_2$ of the oblique parallelepiped built on the vectors $\overline{a}, \overline{b}, \overline{A_1A_2}$.

Therefore, the *distance formula between two straight lines* is [6]:

$$\rho\left(d_1, d_2\right) = \frac{\mathrm{V}_{parallelepiped}}{\mathrm{A}_{base}} = \frac{\left|\overline{a} \cdot \left(\overline{b} \times \overline{A_1A_2}\right)\right|}{\left\|\overline{a} \times \overline{b}\right\|}. \tag{2.37}$$

**Definition 2.8** (see [6]). The support straight line corresponding to the segment which represents the distance between two straight lines is called the **common perpendicular** of the two straight lines.

Let $\delta$ be the straight line which represents the common perpendicular. To determine the equations of the straight line $\delta$:

1. find the direction of the the common perpendicular $\overline{n} = \overline{a} \times \overline{b}$, $\overline{a}$ and $\overline{b}$ being the direction vectors of the two straight lines;
2. write the equation of a plane $\pi_1$, which passes through $d_1$ and contains $\overline{n}$;
3. write the equation of a plane $\pi_2$, which passes through $d_2$ and contains $\overline{n}$;
4. $\delta = \pi_1 \cap \pi_2$ is the common perpendicular searched by us.

If

- $d_1 = \left(A_1, \overline{a}\right),\ A_1\left(x_1, y_1, z_1\right)$,
- $d_2 = \left(A_2, \overline{a}\right),\ A_2\left(x_2, y_2, z_2\right)$,

- $\overline{n} = n_1\overline{i} + n_2\overline{j} + n_3\overline{k}$,
- $\overline{a} = a_1\overline{i} + a_2\overline{j} + a_3\overline{k}$,
- $\overline{b} = b_1\overline{i} + b_2\overline{j} + b_3\overline{k}$,

then the *equations of the common perpendicular* $\delta$ are [6]:

$$\delta : \begin{cases} \begin{vmatrix} x - x_1 & y - y_1 & z - z_1 \\ a_1 & a_2 & a_3 \\ n_1 & n_2 & n_3 \end{vmatrix} = 0 \\ \begin{vmatrix} x - x_2 & y - y_2 & z - z_2 \\ b_1 & b_2 & b_3 \\ n_1 & n_2 & n_3 \end{vmatrix} = 0. \end{cases} \tag{2.38}$$

**Example 2.9** Let be the straight lines

$$d_1 : \begin{cases} x = 1 + 2r \\ y = 3 + r \\ z = -2 + r \end{cases} \text{ and } d_2 : \begin{cases} x = 1 + s \\ y = -2 - 4s, r, s \in \mathbb{R}. \\ z = 9 + 2s \end{cases}$$

Find:

(a) the angle between these straight lines;
(b) the equation of the common perpendicular;
(c) the distance between the two straight lines.

**Solution**

(a) We have

$$\cos \prec (d_1, d_2) = \cos \prec \left(\overline{a}, \overline{b}\right) = \frac{\overline{a} \cdot \overline{b}}{\|\overline{a}\| \left\|\overline{b}\right\|},$$

where

- $d_1 = (A, \overline{a})$, $A(1, 3, -2)$, $\overline{a} = 2\overline{i} + \overline{j} + \overline{k}$,
- $d_2 = \left(B, \overline{b}\right)$, $B(1, -2, 9)$, $\overline{b} = \overline{i} - 4\overline{j} + 2\overline{k}$.

We obtain

$$\cos \prec (d_1, d_2) = 0 \Rightarrow \prec (d_1, d_2) = \frac{\pi}{2}.$$

(b) The direction of the common perpendicular is

$$\overline{n} = \overline{a} \times \overline{b},$$

namely

$$\overline{n} = \begin{vmatrix} \overline{i} & \overline{j} & \overline{k} \\ 2 & 1 & 1 \\ 1 & -4 & 2 \end{vmatrix} = 6\overline{i} - 3\overline{j} - 9\overline{k}.$$

As

- the equation of the plane which passes through $d_1$ and contains $\overline{n}$ is

$$\pi_1 : \begin{vmatrix} x-1 & y-3 & z+2 \\ 2 & 1 & 1 \\ 6 & -3 & -9 \end{vmatrix} = 0 \Leftrightarrow \pi_1 : x - 4y + 2z + 15 = 0;$$

- the equation of the plane which passes through $d_2$ and contains $\overline{n}$ is

$$\pi_2 : \begin{vmatrix} x-1 & y+2 & z-9 \\ 1 & -4 & 2 \\ 6 & -3 & -9 \end{vmatrix} = 0 \Leftrightarrow \pi_2 : 2x + y + z - 9 = 0.$$

The equations of the common perpendicular will be:

$$\begin{cases} x - 4y + 2z + 15 = 0 \\ 2x + y + z - 9 = 0. \end{cases}$$

(c) Using (2.37), we have

$$\rho(d_1, d_2) = \frac{\left|\overline{a} \cdot \left(\overline{b} \times \overline{AB}\right)\right|}{\left\|\overline{a} \times \overline{b}\right\|},$$

where

- $\overline{AB} = (x_B - x_A)\,\overline{i} + (y_B - y_A)\,\overline{j} + (z_B - z_A)\,\overline{k} = -5\overline{j} + 11\overline{k},$
- $\overline{a} \cdot \left(\overline{b} \times \overline{AB}\right) = \begin{vmatrix} 2 & 1 & 1 \\ 1 & -4 & 2 \\ 0 & -5 & 11 \end{vmatrix} = -84,$
- $\left\|\overline{a} \times \overline{b}\right\| = \sqrt{6^2 + (-3)^2 + (-9)^2} = \sqrt{126};$

therefore

$$\rho(d_1, d_2) = \frac{84}{\sqrt{126}} = 2\sqrt{14}.$$

We shall solve this problem using Sage:

```
sage: a=vector([2,1,1]);b=vector([1,-4,2]);var("x y z")
(x, y, z)
sage: a.dot_product(b)
0
sage: n=a.cross_product(b); n
(6, -3, -9)
sage: A=vector([1,3,-2]);B=vector([1,-2,9]);
sage: p1=matrix(SR,3,[x-A[0],y-A[1],z-A[2],a[0],a[1],a[2],n[0],n[1],n[2]]);
sage: pp1=p1.determinant();pp1;pp1/(-6)==0
-6*x + 24*y - 12*z - 90
x - 4*y + 2*z + 15 == 0
sage: p2=matrix(SR,3,[x-B[0],y-B[1],z-B[2],b[0],b[1],b[2],n[0],n[1],n[2]]);
sage: pp2=p2.determinant();pp2;pp2/21==0
42*x + 21*y + 21*z - 189
2*x + y + z - 9 == 0
sage: AB=B-A;
sage: u=a.dot_product(b.cross_product(AB))
sage: rho=u.abs()/n.norm();rho
2*sqrt(14)
```

## 2.5 Angles in E3

### 2.5.1 Angle Between Two Straight Lines

**Definition 2.11** (see [3], p. 112). Let $d_1, d_2$ be two straight lines, which have the direction vectors $\overline{a} = a_1\overline{i} + a_2\overline{j} + a_3\overline{k}$ and respectively $\overline{b} = b_1\overline{i} + b_2\overline{j} + b_3\overline{k}$. The **angle between the straight lines** $d_1$ and $d_2$ is the angle between the vectors $\overline{a}$ and $\overline{b}$ (see Fig. 2.15).

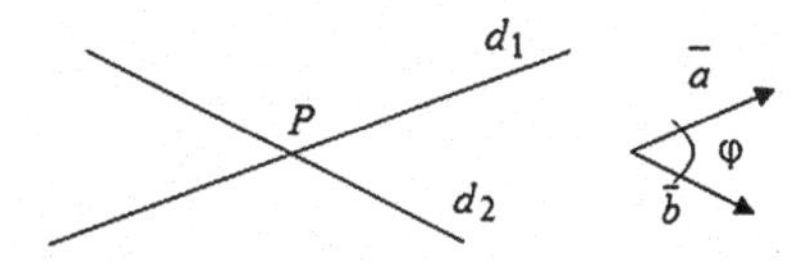


**Fig. 2.15** The angle between two straight lines

Hence

$$\cos\varphi = \frac{\overline{a}\cdot\overline{b}}{\|\overline{a}\|\,\|\overline{b}\|} = \frac{a_1b_1 + a_2b_2 + a_3b_3}{\sqrt{a_1^2 + a_2^2 + a_3^2}\sqrt{b_1^2 + b_2^2 + b_3^2}}, \ \varphi \in [0, \pi]. \tag{2.39}$$

**Remark 2.12** (see [3], p. 112).

(1) $\varphi = \frac{\pi}{2} \Rightarrow$the straight lines are perpendicular $\Leftrightarrow$

$$\overline{a}\cdot\overline{b} = 0 \Leftrightarrow a_1b_1 + a_2b_2 + a_3b_3 = 0.$$

(2) $\varphi = 0 \Rightarrow$the straight lines are parallel$\Leftrightarrow \overline{a}\times\overline{b} = \overline{0} \Leftrightarrow$

$$\begin{vmatrix} \overline{i} & \overline{j} & \overline{k} \\ a_1 & a_2 & a_3 \\ b_1 & b_2 & b_3 \end{vmatrix} = \overline{0} \Leftrightarrow (a_2b_3 - a_3b_2)\,\overline{i} + (a_3b_1 - a_1b_2)\,\overline{j} + (a_1b_2 - a_2b_1)\,\overline{k} = \overline{0} \Leftrightarrow$$

$$\begin{cases} a_2b_3 - a_3b_2 = 0 \\ a_3b_1 - a_1b_2 = 0 \\ a_1b_2 - a_2b_1 = 0 \end{cases} \Leftrightarrow \frac{a_1}{b_1} = \frac{a_2}{b_2} = \frac{a_3}{b_3}. \tag{2.40}$$

Therefore $d_1 || d_2$ the relation (2.40) occurs.

### 2.5.2 Angle Between Two Planes

Let be

- $\pi_1 : a_1x + b_1y + c_1z + d_1 = 0$ and the normal vector $\overline{n}_1 = a_1\overline{i} + b_1\overline{j} + c_1\overline{k}$,

- $\pi_2 : a_2x + b_2y + c_2z + d_2 = 0$ and the normal vector $\overline{n}_2 = a_2\overline{i} + b_2\overline{j} + c_2\overline{k}$.

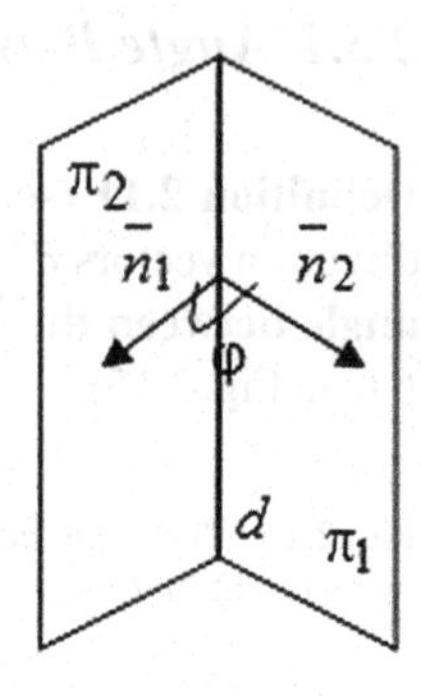


**Fig. 2.16** The angle between two planes

**Definition 2.13** (see [3], p. 113). The **angle** $\varphi$ **between the planes** $\pi_1$ and $\pi_2$ is the angle between the vectors $\overline{n}_1$ and $\overline{n}_2$ (see Fig. 2.16).

Hence

$$\cos\varphi = \frac{\overline{n}_1 \cdot \overline{n}_2}{\|\overline{n}_1\|\,\|\overline{n}_2\|} = \frac{a_1a_2 + b_1b_2 + c_1c_2}{\sqrt{a_1^2 + b_1^2 + c_1^2}\sqrt{a_2^2 + b_2^2 + c_2^2}}, \ \varphi \in [0, \pi]. \tag{2.41}$$

**Remark 2.14** (see [3], p. 113).

(1) $\pi_1 || \pi_2 \Longleftrightarrow \overline{n}_1$ and $\overline{n}_2$ are collinear $\Longleftrightarrow \overline{n}_1 = t\overline{n}_2, t \in \mathbb{R}$; therefore

$$\pi_1 || \pi_2 \Longleftrightarrow \overline{a}_1 = t\overline{a}_2, \overline{b}_1 = t\overline{b}_2, \overline{c}_1 = t\overline{c}_2.$$

(2) $\pi_1 \perp \pi_2 \Longleftrightarrow \overline{n}_1 \cdot \overline{n}_2 = 0 \Longleftrightarrow a_1a_2 + b_1b_2 + c_1c_2 = 0.$

### *2.5.3 Angle Between a Straight Line and a Plane*

Let $d$ be the straight line with the direction vector $\overline{a} = a_1\overline{i} + a_2\overline{j} + a_3\overline{k}$ and the plane $\pi$ having the normal vector $\overline{n} = n_1\overline{i} + n_2\overline{j} + n_3\overline{k}$.

**Definition 2.15** (see [3], p. 113). The angle $\varphi$ between the straight line $d$ and the plane $\pi$ is the angle between the straight line $d$ and the projection of this straight line on the plane $\pi$ (see Fig. 2.17).

**Fig. 2.17** The angle between a straight line and a plane

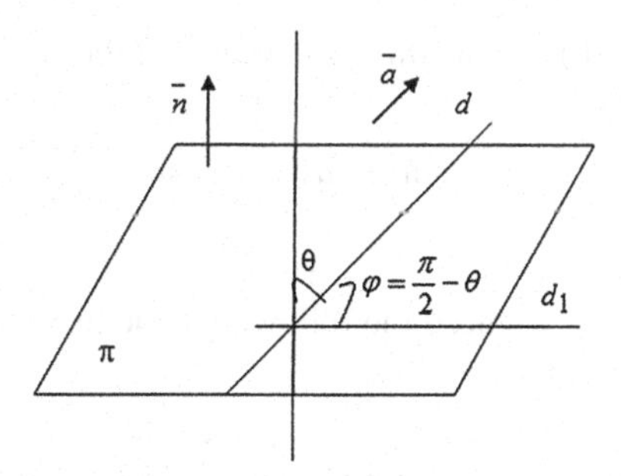


The angle between the straight line $d$ and the plane $\pi$ is related to the angle $\theta$, the angle of the vectors $\overline{a}$ and $\overline{n}$, through the relations: $\theta = \frac{\pi}{2} \pm \varphi$ as the vectors are on the same side of the or in different parts. Hence:

$$\cos\theta = \cos\left(\frac{\pi}{2} \pm \varphi\right) = \pm\sin\varphi,\ \theta \in [0, \pi] \Rightarrow \varphi \in \left[0, \frac{\pi}{2}\right].$$

As

$$\cos\theta = \frac{\overline{n}\cdot\overline{a}}{\|\overline{n}\|\ \|\overline{a}\|},\ \theta \in [0, \pi]$$

it results that

$$\sin\varphi = \frac{|a_1n_1 + a_2n_2 + a_3n_3|}{\sqrt{n_1^2 + n_2^2 + n_3^2}\sqrt{a_1^2 + a_2^2 + a_3^2}},\ \varphi \in \left[0, \frac{\pi}{2}\right]. \tag{2.42}$$

**Remark 2.16** (see [3], p. 113).

(1) $d||\pi \Leftrightarrow \overline{n}\cdot\overline{a} = 0 \Leftrightarrow a_1n_1 + a_2n_2 + a_3n_3 = 0.$
(2) $d\perp\pi \Rightarrow \varphi = \frac{\pi}{2} \Rightarrow \theta = 0 \Rightarrow \overline{n}||\overline{a} \overset{(2.40)}{\Longrightarrow} \frac{n_1}{a_1} = \frac{n_2}{a_2} = \frac{n_3}{a_3}.$

**Example 2.17** Are given

- the planes $\pi_1 : 2x - y + 7 = 0$ and $\pi_2 : x - 5y + 3z = 0$,
- the straight line $d : \frac{x-1}{2} = \frac{y+3}{-1} = \frac{z}{5}$.

Compute:

(a) the angle of these planes;
(b) the angle between the straight line and the plane $\pi_1$.

**Solution**

(a) We have $\overline{n}_1 = (2, -1, 0)$, $\overline{n}_2 = (1, -5, 3)$; hence

$$\cos \prec (\pi_1, \pi_2) = \cos\varphi \overset{(2.41)}{=} \frac{\overline{n}_1\cdot\overline{n}_2}{\|\overline{n}_1\|\ \|\overline{n}_2\|} = \frac{\sqrt{7}}{5} = 0.529.$$

(b) As the direction vector of the straight line $d$ is $\overline{a} = (2, -1, 5)$ we obtain

$$\sin \prec (d, \pi_1) = \sin \varphi \stackrel{(2.42)}{=} \frac{|a_1 n_1 + a_2 n_2 + a_3 n_3|}{\sqrt{n_1^2 + n_2^2 + n_3^2}\sqrt{a_1^2 + a_2^2 + a_3^2}} = \frac{1}{\sqrt{6}}.$$

We need the following code in Sage to solve this problem:

```
sage: var("x y z")
(x, y, z)
sage: pl1=implicit_plot3d(2*x-y+7==0, (x, -10, 10), (y, -10, 10),(z, -10, 10),rgbcolor="green")
sage: pl2=plot3d((5*y-x)/3, (x, -10, 0), (y, -10, 10),rgbcolor="purple")
sage: l=line3d([(7,-6,15), (-15,5,-40)], color='blue')
sage: pl1+pl2+l
```

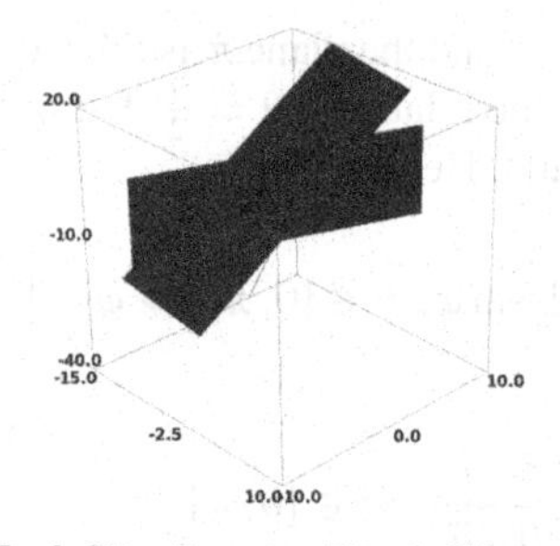


```
sage: n1=vector([2,-1,0]);n2=vector([1,-5,3]);
sage: (n1.dot_product(n2)/(n1.norm()*n2.norm())).n(digits=3)
0.529
sage: a=vector([2,-1,5]);
sage: (a.dot_product(n1)).abs()/(n1.norm()*a.norm()).n(digits=3)
0.408
```

## 2.6 Problems

1. Check if the points $M_1\,(3, 0, 1)$, $M_2\,(0, 2, 4)$, $M_3\left(1, \frac{4}{3}, 3\right)$ are collineare.

**Solution**

Using Sage we shall have:

```
sage: var("x y z")
(x, y, z)
sage: M1=vector([3,0,1]);M2=vector([0,2,4]);M3=vector([1,4/3,3])
sage: (M3[0]-M1[0])/(M2[0]-M1[0])==(M3[1]-M1[1])/(M2[1]-M1[1])==(M3[2]-M1[2])/(M2[2]-M1[2])
True
sage: po1=point3d((3,0,1),size=20,color='red')
sage: po2=point3d((0,2,4),size=20,color='blue')
sage: po3=point3d((1,4/3,3),size=20,color='green')
sage: l=line3d([(3,0,1),(0,2,4)])
sage: po1+po2+po3+l
```

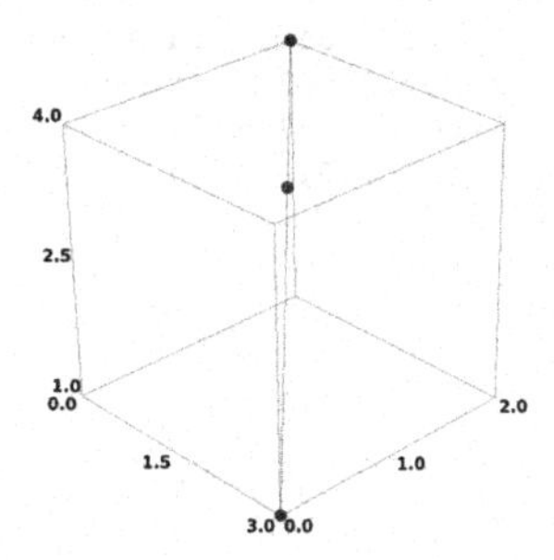


2. Write the equation of the plane determined by the points: $M_1\,(3, 1, 0)$, $M_2\,(0, 7, 2)$, $M_3\,(4, 1, 5)$.

**Solution**

We shall give a solution in Sage:

```
sage: var("x y z")
(x, y, z)
sage: M1=vector([3,1,0]);M2=vector([0,7,2]);M3=vector([4,1,5])
sage: M=matrix(SR,3,[x-M1[0],y-M1[1],z-M1[2],M2[0]-M1[0],M2[1]-M1[1],M2[2]-M1[2],M3[0]-M1[0],M3[1]-M1[1],M3[2]-M1[2]])
sage: M.determinant()==0
30*x + 17*y - 6*z - 107 == 0
sage: po1=point3d((3,1,0),size=20,color='red')
sage: po2=point3d((0,7,2),size=20,color='blue')
sage: po3=point3d((4,1,5),size=20,color='green')
sage: pl=plot3d((30*x+17*y-107)/6, (x, -1, 5), (y, -1, 8),rgbcolor="lightblue")
sage: po1+po2+po3+pl
```

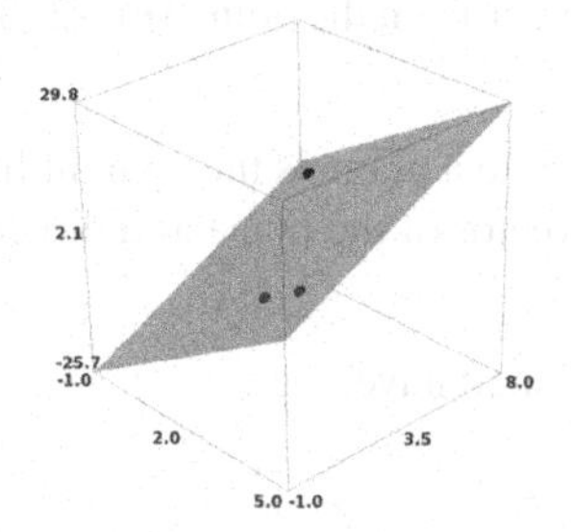


3. Write the equation of a plane perpendicular on the planes $\pi_1 : x - y + z - 1 = 0$ and $\pi_2 : 2x + y + z + 1 = 0$ and which passes through the point $M_0\,(1, -1, 1)$.
4. Write the equation of a plane which passes through the points $M_1\,(3, 1, 2)$, $M_2\,(4, 6, 5)$ and is parallel with the vector $\overline{v} = \overline{i} + 2\overline{j} + 3\overline{k}$.

**Solution**

Solving this problem in Sage we have:

```
sage: var("x y z")
(x, y, z)
sage: M1=vector([2,3,4]);M2=vector([4,6,5])
sage: M1M2=M2-M1;M1M2
(2, 3, 1)
sage: r=vector([x,y,z]);v=vector([1,2,3])
sage: u=M1M2.cross_product(v);u
(7, -5, 1)
sage: w=r-M1;w
(x - 2, y - 3, z - 4)
sage: u.dot_product(w)==0
7*x - 5*y + z - 3 == 0
```

```
sage: a=arrow3d((0,0,0), (1,2,3),color='blue');
sage: po1=point3d((2,3,4),size=20,color='red')
sage: po2=point3d((4,6,5),size=20,color='blue')
sage: pl=plot3d(-7*x+5*y+3, (x, 1.5, 7), (y, 0, 7),rgbcolor="lightblue")
sage: a+pl+po1+po2
```

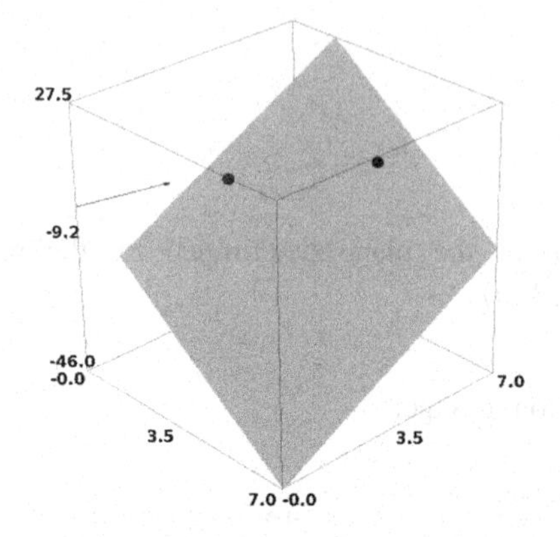


5. Let $d$ be the straight line determined by the point $P_0\,(2, 0, -1)$ and the direction vector $\overline{v} = \overline{i} - \overline{j}$. Compute the distance from the point $P\,(1, 3, -2)$ to the straight line $d$.
6. Write the ecuation of the perpendicular from the point $M\,(-2, 0, 3)$ on the plane $\pi : 7x - 5y + z - 11 = 0$.

*Hint*. The perpendicular from a point to a plane is the straight line which passes through that point and has the normal vector of the plane as a direction vector.

**Solution**

Solving this problem with Sage, we shall have:

```
sage: var("x y z")
(x, y, z)
sage: n=vector([7,-5,1]);M=vector([-2,0,3])
sage: eq1=[(x-M[0])/n[0]==(y-M[1])/n[1]];eq2=[(y-M[1])/n[1]==(z-M[2])/n[2]];eq1+eq2
[1/7*x + 2/7 == -1/5*y, -1/5*y == z - 3]
sage: l=line3d([(-2,0,3),(-9,5,2),(5,-5,4)],color="purple")
sage: po1=point3d((-2,0,3),size=20,color='red')
sage: pl=plot3d(11-7*x+5*y, (x, -1, 1), (y, -1, -0.5),rgbcolor="blue")
sage: po1+pl+l
```

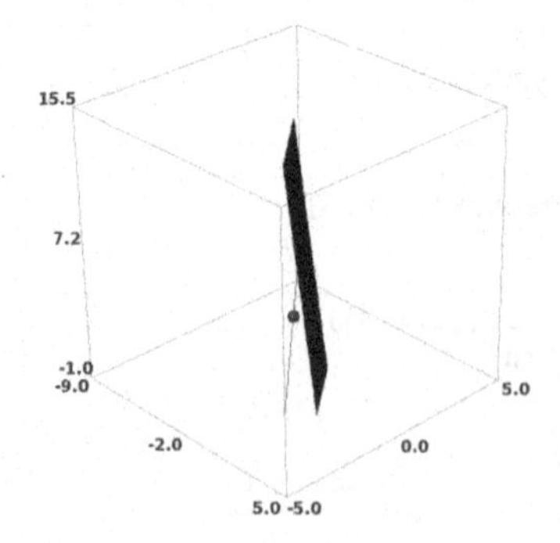

7. It gives a tetrahedron $ABCD$ defined by the points $A\,(3,0,0)$, $B\,(2,4,0)$, $C\,(-3,-1,0)$, $D\,(0,0,5)$. Write the equations of its faces, the edge equations and the equations corresponding to the heights of the tetrahedron $ABCD$.
8. Write the equation of the plane which passes through Oz and is perpendicular on the plane $\pi : 8x + y + 2z - 1 = 0$.

*Hint*. The equation of the plane which passes through Oz is: $ax + by = 0$.

**Solution**

We shall present the solution in Sage:

```
sage: var("a b x y z")
(a, b, x, y, z)
sage: n1=vector([a,b,0]);n2=vector([8,1,2])
sage: eqn=[n1.dot_product(n2)==0];
sage: s=solve(eqn,a,b);s
([a == -1/8*b], [1])
sage: n1=vector([s[0][0].subs(b==1),1,0]);n1
(a == (-1/8), 1, 0)
sage: l=line3d([(0,0,1),(0,0,-1),(0,0,-11)],color="purple")
sage: p1=implicit_plot3d(-x/8+y==0, (x, -10, 10), (y, -10, 10),(z, -10, 0),rgbcolor="lightblue")
sage: p2=plot3d((1-8*x-y)/2, (x, -1, 0), (y, -1, 2),rgbcolor="red")
```

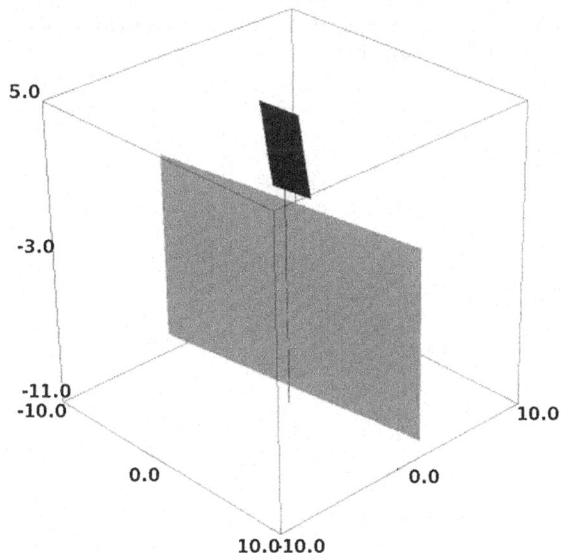


9. Determine the projection equation of the straight line having the equations

$$d : \begin{cases} x - 3z + 1 = 0 \\ y - 2z - 3 = 0 \end{cases}$$

on the plane $\pi : x - y + 2z - 1 = 0$.

10. Let be the straight lines:

$$d_1 : \begin{cases} x - 2y + z + 1 = 0 \\ y - z = 0 \end{cases}$$

and

$$d_2 : \frac{x-1}{2} = \frac{y+3}{1} = \frac{z}{8}.$$

(a) Find the equation of the common perpendicular.
(b) Compute the distance between the two straight lines.

## References

1. S. Chiriţă (ed.), *Probleme de matematici superioare* (Didactică şi Pedagogică, Bucureşti, 1989)
2. V. Postelnicu, S. Coatu (eds.), *Mică enciclopedie matematică* (Tehnică, Bucureşti, 1980)
3. C. Udrişte, *Algebră liniară, geometrie analitică* (Geometry Balkan Press, Bucureşti, 2005)
4. I. Vladimirescu, M. Popescu (eds.), *Algebră liniară şi geometrie n- dimensională* (Radical, Craiova, 1996)
5. I. Vladimirescu, M. Popescu, *Algebră liniară şi geometrie analitică*, ed. Universitaria, Craiova,1993
6. I. Vladimirescu, M. Popescu, M. Sterpu, *Algebră liniară şi geometrie analitică*, Note de curs şi aplica ţii, Universitatea din Craiova, 1993

# Chapter 3
# Linear Transformations

## 3.1 Linear Transformations

The *linear transformations* should be studied [1] because they are compatible with the operations defined in a vector space and allow the transfer of some algebraic situations or problems from a space to another. Matrix operations clearly reflect their similarity with the operations with linear transformations; hence the matrices can be used for numerical representation of the linear transformations. The matrix representation of linear transformations is [1] analogous to the representation of the vectors through $n$ coordinates, relative to a basis.

**Definition 3.1** (see [2], p. 41). Let $U$ and $V$ two vector spaces over the field $K$. The mapping $T : U \to V$ is called a **Linear transformation** (or a **linear mapping**) if the following conditions are satisfied:

(1) $T(\overline{x}+\overline{y}) = T(\overline{x}) + T(\overline{y}), (\forall)\,\overline{x},\overline{y} \in U$, namely $T$ is *additive*;
(2) $T(\alpha\overline{x}) = \alpha T(\overline{x}), (\forall)\,\overline{x} \in U$, namely $T$ is *homogeneous*.

The two properties of the linear maps can be formulated in a single.

**Proposition 3.2** (see [2], p. 41). The mapping $T : U \to V$ is linear if and only if

$$T(\alpha\overline{x}+\beta\overline{y}) = \alpha T(\overline{x}) + \beta T(\overline{y}), (\forall)\,\alpha,\beta \in K, (\forall)\,\overline{x},\overline{y} \in U. \tag{3.1}$$

**Examples of linear transformations**

1. $T : \mathbb{R} \to \mathbb{R}, T(x) = \lambda x, \lambda \in \mathbb{R}$;
2. $\Im : V \to V, \Im(\overline{x}) = \overline{x}, (\forall)\,\overline{x} \in V$ *identity mapping*;
3. $\emptyset : U \to V, \emptyset(\overline{x}_U) = \overline{0}_V$ *zero mapping*;
4. $T : \mathbb{R}^n \to \mathbb{R}^m, T(\overline{x}) = A\overline{x}, (\forall)\,\overline{x} = \left(x^{(1)},\ldots,x^{(n)}\right)^t \in \mathbb{R}^n$, $A \in \mathrm{M}_{m,n}(\mathbb{R})$ is given;
5. $T : \mathrm{M}_n(K) \to K, T(A) = \mathrm{trace}(A) = \sum_{i=1}^{n} a_{ii}$;
6. $T : \mathrm{M}_{m,n}(K) \to \mathrm{M}_{n,m}(K), T(A) = A^t$;
7. $T : V_3 \to \mathbb{R}^3, T(\overline{v}) = \left(v^{(1)}, v^{(2)}, v^{(3)}\right), \overline{v} = v^{(1)}\overline{i} + v^{(2)}\overline{j} + v^{(3)}\overline{k}$.

G. A. Anastassiou and I. F. Iatan, *Intelligent Routines II*,
Intelligent Systems Reference Library 58, DOI: 10.1007/978-3-319-01967-3_3,

We shall denote by $\mathrm{L}(U, V)$ the set of all linear transformations defined on $U$ with their values in $V$.

**Definition 3.3** (see [2], p. 42). Any linear mapping $T : V \to V$ is called an **endomorphism** of the vector space $V$. We shall denote by End $(V)$ the set of the all endomorphisms corresponding to the vector space $V$.

**Definition 3.4** (see [2], p. 42). A linear mapping $T : V \to K$ is called a **linear form**.
**Definition 3.5** (see [2], p. 42). Let be $T \in \mathrm{L}(U, V)$.

(a) The set

$$\mathrm{Ker}\, T = \left\{\overline{x} \in U | T(\overline{x}) = \overline{0}_V\right\} \tag{3.2}$$

is called the **kernel** of the linear transformation $T$.

(b) The set

$$\mathrm{Im}\, T = T(U) = \{\overline{v} \in V |\ (\exists) \overline{u} \in U, T(\overline{u}) = \overline{v}\} \tag{3.3}$$

is called the **image** of the linear transformation $T$.

**Definition 3.6** (see [2], p. 42). A linear transformation $T : U \to V$ is called:

(1) **injective** if and only if $\mathrm{Ker}\, T = \left\{\overline{0}_V\right\}$;
(2) **surjective** if and only if $\mathrm{Im}\, T = V$.

**Definition 3.7** (see [3], p. 57). Any bijective endomorphism of the vector space $V$ is called the **automorphism** of the vector space.

We shall denote by $\mathrm{Aut}(V)$ the set of all automorphisms of the vector space $V$.

**Proposition 3.8** (see [4], p. 31). Let $T : U \to V$ be a bijective linear transformation. If $\mathrm{B}_1 = \{\overline{e}_1, \ldots, \overline{e}_n\}$ is a basis in $U$, then $\mathrm{B}_2 = \{T(\overline{e}_1), \ldots, T(\overline{e}_n)\}$ is a basis in $V$.

**Theorem 3.9** (see [5], p. 45). Let $T : U_n \to V_n$ be a linear mapping between two vector spaces with the same dimension. Then the following statements are equivalent:

(i) $T$ is injective;
(ii) $T$ is surjective;
(iii) $T$ is bijective.

**Definition 3.10** (see [2], p. 42). A linear mapping $T : U \to V$ which is injective and surjective is called **isomorphism** between the vector spaces $U$ and $V$.

**Theorem 3.11** (see [3], p. 29). Let $U$ and $V$ be two finite dimensional vector spaces over the field $K$. Then they are isomorphic if and only if $\dim U = \dim V$.

**Definition 3.12** (see [4], p. 26). Let $S, T \in \mathrm{L}(U, V)$. The **sum of the two linear transformations** is the linear transformation $R \in L(U, V)$,

$$R(\overline{x}) = (S + T)(\overline{x}) = S(\overline{x}) + T(\overline{x}), (\forall)\, \overline{x} \in U. \tag{3.4}$$

**Definition 3.13** (see [4], p. 26). Let $T \in \mathrm{L}(U, V)$. The **scalar multiples of the linear transformation** $T$, denoted $\alpha T \in L(U, V)$ is defined as:

$$(\alpha T)(\overline{x}) = \alpha T(\overline{x}), (\forall)\alpha \in U, (\forall)\overline{x} \in U. \tag{3.5}$$

**Definition 3.14** (see [5], p. 41). The composition of the two linear transformations is called the **product** (or the **multiplication**) and is defined as in the case of the functions.

**Remark 3.15** (see [5], p. 41). The composition isn't commutative but it is associative.

**Proposition 3.16** (see [4], p. 27). If $U, V, W$ are some vector spaces over $K$ and $T : U \to V$, $S : V \to W$ are two linear transformations, then the mapping $S \circ T : U \to V$,

$$(S \circ T)(\overline{x}) = S(T(\overline{x})), (\forall)\ \overline{x} \in U \tag{3.6}$$

is linear.

**Example 3.17** (see [6], p. 36). Let be the linear transformations

$$f : \mathbb{R}^2 \to \mathbb{R}^2, f(\overline{x}) = f(x_1, x_2) = (2x_1 - x_2, 3x_2)$$
$$g : \mathbb{R}^2 \to \mathbb{R}^2, g(\overline{x}) = g(x_1, x_2) = (x_2 - x_1, 4x_1).$$

(a) Show that they are linear maps.
(b) Determine $f \circ g$, $g \circ f$ and check if they are linear maps.

**Solution**

(a) $f$ is a linear mapping$\Leftrightarrow$

$$f(\alpha\overline{x} + \beta\overline{y}) = \alpha f(\overline{x}) + \beta f(\overline{y}), (\forall)\alpha, \beta \in \mathbb{R}, (\forall)\overline{x}, \overline{y} \in \mathbb{R}^2.$$

Let $\alpha, \beta \in \mathbb{R}$ and $(\forall)\ \overline{x}, \overline{y} \in \mathbb{R}^2, \overline{x} = (x_1, x_2), \overline{y} = (y_1, y_2)$. We have:

$$\begin{aligned}
f(\alpha\overline{x} + \beta\overline{y}) &= f(\alpha x_1 + \beta y_1, \alpha x_2 + \beta y_2) \\
&= (2(\alpha x_1 + \beta y_1) - (\alpha x_2 + \beta y_2), 3(\alpha x_2 + \beta y_2)) \\
&= (\alpha(2x_1 - x_2) + \beta(2y_1 - y_2), \alpha \cdot 3x_2 + \beta \cdot 3y_2) \\
&= (\alpha(2x_1 - x_2), \alpha \cdot 3x_2) + (\beta(2y_1 - y_2), \beta \cdot 3y_2) \\
&= \alpha(2x_1 - x_2, 3x_2) + \beta(2y_1 - y_2, 3y_2) = \alpha f(\overline{x}) + \beta f(\overline{y}).
\end{aligned}$$

Therefore, $f$ is a linear mapping. Similarly, one shows that $g$ is also a linear mapping.

(b) We obtain

$$\begin{aligned}
(f \circ g)(\overline{x}) &= f(g(\overline{x})) = f(x_2 - x_1, 4x_1) \\
&= (2(x_2 - x_1) - 4x_1, 12x_1) = (2x_2 - 6x_1, 12x_1);
\end{aligned}$$

$$(g \circ f)(\overline{x}) = g(f(\overline{x})) = g(2x_1 - x_2, 3x_2) = (3x_2 - (2x_1 - x_2), 4(2x_1 - x_2))$$
$$= (4x_2 - 2x_1, 8x_1 - 4x_2).$$

Using the Proposition 3.16 it results that the transformations $f \circ g$ and $g \circ f$ are linear.

We can check this result in Sage, too:

```
sage: F=matrix(RR,[[2,-1],[0,3]])
sage: f=linear_transformation(RR^2, RR^2, F,side='right');f
Vector space morphism represented by the matrix:
[ 2.00000000000000 0.000000000000000]
[-1.00000000000000  3.00000000000000]
Domain: Vector space of dimension 2 over Real Field with 53 bits of precision
Codomain: Vector space of dimension 2 over Real Field with 53 bits of precision
sage: G=matrix(RR,[[-1,1],[4,0]])
sage: g = linear_transformation(RR^2, RR^2, G,side='right');g
Vector space morphism represented by the matrix:
[-1.00000000000000  4.00000000000000]
[ 1.00000000000000 0.000000000000000]
Domain: Vector space of dimension 2 over Real Field with 53 bits of precision
Codomain: Vector space of dimension 2 over Real Field with 53 bits of precision
sage: f*g
Vector space morphism represented by the matrix:
[-6.00000000000000  12.0000000000000]
[ 2.00000000000000 0.000000000000000]
Domain: Vector space of dimension 2 over Real Field with 53 bits of precision
Codomain: Vector space of dimension 2 over Real Field with 53 bits of precision
sage: g*f
Vector space morphism represented by the matrix:
[-2.00000000000000  8.00000000000000]
[ 4.00000000000000 -4.00000000000000]
Domain: Vector space of dimension 2 over Real Field with 53 bits of precision
Codomain: Vector space of dimension 2 over Real Field with 53 bits of precision
```

**Definition 3.18** (see [5], p. 42). The **natural powers** of an endomorphism $T : V \to V$ are defined inductive as:

$$\begin{cases} T^0 = \mathfrak{I} \\ T^n = T \cdot T^{n-1}, n = 1, 2, \ldots \end{cases}$$

**Definition 3.19** (see [5], p. 42). Let $T \in \mathrm{L}(U, V)$ be a bijective linear mapping (hence it is invertible). Its **inverse,** $T^{-1} \in \mathrm{L}(U, V)$ is a linear mapping.

**Example 3.20** (see [6], p. 36). Let be the linear mapping

$$f : \mathbb{R}^3 \to \mathbb{R}^3, f(\overline{x}) = f(x_1, x_2, x_3) = (x_1 - 2x_2 + x_3, 2x_1 + x_2 - x_3, x_2 - 3x_3).$$

(a) Show that $f$ is invertible and determine its inverse $f^{-1} : \mathbb{R}^3 \to \mathbb{R}^3$.
(b) Check if the mapping $f^{-1}$ is linear too.

**Solution**

(a) We know (see the Definition 3.6) is injective if and only if Ker $T = \{\overline{0}_{\mathbb{R}^3}\}$. Let be $\overline{x} \in \mathbb{R}^3$.

$$f(\overline{x}) - \overline{0}_{\mathbb{R}^3} \Leftrightarrow (x_1 - 2x_2 + x_3, 2x_1 + x_2 - x_3, x_2 - 3x_3) = (0, 0, 0)$$
$$\Leftrightarrow \begin{cases} x_1 - 2x_2 + x_3 = 0 \\ 2x_1 + x_2 - x_3 = 0 \\ x_2 - 3x_3 = 0. \end{cases}$$

Since the determinant of the matrix $A$, associated to the previous system is:

$$\Delta = \begin{vmatrix} 1 & -2 & 1 \\ 2 & 1 & -1 \\ 0 & 1 & -3 \end{vmatrix} \neq 0 \Rightarrow \text{rank}(A) = 3 \Rightarrow$$

the homogeneous system is compatible determined, having the unique solution $\overline{x} = (0, 0, 0)$.

Therefore, it results that Ker $T = \{\overline{0}_{\mathbb{R}^3}\}$, i.e. that $f$ is injective. If $f$ is injective then according to the Theorem 3.9 it results that $f$ is bijective, namely (see the Definition 3.19) $f$ is invertible.

Let be $\overline{y} \in \mathbb{R}^3$, $\overline{y} = (y_1, y_2, y_3)$. Then

$$f(\overline{x}) = \overline{y} \Leftrightarrow (x_1 - 2x_2 + x_3, 2x_1 + x_2 - x_3, x_2 - 3x_3) = (y_1, y_2, y_3)$$
$$\Leftrightarrow \begin{cases} x_1 - 2x_2 + x_3 = y_1 \\ 2x_1 + x_2 - x_3 = y_2 \\ x_2 - 3x_3 = y_3. \end{cases}$$

The system solution is:

$$\begin{cases} x_1 = \frac{1}{12}(2y_1 + 5y_2 - y_3) \\ x_2 = \frac{1}{12}(-6y_1 + 3y_2 - 3y_3) \\ x_3 = \frac{1}{12}(-2y_1 + y_2 - 5y_3). \end{cases}$$

We obtain $f^{-1} : \mathbb{R}^3 \to \mathbb{R}^3$,

$$f^{-1}(\overline{x}) = \left(\frac{1}{12}(2x_1 + 5x_2 - x_3), \frac{1}{12}(-6x_1 + 3x_2 - 3x_3), \frac{1}{12}(-2x_1 + x_2 - 5x_3)\right).$$

(b) Using the Definition 3.19 it results that the mapping $f^{-1}$ is linear.

Solving this problem in Sage we shall have:

```
sage: F=matrix(RR,[[1,-2,1],[2,1,-1],[0,1,-3]])
sage: f=linear_transformation(RR^3, RR^3, F,side='right');f
Vector space morphism represented by the matrix:
[ 1.00000000000000  2.00000000000000 0.000000000000000]
[-2.00000000000000  1.00000000000000  1.00000000000000]
[ 1.00000000000000 -1.00000000000000 -3.00000000000000]
Domain: Vector space of dimension 3 over Real Field with 53 bits of precision
Codomain: Vector space of dimension 3 over Real Field with 53 bits of precision
sage: f^(-1)
Vector space morphism represented by the matrix:
[  0.166666666666667  -0.500000000000000  -0.166666666666667]
[  0.416666666666667   0.250000000000000  0.0833333333333333]
[-0.0833333333333334  -0.250000000000000  -0.416666666666667]
Domain: Vector space of dimension 3 over Real Field with 53 bits of precision
Codomain: Vector space of dimension 3 over Real Field with 53 bits of precision
```

**Theorem 3.21** (see [5], p. 41). The set $\mathrm{L}(U, V)$ is a vector space over the field $K$ relative to the addition of the linear transformations and the scalar multiplication of a linear transformation.

**Theorem 3.22** (see [5], p. 43). Let $T \in \mathrm{L}(U, V)$. We have the following properties:

(i) Ker $T$ is a vector subspace of $U$;
(ii) Im $T$ is a vector subspace of $V$.

**Proposition 3.23** (see [5], p. 40 and [3], p. 27). If $T \in L(U, V)$. Then:

(1) $T\left(\overline{0}_U\right) = \overline{0}_V$, i.e. a linear transformation maps the null vector, in the null vector;
(2) $T(-\overline{x}) = -T(\overline{x})$, $(\forall)\, \overline{x} \in U$;
(3) If $W$ is a vector subspace of $U$, then $T(W)$ is a vector subspace of $V$;
(4) If the vectors $\overline{x}_1, \ldots, \overline{x}_n \in U$ are linearly dependent then the vectors $T(\overline{x}_1)$, $\ldots, T(\overline{x}_n) \in V$ are linearly dependent, too;
(5) Being given the vectors $\overline{x}_1, \ldots, \overline{x}_n \in U$, if the vectors $T(\overline{x}_1), \ldots, T(\overline{x}_n) \in V$ are linearly independent then the vectors $\overline{x}_1, \ldots, \overline{x}_n$ are linearly independent, too.

**Proposition 3.24** (see [3], p. 28). Let $T \in L(U, V)$ be an injective linear mapping. If the system of vectors $\{\overline{x}_1, \ldots, \overline{x}_n\} \subset U$ is linearly independent, the system of vectors $T(\overline{x}_1), \ldots, T(\overline{x}_n)$ is linearly independent.

**Proposition 3.25** (see [3], p. 28). Let $U$, $V$ be two vector spaces over the field $K$ and $T \in L(U, V)$. The following statements hold:

(a) If $\{\overline{x}_1, \ldots, \overline{x}_n\} \subset U$ is a system of generators for $U$, then $T(\overline{x}_1), \ldots, T(\overline{x}_n)$ is a system of generators for Im $T$.
(b) If $T$ is surjective and $\{\overline{x}_1, \ldots, \overline{x}_n\} \subset U$ is a system of generators for $U$, then $T(\overline{x}_1), \ldots, T(\overline{x}_n)$ is a system of generators for $V$.

**Theorem 3.26** (see [3], p. 29 and [2], p. 44). Let $U$, $V$ be two vector spaces over the field $K$. If $\dim U < \infty$ and $T \in L(U, V)$, then

$$\dim\ \mathrm{Ker} T + \dim \mathrm{Im} T = \dim U.$$

**Definition 3.27** (see [2], p. 44). Let $T \in L(U, V)$.

(a) The $T$'s kernel dimension is called **defect** of $T$.
(b) The $T$'s image dimension is called the **rank** of $T$.

Let $U$, $V$ be two finite dimensional vector spaces over the field $K$, $\dim U = n$, $\dim V = m$ and $T \in L(U, V)$. Let $B_1 = \{\overline{e}_1, \ldots, \overline{e}_n\}$ and $B_2 = \{\overline{f}_1, \ldots, \overline{f}_m\}$ be two bases in $U$ and respectively in $V$. We consider the expressions of the vectors $T(\overline{e}_1), \ldots, T(\overline{e}_n) \in V$ in the basis $B_2$:

$$\begin{cases} T(\overline{e}_1) = \alpha_1^{(1)}\overline{f}_1 + \alpha_1^{(2)}\overline{f}_2 + \cdots + \alpha_1^{(m)}\overline{f}_m \\ T(\overline{e}_2) = \alpha_2^{(1)}\overline{f}_1 + \alpha_1^{(2)}\overline{f}_2 + \cdots + \alpha_2^{(m)}\overline{f}_m \\ \vdots \\ T(\overline{e}_n) = \alpha_n^{(1)}\overline{f}_1 + \alpha_n^{(2)}\overline{f}_2 + \cdots + \alpha_n^{(m)}\overline{f}_m. \end{cases}$$

We shall denote by

$$\widetilde{T}_{(B_1,B_2)} = \begin{pmatrix} \alpha_1^{(1)} & \ldots & \alpha_n^{(1)} \\ \vdots & & \vdots \\ \alpha_1^{(m)} & \ldots & \alpha_n^{(m)} \end{pmatrix} \tag{3.7}$$

the $m \times m$ matrix, where the column by index $i$ contains the coordinates of the vector $T(\overline{e}_i)$.

**Definition 3.28** (see [2], p. 43). The matrix $\widetilde{T}_{(B_1,B_2)}$ is called the **associated matrix of the linear transformation** $T$ relative to the bases $B_1$ and $B_2$, fixed in the vector spaces $U$ and $V$.

**Example 3.29** (see [9]). Let be

$$f : \mathbb{R}^3 \to \mathbb{R}^2,\ f(\overline{x}) = f(x_1, x_2, x_3) = (2x_1 + x_2 - x_3, x_2 + 7x_3).$$

(a) Show that $f$ is a linear transformation.
(b) Write the associated matrix of $f$ relative to the canonical bases of the two spaces $\mathbb{R}^3$ and $\mathbb{R}^2$.
(c) Determine $\operatorname{Ker} f$ and $\operatorname{Im} f$.
(d) Is $f$ surjective?

**Solution**

(a) Let $\alpha, \beta \in \mathbb{R}$ and $(\forall)\ \overline{x}, \overline{y} \in \mathbb{R}^3$, $\overline{x} = (x_1, x_2, x_3)$, $\overline{y} = (y_1, y_2, y_3)$. We shall have:

$$\begin{aligned} f(\alpha\overline{x} + \beta\overline{y}) &= f(\alpha x_1 + \beta y_1, \alpha x_2 + \beta y_2, \alpha x_3 + \beta y_3) \\ &= (2(\alpha x_1 + \beta y_1) + \alpha x_2 + \beta y_2 - (\alpha x_3 + \beta y_3), \alpha x_2 + \beta y_2 + 7(\alpha x_3 + \beta y_3)) \\ &= (\alpha(2x_1 + x_2 - x_3) + \beta(2y_1 - y_2 - y_3), \alpha(x_2 + 7x_3) + \beta(y_2 + 7y_3)) \\ &= (\alpha(2x_1 + x_2 - x_3), \alpha(x_2 + 7x_3)) + (\beta(2y_1 - y_2 - y_3), \beta(y_2 + 7y_3)) \\ &= \alpha(2x_1 + x_2 - x_3, x_2 + 7x_3) + \beta(2y_1 - y_2 - y_3, y_2 + 7y_3) \\ &= \alpha f(\overline{x}) + \beta f(\overline{y}); \end{aligned}$$

therefore using Proposition 3.2 it results that $f$ is a linear mapping.

(b) As $B_1 = \{\overline{e}_1 = (1,0,0), \overline{e}_2 = (0,1,0), \overline{e}_3 = (0,0,1)\}$ is a basis in $\mathbb{R}^3$ and $B_2 = \{\overline{f}_1 = (1,0), \overline{f}_2 = (0,1)\}$ is a basis from $\mathbb{R}^2$ it results that

$$\begin{aligned} f(\overline{e}_1) &= f(1,0,0) = (2,0) = 2 \cdot \overline{f}_1 + 0 \cdot \overline{f}_2 \\ f(\overline{e}_2) &= f(0,1,0) = (1,1) = 1 \cdot \overline{f}_1 + 1 \cdot \overline{f}_2 \\ f(\overline{e}_3) &= f(0,0,1) = (-1,7) = (-1) \cdot \overline{f}_1 + 7 \cdot \overline{f}_2. \end{aligned}$$

We shall obtain

$$\widetilde{f}_{(B_1,B_2)} = \begin{pmatrix} 2 & 1 & -1 \\ 0 & 1 & 7 \end{pmatrix}.$$

(c) We have (based on the Definition 3.5)

$$\operatorname{Ker} f = \left\{ \overline{x} \in \mathbb{R}^3 | f(\overline{x}) = \overline{0}_{\mathbb{R}^2} \right\}.$$

Let be $\overline{x} \in \mathbb{R}^3$.

$$\begin{aligned} f(\overline{x}) = \overline{0}_{\mathbb{R}^2} &\Leftrightarrow (2x_1 + x_2 - x_3, x_2 + 7x_3) = (0,0) \\ &\Leftrightarrow \begin{cases} 2x_1 + x_2 - x_3 = 0 \\ x_2 + 7x_3 = 0. \end{cases} \end{aligned}$$

The kernel of $f$ is the set of solutions corresponding to the system:

$$\begin{cases} 2x_1 + x_2 = x_3 \\ x_2 = -7x_3 \end{cases} \Rightarrow 2x_1 = 8x_3 \Rightarrow x_1 = 4x_3.$$

Hence, $\overline{x} \in \operatorname{Ker} f \Leftrightarrow \overline{x} = (4x_3, -7x_3, x_3), \ x_3 \in \mathbb{R}^3$.
As, within the Proposition 1.67

$$\dim \operatorname{Ker} f = n - r = 3 - 2 = 1,$$

taking into account that (see the Theorem 3.26)

$$\dim \mathbb{R}^3 = \dim \operatorname{Ker} f + \dim \operatorname{Im} f$$

we obtain

$$\dim \operatorname{Im} f = 2.$$

(d) We know that $\operatorname{Im} f \subseteq \mathbb{R}^2$ i.e. $\operatorname{Im} f$ is a vector space of $\mathbb{R}^2$ (see the Theorem 3.22).

As

$$\left.\begin{array}{c}\mathrm{Im}f \subseteq \mathbb{R}^2\\ \dim \mathrm{Im}\, f = 2\\ \dim \mathbb{R}^2 = 2\end{array}\right\} \Rightarrow \mathrm{Im} f = \mathbb{R}^2 \overset{\text{Definition 3.6}}{\Longrightarrow} f \text{ is surjective.}$$

We can solve this problem in Sage, too:

```
sage: x1,x2,x3 = var('x1,x2,x3')
sage: f_symbolic(x1,x2,x3)=[2*x1+x2-x3,x2+7*x3]
sage: f=linear_transformation(RR^3, RR^2, f_symbolic)
sage: print "f.matrix(side='right')\n",f.matrix(side='right').n(digits=3)
f.matrix(side='right')
[ 2.00  1.00 -1.00]
[0.000  1.00  7.00]
sage: f.kernel()
Vector space of degree 3 and dimension 1 over Real Field with 53 bits of precision
Basis matrix:
[ 1.00000000000000 -1.75000000000000 0.250000000000000]
sage: f.image()
Vector space of degree 2 and dimension 2 over Real Field with 53 bits of precision
Basis matrix:
[ 1.00000000000000 0.000000000000000]
[0.000000000000000  1.00000000000000]
sage: print "is f.image() == RR^2?",f.image() == RR^2
is f.image() == RR^2? True
```

**Example 3.30** (see [9]). Let $V$ be a vector space, $\dim V = 3$ and $f : V \to V$,

$$\widetilde{f}_{(B)} = \begin{pmatrix} 1 & 1 & 1\\ 1 & 1 & 1\\ 1 & 1 & 1\end{pmatrix}, \tag{3.8}$$

where $B = \{\overline{e}_1, \overline{e}_2, \overline{e}_3\}$. Determine $\dim \mathrm{Ker}\, f$ and $\dim \mathrm{Im}\, f$.

**Solution**

Let be

$$\overline{x} \in V \Rightarrow \overline{x} = \alpha_1\overline{e}_1 + \alpha_2\overline{e}_2 + \alpha_3\overline{e}_3;$$

hence

$$f(\overline{x}) = \alpha_1 f(\overline{e}_1) + \alpha_2 f(\overline{e}_2) + \alpha_3 f(\overline{e}_3).$$

From (3.8) it results

$$\begin{cases} f(\overline{e}_1) = \overline{e}_1 + \overline{e}_2 + \overline{e}_3\\ f(\overline{e}_2) = \overline{e}_1 + \overline{e}_2 + \overline{e}_3\\ f(\overline{e}_3) = \overline{e}_1 + \overline{e}_2 + \overline{e}_3.\end{cases}$$

We obtain

$$\mathrm{Ker}\, f = \left\{\overline{x} \in V | f(\overline{x}) = \overline{0}_V\right\} = \{\overline{x} \in V | \alpha_1 + \alpha_2 + \alpha_3 = 0\};$$

therefore

$$\dim \mathrm{Ker}\, f = n - r = 3 - 1 = 2.$$

Whereas

$$\alpha_1 = -\alpha_2 - \alpha_3$$

we deduce

$$\overline{x} = (-\alpha_2 - \alpha_3)\,\overline{e}_1 + \alpha_2 \overline{e}_2 + \alpha_3 \overline{e}_3 = \alpha_2 \underbrace{(\overline{e}_2 - \overline{e}_1)}_{\overline{c}_1} + \alpha_3 \underbrace{(\overline{e}_3 - \overline{e}_1)}_{\overline{c}_2} \Rightarrow$$

$\{\overline{c}_1, \overline{c}_2\}$ is a system of generators for Ker $f$.

However, as dimKer $f = 2$ it results that $\{\overline{c}_1, \overline{c}_2\}$ is a basis in Ker $f$.

We have

$$\dim\ V = \dim\ \mathrm{Ker}\ f + \dim\ \mathrm{Im}\ f \Rightarrow 3 = 2 + \dim \mathrm{Im} f \Rightarrow \dim\ \mathrm{Im} f = 1.$$

We know (see the Definition 3.5) that

$$\mathrm{Im}\ f = \{ f(\overline{x}) \,|\, \overline{x} \in V \}.$$

We shall obtain

$$f(\overline{x}) = (\alpha_1 + \alpha_2 + \alpha_3) \underbrace{(\overline{e}_1 + \overline{e}_2 + \overline{e}_3)}_{\overline{c}_3} \Rightarrow$$

$\{\overline{c}_3\}$ is a system of generators for Im $f$.

However, since dim Im $f = 1$ it results that $\{\overline{c}_3\}$ is a basis in Im $f$. We shall present the solution of this problem in Sage:

```
sage: F=matrix(SR,[[1,1,1],[1,1,1],[1,1,1]])
sage: f=linear_transformation(SR^3, SR^3, F,side='right');f
Vector space morphism represented by the matrix:
[1 1 1]
[1 1 1]
[1 1 1]
Domain: Vector space of dimension 3 over Symbolic Ring
Codomain: Vector space of dimension 3 over Symbolic Ring
sage: f.kernel()
Vector space of degree 3 and dimension 2 over Symbolic Ring
Basis matrix:
[ 1  0 -1]
[ 0  1 -1]
sage: f.image()
Vector space of degree 3 and dimension 1 over Symbolic Ring
Basis matrix:
[1 1 1]
```

**Example 3.31** (see [9]). Let $T : \mathrm{M}_2(\mathbb{R}) \to \mathrm{M}_2(\mathbb{R})$ be a linear mapping, defined by

$$T(A) = \begin{pmatrix} 1 & -2 \\ 0 & 1 \end{pmatrix} A \begin{pmatrix} 1 & 0 \\ 1 & 1 \end{pmatrix}, (\forall)\, A \in \mathrm{M}_2(\mathbb{R}).$$

Build the matrix of the linear mapping $T$ in the canonical basis of that space.

**Solution**

We know that

$$B = \{E_{11}, E_{12}, E_{21}, E_{22}\} = \left\{ \begin{pmatrix} 1 & 0 \\ 0 & 0 \end{pmatrix}, \begin{pmatrix} 0 & 1 \\ 0 & 0 \end{pmatrix}, \begin{pmatrix} 0 & 0 \\ 1 & 0 \end{pmatrix}, \begin{pmatrix} 0 & 0 \\ 0 & 1 \end{pmatrix} \right\}$$

is a canonical basis in $M_2(\mathbb{R})$. We compute:

$$T(E_{11}) = \begin{pmatrix} 1 & -2 \\ 0 & 1 \end{pmatrix} \cdot \begin{pmatrix} 1 & 0 \\ 0 & 0 \end{pmatrix} \cdot \begin{pmatrix} 1 & 0 \\ 1 & 1 \end{pmatrix} = \begin{pmatrix} 1 & 0 \\ 0 & 0 \end{pmatrix} = E_{11}$$

$$T(E_{12}) = \begin{pmatrix} 1 & -2 \\ 0 & 1 \end{pmatrix} \cdot \begin{pmatrix} 0 & 1 \\ 0 & 0 \end{pmatrix} \cdot \begin{pmatrix} 1 & 0 \\ 1 & 1 \end{pmatrix} = \begin{pmatrix} 1 & 1 \\ 0 & 0 \end{pmatrix} = E_{11} + E_{12}$$

$$T(E_{21}) = \begin{pmatrix} 1 & -2 \\ 0 & 1 \end{pmatrix} \cdot \begin{pmatrix} 0 & 0 \\ 1 & 0 \end{pmatrix} \cdot \begin{pmatrix} 1 & 0 \\ 1 & 1 \end{pmatrix} = \begin{pmatrix} -2 & 0 \\ 1 & 0 \end{pmatrix} = -2E_{11} + E_{21}$$

$$T(E_{22}) = \begin{pmatrix} 1 & -2 \\ 0 & 1 \end{pmatrix} \cdot \begin{pmatrix} 0 & 0 \\ 0 & 1 \end{pmatrix} \cdot \begin{pmatrix} 1 & 0 \\ 1 & 1 \end{pmatrix} = \begin{pmatrix} -2 & -2 \\ 1 & 1 \end{pmatrix}$$
$$= -2E_{11} - 2E_{12} + E_{21} + E_{22}.$$

It results

$$\widetilde{T}_{(B)} = \begin{pmatrix} 1 & 1 & -2 & -2 \\ 0 & 1 & 0 & -2 \\ 0 & 0 & 1 & 1 \\ 0 & 0 & 0 & 1 \end{pmatrix}.$$

Using Sage, we achieve:

```
sage: M=MatrixSpace(RR,2)
sage: A1=M([[1,-2],[0,1]]);A2=M([[1,0],[1,1]]);
sage: var("a11 a12 a21 a22")
(a11, a12, a21, a22)
sage: A=matrix([[a11,a12],[a21,a22]]);
sage: T=A1*A*A2
sage: E=M.basis()
sage: T0=T.substitute(a11=E[0][0][0],a12=E[0][0][1],a21=E[0][1][0],a22=E[0][1][1])
sage: T1=T.substitute(a11=E[1][0][0],a12=E[1][0][1],a21=E[1][1][0],a22=E[1][1][1])
sage: T2=T.substitute(a11=E[2][0][0],a12=E[2][0][1],a21=E[2][1][0],a22=E[2][1][1])
sage: T3=T.substitute(a11=E[3][0][0],a12=E[3][0][1],a21=E[3][1][0],a22=E[3][1][1])
sage: U=a11*E[0]+a12*E[1]+a21*E[2]+a22*E[3]
sage: s0=solve([T0[0][0]==U[0][0],T0[0][1]==U[0][1],T0[1][0]==U[1][0],T0[1][1]==U[1][1]],a11,a12,a21,a22);s0
[[a11 == 1, a12 == 0, a21 == 0, a22 == 0]]
sage: s1=solve([T1[0][0]==U[0][0],T1[0][1]==U[0][1],T1[1][0]==U[1][0],T1[1][1]==U[1][1]],a11,a12,a21,a22);s1
[[a11 == 1, a12 == 1, a21 == 0, a22 == 0]]
sage: s2=solve([T2[0][0]==U[0][0],T2[0][1]==U[0][1],T2[1][0]==U[1][0],T2[1][1]==U[1][1]],a11,a12,a21,a22);s2
[[a11 == -2, a12 == 0, a21 == 1, a22 == 0]]
sage: s3=solve([T3[0][0]==U[0][0],T3[0][1]==U[0][1],T3[1][0]==U[1][0],T3[1][1]==U[1][1]],a11,a12,a21,a22);s3
[[a11 == -2, a12 == -2, a21 == 1, a22 == 1]]
sage: v0=vector([1,0,0,0]);v1=vector([1,1,0,0]);
sage: v2=vector([-2,0,1,0]);v3=vector([-2,-2,1,1])
sage: TB=column_matrix([v0,v1,v2,v3]);TB
[ 1  1 -2 -2]
[ 0  1  0 -2]
[ 0  0  1  1]
[ 0  0  0  1]
```

**Example 3.32** (see [9]). Let be $f : R_4[X] \to R_3[X]$ defined by

$$f(P) = P', (\forall) P \in R_4[X],$$

where $P'$ means the formal derivative of the polynomial $P$.

(a) Show that the mapping $f$ is linear.
(b) Write $\widetilde{f}_{(B_1,B_2)}$ namely the matrix of $f$ relative to the canonical bases $B_1 = \{1, X, X^2, X^3, X^4\}$ from $R_4[X]$ and respectively to $B_2 = \{1, X, X^2, X^3\}$ from $R_3[X]$.
(c) Determine Ker $f$ and Im $f$.

**Solution**

(a) Let $\alpha, \beta \in \mathbb{R}$ and $P, Q \in R_4[X]$. We have

$$\begin{aligned} f(\alpha P + \beta Q) &= (\alpha P + \beta Q)' = \alpha P' + \beta Q' \\ &= \alpha f(P) + \beta f(Q), (\forall)\, \alpha, \beta \in \mathbb{R}, (\forall)\, P, Q \in R_4[X], \end{aligned}$$

namely the mapping $f$ is linear.

We can check that in Sage, too:

```
sage: F=SR[x];var("a0 a1 a2 a3 a4 b0 b1 b2 b3 b4 al be")
(a0, a1, a2, a3, a4, b0, b1, b2, b3, b4, al, be)
sage: P=F([a0,a1,a2,a3,a4]);Q=F([b0,b1,b2,b3,b4])
sage: S=al*P+be*Q
sage: expand(al*diff(P,x)+be*diff(Q,x))==expand(diff(S,x))
True
```

(b) We compute

$$\begin{cases} f(1) = 1' = 0 = 0 \cdot 1 + 0 \cdot X + 0 \cdot X^2 + 0 \cdot X^3 \\ f(X) = X' = 1 = 1 \cdot 1 + 0 \cdot X + 0 \cdot X^2 + 0 \cdot X^3 \\ f(X^2) = (X^2)' = 2X = 0 \cdot 1 + 2 \cdot X + 0 \cdot X^2 + 0 \cdot X^3 \\ f(X^3) = (X^3)' = 3X^2 = 0 \cdot 1 + 0 \cdot X + 3 \cdot X^2 + 0 \cdot X^3 \\ f(X^4) = (X^4)' = 4X^3 = 0 \cdot 1 + 0 \cdot X + 0 \cdot X^2 + 4 \cdot X^3. \end{cases}$$

We deduce that

$$\widetilde{f}_{(B_1,B_2)} = \begin{pmatrix} 0 & 1 & 0 & 0 & 0 \\ 0 & 0 & 2 & 0 & 0 \\ 0 & 0 & 0 & 3 & 0 \\ 0 & 0 & 0 & 0 & 4 \end{pmatrix}.$$

We can obtain the same matrix with Sage, too:

```
sage: R.<x>=RR['x'];K=R^5
sage: M=K.span([[1,x,x^2,x^3,x^4]])
sage: U=M([diff(x^0,x),diff(x,x),diff(x^2,x),diff(x^3,x),diff(x^4,x)])
sage: v0=vector([U[0][0],U[0][1],U[0][2],U[0][3]])
sage: v1=vector([U[1][0],U[1][1],U[1][2],U[1][3]])
sage: v2=vector([U[2][0],U[2][1],U[2][2],U[2][3]])
sage: v3=vector([U[3][0],U[3][1],U[3][2],U[3][3]])
sage: v4=vector([U[4][0],U[4][1],U[4][2],U[4][3]])
sage: fB=column_matrix([v0,v1,v2,v3,v4]).n(digits=2);fB
[0.00  1.0 0.00 0.00 0.00]
[0.00 0.00  2.0 0.00 0.00]
[0.00 0.00 0.00  3.0 0.00]
[0.00 0.00 0.00 0.00  4.0]
```

(c) We have

$$\operatorname{Ker} f = \left\{P \in R_4[X] | f(P) = O_{R_3[X]}\right\}.$$

Let $P \in \mathrm{R}_4[X]$, hence

$$P = a_0 + a_1 X + a_2 X^2 + a_3 X^3 + a_4 X^4;$$

then

$$\begin{aligned} f(P) = \mathrm{O}_{\mathrm{R}_3[X]} &\Leftrightarrow P' = \mathrm{O}_{\mathrm{R}_3[X]} \\ &\Leftrightarrow a_1 + 2a_2 X + 3a_3 X^2 + 4a_4 X^3 = 0 + 0 \cdot X + 0 \cdot X^2 + 0 \cdot X^3. \end{aligned}$$

It results that: $a_1 = \ldots = a_4 = 0$, i.e. $P = a_0$. Therefore

$$\operatorname{Ker} f = \{P \in \mathrm{R}_4[X] | P = a_0 = a_0 \cdot 1\},$$

i.e. $B = \{1\}$ is a basis for $\operatorname{Ker} f \Rightarrow \dim \operatorname{Ker} f = 1$.

As

$$\dim \mathrm{R}_4[X] = \dim \operatorname{Ker} f + \dim \operatorname{Im} f \Rightarrow 5 = 1 + \dim \operatorname{Im} f \Rightarrow \dim \operatorname{Im} f = 4.$$

However

$$\left.\begin{array}{r} \operatorname{Im} f \subseteq \mathrm{R}_3[X] \\ \dim \operatorname{Im} f = 4 \\ \dim \mathrm{R}_3[X] = 4 \end{array}\right\} \Rightarrow \operatorname{Im} f = \mathrm{R}_3[X].$$

## 3.2 Matrix as a Linear Mapping

We propose to define the matrix operations, starting from the corresponding operations of linear maps.

Let be $S,T \in \mathrm{L}(U, V)$ and $B_1 = \{\overline{e}_1, \ldots, \overline{e}_n\}$, $B_2 = \{\overline{f}_1, \ldots, \overline{f}_m\}$ be two bases in $U$ and respectively in $V$. Let $A$, $B$ be the associated matrices of $S$ and $T$ relative to the two bases: $A = \widetilde{S}_{(B_1,B_2)}$, $A = (a_{ij})_{\substack{1\le i\le m\\ 1\le j\le n}}$ and $B = \widetilde{T}_{(B_1,B_2)}$, $B = (b_{ij})_{\substack{1\le i\le m\\ 1\le j\le n}}$.

We have:

$$\begin{cases} S(\overline{e}_1) = a_{11}\overline{f}_1 + \cdots + a_{m1}\overline{f}_m \\ \vdots \\ S(\overline{e}_j) = a_{1j}\overline{f}_1 + \cdots + a_{mj}\overline{f}_m \\ \vdots \\ S(\overline{e}_n) = a_{1n}\overline{f}_1 + \cdots + a_{mn}\overline{f}_m, \end{cases}$$

$$\begin{cases} T(\overline{e}_1) = b_{11}\overline{f}_1 + \cdots + b_{m1}\overline{f}_m \\ \vdots \\ T(\overline{e}_j) = b_{1j}\overline{f}_1 + \cdots + b_{mj}\overline{f}_m \\ \vdots \\ T(\overline{e}_n) = b_{1n}\overline{f}_1 + \cdots + b_{mn}\overline{f}_m. \end{cases}$$

(1) *Matrix equality*

$S$ and $T$ are equal$\Leftrightarrow S(\overline{e}_j) = T(\overline{e}_j)$, $(\forall)\ j = \overline{1,n} \Leftrightarrow$

$$a_{1j}\overline{f}_1 + \cdots a_{ij}\overline{f}_i + \cdots + a_{mj}\overline{f}_m = b_{1j}\overline{f}_1 + \cdots + b_{ij}\overline{f}_i + \cdots + b_{mj}\overline{f}_m \quad (3.9)$$

As $\{\overline{f}_1, \ldots, \overline{f}_m\}$ is a basis and we know that the writing of a vector into a basis is unique, from (3.9) it results that $a_{ij} = b_{ij}$, $(\forall)\ i = \overline{1,m}$, $(\forall)\ j = \overline{1,n}$.

**Definition 3.33** (see [8], p. 60). The matrices $A = (a_{ij})_{\substack{1\le i\le m\\ 1\le j\le n}}$ and $B = (b_{ij})_{\substack{1\le i\le m\\ 1\le j\le n}}$ are **equal** if and only if

$$a_{ij} = b_{ij},\ (\forall)\ i = \overline{1,m},\ (\forall)\ j = \overline{1,n}.$$

(2) *Matrix addition*

We denote by $C$ the associated matrix of the linear mapping $S + T$. We have

$$\begin{aligned}(S+T)(\overline{e}_j) &\stackrel{def}{=} S(\overline{e}_j) + T(\overline{e}_j) \\ &= a_{1j}\overline{f}_1 + \cdots a_{ij}\overline{f}_i + \cdots + a_{mj}\overline{f}_m + b_{1j}\overline{f}_1 + \cdots + b_{ij}\overline{f}_i + \cdots + b_{mj}\overline{f}_m,\end{aligned}$$

namely

$$(S+T)(\overline{e}_j) = (a_{1j} + b_{1j})\overline{f}_1 + \cdots (a_{ij} + b_{ij})\overline{f}_i + \cdots + (a_{mj} + b_{mj})\overline{f}_m. \quad (3.10)$$

However

$$(S+T)\left(\overline{e}_j\right)=c_{1j}\overline{f}_1+\cdots+c_{ij}\overline{f}_i+\cdots+c_{mj}\overline{f}_m. \quad (3.11)$$

From (3.10) and (3.11) it results

$$c_{ij}=a_{ij}+b_{ij},\ (\forall)\, i=\overline{1,m},\ (\forall)\, j=\overline{1,n}.$$

**Definition 3.34** (see [8], p. 60). The **sum of matrices** $A=\left(a_{ij}\right)_{\substack{1\le i\le m\\ 1\le j\le n}}$ and $B=\left(b_{ij}\right)_{\substack{1\le i\le m\\ 1\le j\le n}}$ is the matrix $C=\left(c_{ij}\right)_{\substack{1\le i\le m\\ 1\le j\le n}}$, $c_{ij}=a_{ij}+b_{ij}$, $(\forall)\, i=\overline{1,m}$, $(\forall)\, j=\overline{1,n}$. We denote $C=A+B$.

(3) *Scalar multiplication of the matrices*

We shall denote by $C$ the associated matrix of the linear mapping $\alpha S$, $\alpha\in K$.

$$(\alpha S)\left(\overline{e}_j\right)=\alpha a_{1j}\overline{f}_1+\cdots+\alpha a_{ij}\overline{f}_i+\cdots+\alpha a_{mj}\overline{f}_m. \quad (3.12)$$

However

$$(\alpha S)\left(\overline{e}_j\right)=c_{1j}\overline{f}_1+\cdots+c_{ij}\overline{f}_i+\cdots+c_{mj}\overline{f}_m. \quad (3.13)$$

From (3.12) and (3.13) it results

$$c_{ij}=\alpha a_{ij},\ (\forall)\, i=\overline{1,m},\ (\forall)\, j=\overline{1,n}.$$

**Definition 3.35** (see [8], p. 60). By **multiplying** a matrix $A=\left(a_{ij}\right)_{\substack{1\le i\le m\\ 1\le j\le n}}$ with a scalar $\alpha\in K$ it results the matrix $\alpha A$, whose elements are obtained by multiplying the all elements of $A$ with $\alpha$.

(4) *Matrix multiplication*

Let $S\in \mathrm{L}(U,V)$, $T\in \mathrm{L}(V,W)$ and $B_1=\{\overline{e}_1,\ldots,\overline{e}_n\}$ be a basis in $U$, $B_2=\left\{\overline{f}_1,\ldots,\overline{f}_m\right\}$ be a basis in $V$, $B_3=\left\{\overline{g}_1,\ldots,\overline{g}_p\right\}$ be a basis in $W$. We denote by:

- $A=\left(a_{ij}\right)_{\substack{1\le i\le m\\ 1\le j\le n}}$ the associated matrix of $S$ relative to the bases $B_1$ and $B_2$;
- $B=\left(b_{ij}\right)_{\substack{1\le i\le p\\ 1\le j\le m}}$ the associated matrix of $T$ relative to the bases $B_2$ and $B_3$.

We have:

$$\begin{aligned}(T\circ S)\left(\overline{e}_j\right)&=T\left(S\left(\overline{e}_j\right)\right)=T\left(a_{1j}\overline{f}_1+\cdots a_{kj}\overline{f}_k+\cdots+a_{mj}\overline{f}_m\right)\\&=a_{1j}T\left(\overline{f}_1\right)+\cdots+a_{kj}T\left(\overline{f}_k\right)+\cdots+a_{mj}T\left(\overline{f}_m\right),\end{aligned}$$

i.e.

$$\begin{aligned}(T \circ S)\left(\overline{e}_j\right) = {} & a_{1j}\left(b_{11}\overline{g}_1 + \cdots + b_{i1}\overline{g}_i + \cdots + b_{p1}\overline{g}_p\right) + \cdots + \\ & a_{kj}\left(b_{1k}\overline{g}_1 + \cdots + b_{ik}\overline{g}_i + \cdots + b_{pk}\overline{g}_p\right) + \cdots + \\ & a_{mj}\left(b_{1m}\overline{g}_1 + \cdots + b_{im}\overline{g}_i + \cdots + b_{pm}\overline{g}_p\right).\end{aligned} \tag{3.14}$$

We denote by $C = \left(c_{ij}\right)_{\substack{1 \le i \le p \\ 1 \le j \le n}}$ the associated matrix of the linear mapping $T \circ S \in L(U, W)$. It results

$$(T \circ S)\left(\overline{e}_j\right) = c_{1j}\overline{g}_1 + \cdots + c_{ij}\overline{g}_i + \cdots + c_{pj}\overline{g}_p. \tag{3.15}$$

We write (3.14) as

$$\begin{aligned}(T \circ S)\left(\overline{e}_j\right) = {} & \left(b_{11}a_{1j} + \cdots + b_{1k}a_{kj} + \cdots + b_{1m}a_{mj}\right)\overline{g}_1 \\ & + \cdots + \left(b_{i1}a_{1j} + \cdots + b_{ik}a_{kj} + \cdots + b_{im}a_{mj}\right)\overline{g}_i \\ & + \cdots + \left(b_{p1}a_{1j} + \cdots + b_{pk}a_{kj} + \cdots + b_{pm}a_{mj}\right)\overline{g}_p.\end{aligned} \tag{3.16}$$

From (3.15) and (3.16) it results

$$c_{ij} = b_{i1}a_{1j} + \cdots + b_{ik}a_{kj} + \cdots + b_{im}a_{mj}, (\forall)\, i = \overline{1, p}, (\forall)\, j = \overline{1, n}. \tag{3.17}$$

**Definition 3.36** (see [8], p. 60). The **product** of the matrices $B = \left(b_{ij}\right)_{\substack{1 \le i \le p \\ 1 \le j \le m}}$ and $A = \left(a_{ij}\right)_{\substack{1 \le i \le m \\ 1 \le j \le n}}$ is the matrix $C = \left(c_{ij}\right)_{\substack{1 \le i \le p \\ 1 \le j \le n}}$, $c_{ij}$, $(\forall)\, i = \overline{1, p}$, $(\forall)\, j = \overline{1, n}$ being defined in (3.17). We denote $C = B \cdot A$.

**Remark 3.37** (see [8], p. 61). The product $B \cdot A$ is defined if and only if the number of columns of $B$ is equal to the number of lines of $A$.

**Proposition 3.38** (see [8], p. 60). If $A, B, C$ are matrices with proper dimensions, such that the following products are defined and $\lambda \in K$, then:

(a) $A(BC) = (AB)C$
(b) $A(B + C) = AB + AC$
(c) $(B + C)A = BA + CA$
(d) $A(\lambda B) = (\lambda A)B = \lambda(AB)$.

**Remark 3.39** (see [8], p. 61). In general, the matrix multiplication is not commutative.

(5) *Matrix inversion*

**Definition 3.40** (see [8], p. 61). The matrix $A \in M_n(K)$ is **invertible** if there is a unique matrix $B \in M_n(K)$ such that

$$AB = BA = I_n.$$

The inverse of $A$ is denoted with $A^{-1}$.

(6) *Rank of a matrix*

**Theorem 3.41** (see [8], p. 61). Let $U$, $V$ be two vector spaces over the field $K$ and $T \in \mathrm{L}(U, V)$. If $B_1 = \{\overline{e}_1, \cdots, \overline{e}_n\}$ is a basis in $U$ and $B_2 = \{\overline{f}_1, \cdots, \overline{f}_n\}$ is a basis in $V$ and $A$ is the associated matrix of the linear mapping $T$ relative to the bases $B_1$ and $B_2$, then

$$\text{rank } T = \text{rank } A.$$

**Proposition 3.42** (see [8], p. 62). A square matrix is invertible if and only if it is nonsingular.

## 3.3 Changing the Associated Matrix to the Change of Basis

Let $\overline{x} \in U$. We can write

$$\overline{x} = x^{(1)}\overline{e}_1 + \cdots + x^{(n)}\overline{e}_n.$$

We shall have

$$\begin{aligned} T(\overline{x}) &= x^{(1)}T(\overline{e}_1) + \cdots + x^{(n)}T(\overline{e}_n) \\ &= x^{(1)}\left(\alpha_1^{(1)}\overline{f}_1 + \alpha_1^{(2)}\overline{f}_2 + \cdots + \alpha_1^{(m)}\overline{f}_m\right) \\ &\quad + \cdots + x^{(n)}\left(\alpha_n^{(1)}\overline{f}_1 + \alpha_n^{(2)}\overline{f}_2 + \cdots + \alpha_n^{(m)}\overline{f}_m\right), \end{aligned}$$

namely

$$T(\overline{x}) = \left(x^{(1)}\alpha_1^{(1)} + \cdots + x^{(n)}\alpha_n^{(1)}\right)\overline{f}_1 + \cdots + \left(x^{(1)}\alpha_1^{(m)} + \cdots + x^{(n)}\alpha_n^{(m)}\right)\overline{f}_m. \tag{3.18}$$

As $T(\overline{x}) \in V$ it results

$$T(\overline{x}) = y^{(1)}\overline{f}_1 + \cdots + y^{(m)}\overline{f}_m. \tag{3.19}$$

From (3.18) and (3.19) we deduce

$$\begin{cases} y^{(1)} = x^{(1)}\alpha_1^{(1)} + \cdots + x^{(n)}\alpha_n^{(1)} \\ \qquad\vdots \\ y^{(m)} = x^{(1)}\alpha_1^{(m)} + \cdots + x^{(n)}\alpha_n^{(m)}; \end{cases}$$

therefore

$$(T(\overline{x}))_{B_2} = \widetilde{T}_{(B_1,B_2)} \cdot \overline{x}_{B_1}. \tag{3.20}$$

Let $B_1'$ be another basis of $U$ and $B_2'$ be another basis of $V$. Let $C$ be the transition matrix from the basis $B_1$ to the basis $B_1'$ and $D$ the transition matrix from the basis $B_2$ to the basis $B_2'$.

Within the relation (3.20) we have

$$(T(\overline{x}))_{B_2'} = \widetilde{T}_{(B_1',B_2')} \cdot \overline{x}_{B_1'}. \tag{3.21}$$

We know that

$$\overline{x}_{B_1} = C_{(B_1,B_1')} \cdot \overline{x}_{B_1'} \tag{3.22}$$

and

$$(T(\overline{x}))_{B_2} = D_{(B_2,B_2')} \cdot (T(\overline{x}))_{B_2'}. \tag{3.23}$$

Equating (3.20) and (3.23) it results

$$D_{(B_2,B_2')} \cdot (T(\overline{x}))_{B_2'} = \widetilde{T}_{(B_1,B_2)} \cdot \overline{x}_{B_1}. \tag{3.24}$$

Substituting (3.24) into (3.22) it results

$$D_{(B_2,B_2')} \cdot (T(\overline{x}))_{B_2'} = \widetilde{T}_{(B_1,B_2)} \cdot C_{(B_1,B_1')} \cdot \overline{x}_{B_1'}. \tag{3.25}$$

If in the relation (3.25) we multiply at the left, the both members with the inverse of $D$, we obtain

$$(T(\overline{x}))_{B_2'} = D^{-1}_{(B_2,B_2')} \cdot \widetilde{T}_{(B_1,B_2)} \cdot C_{(B_1,B_1')} \cdot \overline{x}_{B_1'}. \tag{3.26}$$

From (3.21) and (3.26) it results

$$\widetilde{T}_{(B_1',B_2')} \cdot \overline{x}_{B_1'} = D^{-1}_{(B_2,B_2')} \cdot \widetilde{T}_{(B_1,B_2)} \cdot C_{(B_1,B_1')} \cdot \overline{x}_{B_1'},$$

i.e.

$$\widetilde{T}_{(B_1',B_2')} = D^{-1}_{(B_2,B_2')} \cdot \widetilde{T}_{(B_1,B_2)} \cdot C_{(B_1,B_1')}. \tag{3.27}$$

The formula (3.27) constitutes [9] the *changing formula of the associated matrix of a linear mapping when one changes the bases in the two vector spaces* $U$ and $V$.

**Example 3.43** (see [5], p. 48). Let $T_1, T_2 \in \mathrm{End}\left(\mathbb{R}^3\right)$, be defined as

$$\begin{aligned}
&T_1(\overline{x}) = \left(5x^{(1)} - x^{(2)} - 5x^{(3)}, 20x^{(1)} - 15x^{(2)} + 8x^{(3)}, 3x^{(1)} - 2x^{(2)} + x^{(3)}\right) \\
&T_2(\overline{x}) = \left(10x^{(1)} - 10x^{(2)} + 10x^{(3)}, 0, 5x^{(1)} - 5x^{(2)} + 5x^{(3)}\right), \\
&(\forall)\, \overline{x} = \left(x^{(1)}, x^{(2)}, x^{(3)}\right) \in \mathbb{R}^3.
\end{aligned}$$

Find the sum of the two endomorfisme matrix $T = T_1 + T_2$ relative to the basis $B' = \{\overline{v}_1 = (2,3,1), \overline{v}_2 = (3,4,1), \overline{v}_3 = (1,2,2)\} \subset \mathbb{R}^3$.

**Solution**

We have:

$$\begin{aligned} T(\overline{x}) &= (T_1 + T_2)(\overline{x}) = T_1(\overline{x}) + T_2(\overline{x}) \\ &= \left(15x^{(1)} - 11x^{(2)} + 5x^{(3)}, 20x^{(1)} - 15x^{(2)} + 8x^{(3)}, 8x^{(1)} - 7x^{(2)} + 6x^{(3)}\right). \end{aligned}$$

Let $B_1 = \{\overline{e}_1 = (1,0,0), \overline{e}_2 = (0,1,0), \overline{e}_3 = (0,0,1)\}$ be the canonical base of the space $\mathbb{R}^3$.

Computing

$$\begin{cases} T(\overline{e}_1) = (15, 20, 8) \\ T(\overline{e}_2) = (-11, -15, -7) \\ T(\overline{e}_3) = (5, 8, 6) \end{cases}$$

we shall obtain

$$\widetilde{T}_{(B)} = \begin{pmatrix} 15 & -11 & 5 \\ 20 & -15 & 8 \\ 8 & -7 & 6 \end{pmatrix}.$$

We denote by $C$ the transition matrix from the basis $B$ to the basis $B'$.

As

$$\begin{cases} \overline{v}_1 = (2,3,1) = 2\overline{e}_1 + 3\overline{e}_2 + \overline{e}_3 \\ \overline{v}_2 = (3,4,1) = 3\overline{e}_1 + 4\overline{e}_2 + \overline{e}_3 \\ \overline{v}_3 = (1,2,2) = \overline{e}_1 + 2\overline{e}_2 + 2\overline{e}_3 \end{cases}$$

we have

$$C = \begin{pmatrix} 2 & 3 & 1 \\ 3 & 4 & 2 \\ 1 & 1 & 2 \end{pmatrix}.$$

Hence

$$\widetilde{T}_{(B')} = C^{-1} \cdot \widetilde{T}_{(B)} \cdot C = \begin{pmatrix} 1 & 0 & 0 \\ 0 & 2 & 0 \\ 0 & 0 & 3 \end{pmatrix}.$$

We can check this result using Sage:

```
sage: V=RR^3
sage: v1=vector(RR,[2,3,1]);v2=vector(RR,[3,4,1]);v3=vector(RR,[1,2,2]);
sage: W=V.subspace_with_basis([v1,v2,v3])
sage: var("x1 x2 x3")
(x1, x2, x3)
sage: T_symbolic(x1,x2,x3) = [5*x1-x2-5*x3,20*x1-15*x2+8*x3,3*x1-2*x2+x3]
sage: T1=linear_transformation(W, W, T_symbolic)
sage: Tt_symbolic(x1,x2,x3) = [10*x1-10*x2+10*x3,0,5*x1-5*x2+5*x3]
sage: T2=linear_transformation(W, W, Tt_symbolic)
```

```
sage: print "(T1+T2).matrix(side='right')\n", (T1+T2).matrix(side='right').n(digits=3)
(T1+T2).matrix(side='right')
[ 1.00 0.000 0.000]
[0.000  2.00 0.000]
[0.000 0.000  3.00]
```

## 3.4 Eigenvalues and Eigenvectors

The eigenvalue problems are of great importance in many branches of physics. They make it possible to find some coordinate systems in which the linear transformations take the simplest forms.

For example, in mechanics the main moments of a solid body one finds with the eigenvalues of a symmetric matrix representing the vector tensor. The situation is similar in continuous mechanics, where a body rotations and deformations in the main directions are found using the eigenvalues of a symmetric matrix.

Eigenvalues have [1] a central importance in quantum mechanics, where the measured values of the observable physical quantities appear as eigenvalues of operators.

Also, the eigenvalues are useful in the study of differential equations and continuous dynamical systems that arise in areas such as physics and chemistry.

**Definition 3.44** (see [2], p. 45). Let $V$ be a vector space over the field $K$ and $T$ $\in$End$(V)$. The subspace vector $W$ of $V$ is called an **invariant subspace** relative to $T$ if from $\overline{x} \in W$ it results $T(\overline{x}) \in W$ or $T(W) \subseteq W$.

**Definition 3.45** (see [2], p. 45). Let $V$ be a vector space over the field $K$ and $T$ $\in$End$(V)$. We say that the scalar $\lambda \in K$ is an **eigenval** for $T$ if there is $\overline{x} \in V \backslash \{\overline{0}\}$ such that

$$T(\overline{x}) = \lambda \overline{x}. \tag{3.28}$$

**Definition 3.46** (see [2], p. 45). The vector $\overline{x} \in V \backslash \{\overline{0}\}$ for which there is $\lambda \in K$ such that $T(\overline{x}) = \lambda \overline{x}$ is called the **eigenvector** corresponding to the eigenvalue $\lambda$.

**Remark 3.47** (see [9]). If $\lambda$ is an eigenval for $T$ then there are an infinite number of eigenvectors corresponding to $\lambda$.

**Proposition 3.48** (see [2], p. 45). Let $V$ be a vector space over the field $K$ and $\lambda \in K$ an eigenvalue for $T$. Then the set

$$V_\lambda = \{\overline{x} \in V | T(\overline{x}) = \lambda \overline{x}\} \tag{3.29}$$

is a vector subspace of $V$, invariant relative to $T$.

**Definition 3.49** (see [2], p. 45). Let $V$ be a vector space over the field $K$. If $T$ $\in$End$(V)$ and $\lambda \in K$ an eigenvalue for $T$, then the vector subspace $V_\lambda$ is called the **eigensubspace** associated to the eigenvalue $\lambda$.

**Proposition 3.50** (see [3], p. 64). Let $T : V \to V$ be an endomorphism of the vector space $V$ and $\overline{a}_1, \ldots \overline{a}_p$ the eigenvectors of the endomorphism $T$ respectively corresponding to the distinct eigenvalues $\lambda_1, \ldots \lambda_p$, i.e.

$$T(\overline{a}_i) = \lambda_i \overline{a}_i, i = \overline{1, p},$$

then $\overline{a}_1, \ldots \overline{a}_p$ are linearly independent.

### *3.4.1 Characteristic Polynomial of an Endomorphism*

Let be the endomorphism $T : V \to V$ and $B = \{\overline{e}_1, \ldots, \overline{e}_n\}$ be a basis of $V$. We consider the expressions of the vectors $T(\overline{e}_1), \ldots, T(\overline{e}_n) \in V$ in the basis $B$ :

$$\begin{cases} T(\overline{e}_1) = \alpha_1^{(1)}\overline{e}_1 + \alpha_1^{(2)}\overline{e}_2 + \cdots + \alpha_1^{(n)}\overline{e}_n \\ T(\overline{e}_2) = \alpha_2^{(1)}\overline{e}_1 + \alpha_1^{(2)}\overline{e}_2 + \cdots + \alpha_2^{(n)}\overline{e}_n \\ \vdots \\ T(\overline{e}_n) = \alpha_n^{(1)}\overline{e}_1 + \alpha_n^{(2)}\overline{e}_2 + \cdots + \alpha_n^{(n)}\overline{e}_n. \end{cases} \tag{3.30}$$

Let $\widetilde{T}_{(B)}$ be the associated matrix of the linear mapping $T$ relative to the basis $B$,

$$\widetilde{T}_{(B)} = \begin{pmatrix} \alpha_1^{(1)} & \ldots & \alpha_n^{(1)} \\ \vdots & & \vdots \\ \alpha_1^{(n)} & \ldots & \alpha_n^{(n)} \end{pmatrix}. \tag{3.31}$$

Let be $\overline{x} \in V$. We can write

$$\overline{x} = x^{(1)}\overline{e}_1 + \cdots + x^{(n)}\overline{e}_n.$$

The relation (3.28) becomes

$$T\left(x^{(1)}\overline{e}_1 + \cdots + x^{(n)}\overline{e}_n\right) = \lambda\left(x^{(1)}\overline{e}_1 + \cdots + x^{(n)}\overline{e}_n\right). \tag{3.32}$$

As $T$ is a linear mapping, from (3.32) it results

$$x^{(1)}T(\overline{e}_1) + \cdots + x^{(n)}T(\overline{e}_n) = \lambda\left(x^{(1)}\overline{e}_1 + \cdots + x^{(n)}\overline{e}_n\right). \tag{3.33}$$

If in (3.33) we take into account the relations (3.30) we obtain

$$x^{(1)}\left(\alpha_1^{(1)}\overline{e}_1 + \alpha_1^{(2)}\overline{e}_2 + \cdots + \alpha_1^{(n)}\overline{e}_n\right) + \cdots + \tag{3.34}$$

$$x^{(n)}\left(\alpha_n^{(1)}\overline{e}_1+\alpha_n^{(2)}\overline{e}_2+\cdots+\alpha_n^{(n)}\overline{e}_n\right)$$
$$=\lambda\left(x^{(1)}\overline{e}_1+\cdots+x^{(n)}\overline{e}_n\right).$$

From (3.34) we deduce

$$\begin{cases}\left(\alpha_1^{(1)}-\lambda\right)x^{(1)}+\alpha_2^{(1)}x^{(2)}+\cdots+\alpha_n^{(1)}x^{(n)}=0\\ \alpha_1^{(2)}x^{(1)}+\left(\alpha_2^{(2)}-\lambda\right)x^{(2)}+\cdots+\alpha_n^{(2)}x^{(n)}=0\\ \vdots\\ \alpha_1^{(n)}x^{(1)}+\alpha_2^{(n)}x^{(2)}+\cdots+\left(\alpha_n^{(n)}-\lambda\right)x^{(n)}=0\end{cases} \tag{3.35}$$

i.e. a linear and homogeneous system in the unknowns $x^{(1)}, x^{(2)}, \ldots, x^{(n)}$.

The scalar $\lambda$ is an eigenvalue of $T$ if and only if the system (3.35) admits the nonbanal solutions, i.e. there is a scalar system $x^{(1)}, x^{(2)}, \ldots, x^{(n)}$, not all null, that verifies the system (3.35).

This is achieved if and only if

$$\Delta=\begin{vmatrix}\alpha_1^{(1)}-\lambda & \alpha_2^{(1)} & \cdots & \alpha_n^{(1)}\\ \alpha_1^{(2)} & \alpha_2^{(2)}-\lambda & \cdots & \alpha_n^{(2)}\\ \cdots & \cdots & \cdots & \cdots\\ \alpha_1^{(n)} & \alpha_2^{(n)} & \cdots & \alpha_n^{(n)}-\lambda\end{vmatrix}=0,$$

i.e.

$$\det\left(\widetilde{T}_{(B)}-\lambda \mathrm{I}_n\right)=0, \tag{3.36}$$

$\mathrm{I}_n$ being the unit matrix of order $n$.

In conclusion, $\lambda$ is an eigenvalue of $T$ if and only if there is a root in $K$ of the Eq. (3.36).

We denote

$$P(\lambda)=\det\left(\widetilde{T}_{(B)}-\lambda \mathrm{I}_n\right). \tag{3.37}$$

**Remark 3.51** (see [3], p. 66). The polynomial $P(\lambda)$ is a polynomial in $\lambda$, of degree $n$:

$$P(\lambda)=(-1)^n\lambda^n+(-1)^{n-1}\delta_1\lambda^{n-1}+(-1)^{n-2}\delta_2\lambda^{n-2}+\cdots+(-1)\,\delta_{n-1}\lambda+\delta_n,$$

where

$$\begin{cases} \delta_1 = \text{trace}\left(\widetilde{T}_{(B)}\right) = \sum_{i=1}^{n} \alpha_i^{(i)} \\ \delta_2 = \sum_{1 \le i < j \le n} \begin{vmatrix} \alpha_i^{(i)} & \alpha_j^{(i)} \\ \alpha_i^{(j)} & \alpha_j^{(j)} \end{vmatrix} \\ \vdots \\ \delta_n = \det\left(\widetilde{T}_{(B)}\right). \end{cases}$$

**Definition 3.52** (see [2], p. 45). The polynomial $P(\lambda)$ defined in (3.37), in the indeterminate $\lambda$, of degree $n$ is called the **characteristic polynomial** associated of the endomorphism $T$.

**Definition 3.53** (see [3], p. 66). The equation $P(\lambda) = 0$ is called the **characteristic equation** associated of the endomorphism $T$.

**Remark 3.54** (see [3], p. 66). The scalar $\lambda$ is an eigenvalue of $T$ if and only if $\lambda$ is the root of the characteristic equation.

**Example 3.55** (see [9]). Let $T$ be be an endomorphism of $\mathbb{R}^3$ such that $T$ has the eigenvalues: $\lambda_1 = 1, \lambda_2 = -1, \lambda_3 = 2$ with the eigenvectors $\overline{x}_1 = (1, 0, 1)$, $\overline{x}_2 = (-1, 2, 1), \overline{x}_3 = (2, 1, -1)$. Write the associated matrix of $T$ in the canonical basis from $\mathbb{R}^3$.

**Solution**

We note that

$$\begin{cases} \overline{x}_1 = \overline{e}_1 + \overline{e}_3 \\ \overline{x}_2 = -\overline{e}_1 + 2\overline{e}_2 + \overline{e}_3 \\ \overline{x}_3 = 2\overline{e}_1 + \overline{e}_2 - \overline{e}_3. \end{cases}$$

As

$$T(\overline{x}_1) = \lambda_1 \overline{x}_1$$

we deduce

$$T(\overline{e}_1 + \overline{e}_3) = \lambda_1 (\overline{e}_1 + \overline{e}_3). \tag{3.38}$$

Taking into account that $T$ is a linear mapping from (3.38) we obtain

$$T(\overline{e}_1) + T(\overline{e}_3) = \lambda_1 (\overline{e}_1 + \overline{e}_3). \tag{3.39}$$

Similarly, because

$$T(\overline{x}_2) = \lambda_2 \overline{x}_2$$

we deduce

$$T(-\overline{e}_1 + 2\overline{e}_2 + \overline{e}_3) = \lambda_2 (-\overline{e}_1 + 2\overline{e}_2 + \overline{e}_3). \tag{3.40}$$

Taking into account that $T$ is a linear mapping from (3.40) we achieve:

$$-T(\overline{e}_1)+2T(\overline{e}_2)+T(\overline{e}_3)=\lambda_2(-\overline{e}_1+2\overline{e}_2+\overline{e}_3). \tag{3.41}$$

Similarly, as

$$T(\overline{x}_3)=\lambda_3\overline{x}_3$$

we shall have

$$T(2\overline{e}_1+\overline{e}_2-\overline{e}_3)=\lambda_3(2\overline{e}_1+\overline{e}_2-\overline{e}_3). \tag{3.42}$$

As $T$ is a linear mapping from (3.42), it results:

$$2T(\overline{e}_1)+T(\overline{e}_2)-T(\overline{e}_3)=\lambda_3(2\overline{e}_1+\overline{e}_2-\overline{e}_3). \tag{3.43}$$

We shall solve the system of equations resulting from the relations (3.39), (3.41) and (3.43) to determine the expression of $T(\overline{e}_i), i=\overline{1,3}$ as a linear combination of the elements of the basis, i.e.:

$$\begin{cases} T(\overline{e}_1)+T(\overline{e}_3)=\overline{e}_1+\overline{e}_3 \\ -T(\overline{e}_1)+2T(\overline{e}_2)+T(\overline{e}_3)=\overline{e}_1-2\overline{e}_2-\overline{e}_3 \\ 2T(\overline{e}_1)+T(\overline{e}_2)-T(\overline{e}_3)=4\overline{e}_1+2\overline{e}_2-2\overline{e}_3. \end{cases} \tag{3.44}$$

Adding the first two equations we deduce:

$$2T(\overline{e}_2)+2T(\overline{e}_3)=2(\overline{e}_1-\overline{e}_2),$$

i.e.

$$T(\overline{e}_2)+T(\overline{e}_3)=\overline{e}_1-\overline{e}_2. \tag{3.45}$$

Adding the last two equations we deduce:

$$T(\overline{e}_1)+3T(\overline{e}_2)=5\overline{e}_1-3\overline{e}_3. \tag{3.46}$$

From (3.45) it results

$$T(\overline{e}_3)=\overline{e}_1-\overline{e}_2-T(\overline{e}_2). \tag{3.47}$$

From (3.46) it results

$$T(\overline{e}_1)=5\overline{e}_1-3\overline{e}_3-3T(\overline{e}_2). \tag{3.48}$$

Substituting (3.47) and (3.48) into the first equation of the system (3.44) we obtain

$$5\overline{e}_1-3\overline{e}_3-3T(\overline{e}_2)+\overline{e}_1-\overline{e}_2-T(\overline{e}_2)=\overline{e}_1+\overline{e}_3,$$

i.e.

$$5\overline{e}_1-\overline{e}_2-4\overline{e}_3=4T(\overline{e}_2);$$

hence

$$T(\overline{e}_2) = \frac{5}{4}\overline{e}_1 - \frac{1}{4}\overline{e}_2 - \overline{e}_3.$$

We shall have

$$T(\overline{e}_1) = 5\overline{e}_1 - 3\overline{e}_3 - 3\left(\frac{5}{4}\overline{e}_1 - \frac{1}{4}\overline{e}_2 - \overline{e}_3\right) = \frac{5}{4}\overline{e}_1 + \frac{3}{4}\overline{e}_2$$

and

$$T(\overline{e}_3) = \overline{e}_1 - \overline{e}_2 - \left(\frac{5}{4}\overline{e}_1 - \frac{1}{4}\overline{e}_2 - \overline{e}_3\right) = -\frac{1}{4}\overline{e}_1 - \frac{3}{4}\overline{e}_2 + \overline{e}_3.$$

We obtain

$$\widetilde{T}_{(B)} = \begin{pmatrix} 5/4 & 5/4 & -1/4 \\ 3/4 & -1/4 & -3/4 \\ 0 & -1 & 1 \end{pmatrix}.$$

The same matrix can be achieved using Sage:

```
sage: la1=1;la2=-1;la3=2;V=VectorSpace(RR,3)
sage: x1=V([1,0,1]);x2=V([-1,2,1]);x3=V([2,1,-1])
sage: var("T1 T2 T3 e1 e2 e3")
(T1, T2, T3, e1, e2, e3)
sage: vv1=la1*(x1[0]*e1+x1[1]*e2+x1[2]*e3)
sage: vv2=la2*(x2[0]*e1+x2[1]*e2+x2[2]*e3)
sage: vv3=la3*(x3[0]*e1+x3[1]*e2+x3[2]*e3)
sage: eq=[T1+T3==vv1,-T1+2*T2+T3==vv2,2*T1+T2-T3==vv3]
sage: s=solve(eq,T1,T2,T3);s
[[T1 == 5/4*e1 + 3/4*e2, T2 == 5/4*e1 - 1/4*e2 - e3, T3 == -1/4*e1 - 3/4*e2 + e3]]
sage: v1=V([5/4,3/4,0]);v2=V([5/4,-1/4,-1]);v3=V([-1/4,-3/4,1])
sage: M=column_matrix([v1,v2,v3]);M.n(digits=3)
[  1.25   1.25 -0.250]
[ 0.750 -0.250 -0.750]
[ 0.000  -1.00   1.00]
```

**Definition 3.56** (see [5], p. 64). The set of the roots of the characteristic equation, that is associated to the endomorphism $T$ is called the **spectrum of the endomorphism** $T$. If all the roots are simple on says that $T$ is an endomorphism with a **simple spectrum**. We denote by $\sigma(T)$ the spectrum of $T$.

**Theorem 3.56 (Hamilton-Cayley**, see [2], p. 49). Let $V$ be a $n$ dimensional vector space over $K, n \geq 1$ and be an endomorphism $T \in \text{End}(V)$. If $P(\lambda)$ is the characteristic polynomial of $A = \widetilde{T}_{(B)}$ (the matrix of the endomorphism $T$ relative to a basis $B$ of $V$, then $P(A) = 0_{\text{End}(V)}$.

**Example 3.57** (see [7], p. 29). Compute

$$P(A) = A^4 - 8A^3 + 24A^2 - 32A + 16I_4,$$

where $A$ is the matrix

$$A = \begin{pmatrix} 2 & 0 & 0 & 0 \\ 1 & 3 & 1 & 1 \\ 0 & 0 & 1 & -1 \\ -1 & -1 & 0 & 2 \end{pmatrix}.$$

**Solution**

The characteristic polynomial, that is associated to the matrix $A$ is

$$P(\lambda) = \begin{vmatrix} 2-\lambda & 0 & 0 & 0 \\ 1 & 3-\lambda & 1 & 1 \\ 0 & 0 & 1-\lambda & -1 \\ -1 & -1 & 0 & 2-\lambda \end{vmatrix} = \lambda^4 - 8\lambda^3 + 24\lambda^2 - 32\lambda + 16.$$

Using the Hamilton- Cayley theorem we obtain $P(A) = O_{M_4(\mathbb{R}^4)}$.
We shall use Sage to check the previous result:

```
sage: A=matrix([[2,0,0,0],[1,3,1,1],[0,0,1,-1],[-1,-1,0,2]])
sage: var("la")
la
sage: I4=identity_matrix(4)
sage: P=(A-la*I4).determinant();P
la^4 - 8*la^3 + 24*la^2 - 32*la + 16
sage: A^4-8*A^3+24*A^2-32*A+16*I4
[0 0 0 0]
[0 0 0 0]
[0 0 0 0]
[0 0 0 0]
```

### *3.4.2 Determining the Eigenvalues and the Eigenvectors for an Endomorphism*

To determine the eigenvalues associated of an endomorphism we proceed [9] as:

- write the characteristic equation;
- solve the characteristic equation;
- achieve the eigenvalues of the endomorphism as roots of the characteristic equation, that are in $K$.

To determine the eigenvectors corresponding to an eigenvalue $\lambda_0$ of $T$ we proceed as:

- rewrite the system (3.35), by replacing $\lambda$ with $\lambda_0$;
- determine the vector subspace $W_{\lambda_0}$ of the solutions of the obtained linear and homogeneous system (we find its dimension and a basis), called the *associated eigensubspace* of the eigenvalue $\lambda_0$;
- all non-zero vectors from the vector subspace are eigenvectors for the associated eigenvalue $\lambda_0$.

**Definition 3.58** (see [3], p. 66). The dimension of the eigensubspace $W_{\lambda_0}$ associated of the eigenvalue $\lambda_0$ is called the **geometric multiplicity** of $\lambda_0$ and it is denoted by $g_{\lambda_0}$.

**Definition 3.59** (see [3], p. 66). The **algebraic multiplicity** of the eigenvalue $\lambda_0$, denoted by $a_{\lambda_0}$ means the multiplicity of $\lambda_0$ as a root of the characteristic polynomial $P(\lambda)$, associated to the endomorphism $T$.

**Proposition 3.60** (see [3], p. 66). The characteristic polynomial $P(\lambda)$ is invariant relative to the basis changing in the vector space $V$.

**Theorem 3.61** (see [3], p. 66). Let $V$ be a vector space over the field $K$, $\dim V = n < \infty$, $T \in \mathrm{End}(V)$ and $\lambda_0$ an eigenvalue of $T$. Then the geometric multiplicity of $\lambda_0$ is not greater than the algebraic multiplicity of $\lambda_0$, i.e. $g_{\lambda_0} \leq a_{\lambda_0}$.

**Proposition 3.62** (see [5], p. 64). Let $V$ be a vector space over the field $K$ and $T \in \mathrm{End}(V)$. Then each eigenvector of $T$ corresponds to a single eigenvalue $\lambda \in \sigma(T)$.

### 3.4.3 Diagonalization Algorithm of an Endomorphism

We suppose that $T \in \mathrm{End}(V)$ has $n$ distinct eigenvalues $\lambda_1, \ldots, \lambda_n$ and $\dim V = n$.

We denote by $\overline{a}_1, \ldots, \overline{a}_n$ the eigenvectors of the endomorphism $T$ corresponding respectively to the eigenvalues $\lambda_1, \ldots, \lambda_n$.

Let $B = \{\overline{a}_1, \ldots, \overline{a}_n\}$ be a basis of eigenvectors corresponding to $T$; therefore

$$\begin{cases} T(\overline{a}_1) = \lambda_1 \overline{a}_1 \\ \quad \vdots \\ T(\overline{a}_n) = \lambda_n \overline{a}_n. \end{cases}$$

The matrix

$$\widetilde{T}_{(B)} = \begin{pmatrix} \lambda_1 & 0 & \cdots & 0 \\ 0 & \lambda_2 & & 0 \\ 0 & 0 & \ddots & 0 \\ 0 & 0 & & \lambda_n \end{pmatrix}$$

is a diagonal matrix.

**Definition 3.63** (see [5], p. 69). Let $V$ be a vector space over the finite dimensional field $K$, $n \geq 1$. We say that the endomorphism $T \in \mathrm{End}(V)$ is **diagonalizable** if there is a basis of $V$ relative to which its matrix is a diagonal matrix.

**Theorem 3.64** (see [3], p. 68). Let $V$ be a vector space over the $n$ finite dimensional field $K$, $n \geq 1$. The necessary and sufficient condition that the endomorphism $T \in \mathrm{End}(V)$ to be diagonalizable is that the characteristic polynomial $P(\lambda)$ to have all the roots in $K$ and the geometric multiplicity of each eigenvalue to be equal to its algebraic multiplicity.

The diagonalization algorithm of an endomorphism $T \in \text{End}(V)$ consists [9] in the following steps:

1. choose a certain basis $B \subseteq V$ and determine the matrix $\widetilde{T}_{(B)}$ associated to the endomorphism $T$ in this basis;
2. find the eigenvalues $\lambda_1, \ldots, \lambda_p$ and of their corresponding algebraic multiplicites: $a_{\lambda_1}, \ldots, a_{\lambda_p}$;
3. achieve the eigensubspaces $W_{\lambda_1}, \ldots, W_{\lambda_p}$ corresponding to the eigenvalues $\lambda_1, \ldots, \lambda_p$;
4. determine the basis $B_i$ of the eigenspace $W_{\lambda_i}$ associated to the eigenvalue $\lambda_i$, $i = \overline{1, p}$ and of the geometric multiplicity $g_{\lambda_i}$, corresponding to the eigenvalue $\lambda_i, i = \overline{1, p}$;
5. check the condition (from the Theorem 3.64) that the endomorphism $T$ to be diagonalizable;
6. obtain the basis $B'$ of the vector space $V$ relative to which the associated matrix of $T$ has the canonical diagonal form:

$$B' = B_1 \cup B_2 \cup \ldots \cup B_p.$$

The associated matrix of $T$ in the basis $B'$ is a diagonal matrix, having on its diagonal the eigenvalues $\lambda_1, \ldots, \lambda_p$, each of them appearing of a number of times equal to its order of multiplicity:

$$\widetilde{T}_{(B')} = \begin{pmatrix} \lambda_1 & & & & & 0 \\ & \ddots & & & & \\ & & \lambda_1 & & & \\ & & & \ddots & & \\ & & & & \lambda_p & \\ & & & & & \ddots \\ 0 & & & & & \lambda_p \end{pmatrix}$$

7. build the transition matrix from the basis $B$ to the basis $B'$, i.e. $M_{(B,B')}$;
8. test the correctness of the calculations using the relation

$$\widetilde{T}_{(B')} = M^{-1}_{(B,B')} \cdot \widetilde{T}_{(B)} \cdot M_{(B,B')}.$$

**Example 3.65** (see [9]). On the vector space of matrices of second order one considers the mapping:

$$T : M_2(\mathbb{R}) \to M_2(\mathbb{R}), T(A) = A^t.$$

(a) Write the associated matrix of $T$ relative to the canonical basis of the space $M_2(\mathbb{R})$.
(b) Determine the eigenvalues and the corresponding eigenspaces.

(c) Determine a basis $B'$ of the vector space $M_2(\mathbb{R})$ relative to which the associated matrix of $T$ has a diagonal form.

**Solution**

(a) We know that (see the Example 3.31)

$$B = \{E_{11}, E_{12}, E_{21}, E_{22}\} = \left\{\begin{pmatrix}1&0\\0&0\end{pmatrix}, \begin{pmatrix}0&1\\0&0\end{pmatrix}, \begin{pmatrix}0&0\\1&0\end{pmatrix}, \begin{pmatrix}0&0\\0&1\end{pmatrix}\right\}$$

is a canonical basis in $M_2(\mathbb{R})$.

We compute

$$\begin{aligned} T(E_{11}) &= E_{11}^t = \begin{pmatrix}1&0\\0&0\end{pmatrix} = E_{11} \\ T(E_{12}) &= E_{12}^t = \begin{pmatrix}0&0\\1&0\end{pmatrix} = E_{21} \\ T(E_{21}) &= E_{21}^t = \begin{pmatrix}0&1\\0&0\end{pmatrix} = E_{12} \\ T(E_{22}) &= E_{22}^t = \begin{pmatrix}0&0\\0&1\end{pmatrix} = E_{22}. \end{aligned}$$

The associated matrix of $T$ relative to the canonical basis of the space $M_2(\mathbb{R})$ will be

$$\widetilde{T}_{(B)} = \begin{pmatrix}1&0&0&0\\0&0&1&0\\0&1&0&0\\0&0&0&1\end{pmatrix}.$$

(b) We determine

$$\begin{aligned} P(\lambda) = \det\left(\widetilde{T}_{(B)} - \lambda I_4\right) &= \begin{vmatrix}1-\lambda&0&0&0\\0&-\lambda&1&0\\0&1&-\lambda&0\\0&0&0&1-\lambda\end{vmatrix} \\ &= (1-\lambda)^2\left(\lambda^2-1\right) = (\lambda-1)^3(\lambda+1). \end{aligned}$$

The roots of the characteristic equation are:

- $\lambda_1 = 1$, having $a_{\lambda_1} = 3$;
- $\lambda_2 = -1$, having $a_{\lambda_2} = 1$.

The associated eigenspace of the eigenvalue $\lambda_1$ is

$$W_{\lambda_1} = \{A \in \mathrm{M}_2(\mathbb{R}) \,|T(A) = \lambda_1 A\}.$$

As $T(A) = A^t$ and $\lambda_1 = 1$ we obtain

$$W_{\lambda_1} = \left\{A \in \mathrm{M}_2(\mathbb{R}) \,|A^t = A\right\}.$$

Let be $A \in \mathrm{M}_2(\mathbb{R})$, $A = \begin{pmatrix} a_{11} & a_{12} \\ a_{21} & a_{22} \end{pmatrix}$. From the condition $A^t = A$ we deduce

$$\begin{pmatrix} a_{11} & a_{21} \\ a_{12} & a_{22} \end{pmatrix} = \begin{pmatrix} a_{11} & a_{12} \\ a_{21} & a_{22} \end{pmatrix} \Leftrightarrow a_{12} = a_{21}.$$

Hence

$$W_{\lambda_1} = \left\{A \in \mathrm{M}_2(\mathbb{R}) \,|A = \begin{pmatrix} a_{11} & a_{12} \\ a_{12} & a_{22} \end{pmatrix}\right\}.$$

We can write

$$\begin{aligned} A &= a_{11} \begin{pmatrix} 1 & 0 \\ 0 & 0 \end{pmatrix} + a_{12} \begin{pmatrix} 0 & 1 \\ 1 & 0 \end{pmatrix} + a_{22} \begin{pmatrix} 0 & 0 \\ 0 & 1 \end{pmatrix} \\ &= a_{11} E_{11} + a_{12} \begin{pmatrix} 0 & 1 \\ 1 & 0 \end{pmatrix} + a_{22} E_{22}; \end{aligned}$$

it results that

$$B_1 = \left\{E_{11}, E_{22}, \begin{pmatrix} 0 & 1 \\ 1 & 0 \end{pmatrix}\right\} \tag{3.49}$$

is a system of generators for $W_{\lambda_1}$.

We note that $B_1$ is also linearly independent as if

$$a_{11} E_{11} + a_{12} \begin{pmatrix} 0 & 1 \\ 1 & 0 \end{pmatrix} + a_{22} E_{22} = \mathrm{O}_{\mathrm{M}_2(\mathbb{R})}$$

it results

$$\begin{pmatrix} a_{11} & a_{12} \\ a_{12} & a_{22} \end{pmatrix} = \begin{pmatrix} 0 & 0 \\ 0 & 0 \end{pmatrix},$$

i.e. $a_{11} = a_{12} = a_{22} = 0$.

Therefore $B_1$ from (3.49) is a basis of $W_{\lambda_1}$ and $g_{\lambda_1} = 3$.

The associated eigenspace of the eigenvalue $\lambda_2$ is

$$W_{\lambda_2} = \{A \in \mathrm{M}_2(\mathbb{R}) \,|T(A) = \lambda_2 A\},$$

i.e.

$$W_{\lambda_2} = \left\{A \in \mathrm{M}_2(\mathbb{R}) \,|\, A^t = -A\right\}.$$

From the condition $A^t = -A$ we deduce

$$\begin{pmatrix} a_{11} & a_{21} \\ a_{12} & a_{22} \end{pmatrix} = \begin{pmatrix} -a_{11} & -a_{12} \\ -a_{21} & -a_{22} \end{pmatrix} \Leftrightarrow \begin{cases} a_{11} = 0 \\ a_{12} = -a_{21} \\ a_{22} = 0. \end{cases}$$

Therefore

$$W_{\lambda_2} = \left\{A \in \mathrm{M}_2(\mathbb{R}) \,|\, A = \begin{pmatrix} 0 & a_{12} \\ -a_{12} & 0 \end{pmatrix}\right\}.$$

We can write

$$A = a_{12} \begin{pmatrix} 0 & 1 \\ -1 & 0 \end{pmatrix}.$$

Similarly, we obtain

$$B_2 = \left\{\begin{pmatrix} 0 & 1 \\ -1 & 0 \end{pmatrix}\right\} \tag{3.50}$$

is a basis of $W_{\lambda_2}$ and $g_{\lambda_2} = 1$.

(c) As

$$\left.\begin{array}{c} a_{\lambda_1} = g_{\lambda_1} \\ a_{\lambda_2} = g_{\lambda_2} \\ \text{the characteristic equation has some real roots} \end{array}\right\} \Rightarrow T \text{ is diagonalizable.}$$

The basis of the vector space $\mathrm{M}_2(\mathbb{R})$ relative to which the associated matrix of $T$ has the canonical diagonal form is

$$B' = B_1 \cup B_2 \overset{(3.49)+(3.50)}{=} \left\{E_{11}, E_{22}, \begin{pmatrix} 0 & 1 \\ 1 & 0 \end{pmatrix}, \begin{pmatrix} 0 & 1 \\ -1 & 0 \end{pmatrix}\right\} = \{F_1, F_2, F_3, F_4\}$$

and

$$\widetilde{T}_{(B')} = \begin{pmatrix} 1 & 0 & 0 & 0 \\ 0 & 1 & 0 & 0 \\ 0 & 0 & 1 & 0 \\ 0 & 0 & 0 & -1 \end{pmatrix}.$$

We shall have:

$$F_1 = E_{11} = 1 \cdot E_{11} + 0 \cdot E_{12} + 0 \cdot E_{21} + 0 \cdot E_{22}$$
$$F_2 = E_{22} = 0 \cdot E_{11} + 0 \cdot E_{12} + 0 \cdot E_{21} + 1 \cdot E_{22}$$

$$F_3 = E_{12} + E_{21} = 0 \cdot E_{11} + 1 \cdot E_{12} + 1 \cdot E_{21} + 0 \cdot E_{22}$$
$$F_4 = E_{12} - E_{21} = 0 \cdot E_{11} + 1 \cdot E_{12} - 1 \cdot E_{21} + 0 \cdot E_{22}.$$

Therefore

$$\mathrm{M}_{(B,B')} = \begin{pmatrix} 1 & 0 & 0 & 0 \\ 0 & 0 & 1 & 1 \\ 0 & 0 & 1 & -1 \\ 0 & 1 & 0 & 0 \end{pmatrix}.$$

It results that

$$\mathrm{M}^{-1}_{(B,B')} \cdot \widetilde{T}_{(B)} \cdot \mathrm{M}_{(B,B')} = \begin{pmatrix} 1 & 0 & 0 & 0 \\ 0 & 1 & 0 & 0 \\ 0 & 0 & 1 & 0 \\ 0 & 0 & 0 & -1 \end{pmatrix} = \widetilde{T}_{(B')}.$$

The same matrix can be achieved in Sage:

```
sage: M=MatrixSpace(RR,2)
sage: var("a11 a12 a21 a22")
(a11, a12, a21, a22)
sage: A=matrix([[a11,a12],[a21,a22]])
sage: T=A.transpose()
sage: E=M.basis()
sage: T0=T.substitute(a11=E[0][0][0],a12=E[0][0][1],a21=E[0][1][0],a22=E[0][1][1])
sage: T1=T.substitute(a11=E[1][0][0],a12=E[1][0][1],a21=E[1][1][0],a22=E[1][1][1])
sage: T2=T.substitute(a11=E[2][0][0],a12=E[2][0][1],a21=E[2][1][0],a22=E[2][1][1])
sage: T3=T.substitute(a11=E[3][0][0],a12=E[3][0][1],a21=E[3][1][0],a22=E[3][1][1])
sage: U=a11*E[0]+a12*E[1]+a21*E[2]+a22*E[3]
sage: s0=solve([T0[0][0]==U[0][0],T0[0][1]==U[0][1],T0[1][0]==U[1][0],T0[1][1]==U[1][1]],a11,a12,a21,a22);s0
[[a11 == 1, a12 == 0, a21 == 0, a22 == 0]]
sage: s1=solve([T1[0][0]==U[0][0],T1[0][1]==U[0][1],T1[1][0]==U[1][0],T1[1][1]==U[1][1]],a11,a12,a21,a22);s1
[[a11 == 0, a12 == 0, a21 == 1, a22 == 0]]
sage: s2=solve([T2[0][0]==U[0][0],T2[0][1]==U[0][1],T2[1][0]==U[1][0],T2[1][1]==U[1][1]],a11,a12,a21,a22);s2
[[a11 == 0, a12 == 1, a21 == 0, a22 == 0]]
sage: s3=solve([T3[0][0]==U[0][0],T3[0][1]==U[0][1],T3[1][0]==U[1][0],T3[1][1]==U[1][1]],a11,a12,a21,a22);s3
[[a11 == 0, a12 == 0, a21 == 0, a22 == 1]]
sage: v0=vector([1,0,0,0]);v1=vector([0,0,1,0]);
sage: v2=vector([0,1,0,0]);v3=vector([0,0,0,1])
sage: TB=column_matrix([v0,v1,v2,v3])
sage: la=TB.eigenvalues();la
[-1, 1, 1, 1]
sage: ss=solve([T[0][0]==A[0][0],T[0][1]==A[0][1],T[1][0]==A[1][0],T[1][1]==A[1][1]],a11,a12,a21,a22);ss
[[a11 == r3, a12 == r2, a21 == r2, a22 == r1]]
sage: si=solve([T[0][0]==-A[0][0],T[0][1]==-A[0][1],T[1][0]==-A[1][0],T[1][1]==-A[1][1]],a11,a12,a21,a22);si
[[a11 == 0, a12 == r4, a21 == -r4, a22 == 0]]
sage: TBp=diagonal_matrix([la[1],la[2],la[3],la[0]]);TBp
[ 1  0  0  0]
[ 0  1  0  0]
[ 0  0  1  0]
[ 0  0  0 -1]
```

### 3.4.4 Jordan Canonical Form

Let $V$ be a $n$ finite dimensional vector space over the field $K$.

If in the previous paragraph we showed the necessary and sufficient conditions that an endomorphism $T \in \mathrm{End}(V)$ to be diagonalizable, in this section we propose to determine a basis of $V$ relative to which the associated matrix of $T$ to have a simpler form called the **Jordan canonical form**.

**Definition 3.66** (see [2], p. 47). A square matrix of the form

$$\begin{pmatrix} \lambda & 1 & 0 & 0 & \cdots & 0 & 0 & 0 \\ 0 & \lambda & 1 & 0 & \cdots & 0 & 0 & 0 \\ \cdots & \cdots & \cdots & \cdots & \cdots & \cdots & \cdots & \cdots \\ 0 & 0 & 0 & 0 & \cdots & 0 & \lambda & 1 \\ 0 & 0 & 0 & 0 & 0 & \cdots & 0 & \lambda \end{pmatrix} \in \mathrm{M}_p(K) \tag{3.51}$$

is called a **Jordan cell** of the order $p$, denoted with $\mathrm{J}_p(\lambda)$.

**Example 3.67** The matrix

$$J_4(\lambda) = \begin{pmatrix} \lambda & 1 & 0 & 0 \\ 0 & \lambda & 1 & 0 \\ 0 & 0 & \lambda & 1 \\ 0 & 0 & 0 & \lambda \end{pmatrix}$$

is a Jordan cell of the fourth order.

This matrix can be generated with Sage:

```
sage: la=var('la')
sage: jordan_block(la, 4)
[la  1  0  0]
[ 0 la  1  0]
[ 0  0 la  1]
[ 0  0  0 la]
```

**Definition 3.68** (see [2], p. 47). A square matrix of the form

$$B_p(\lambda) = \begin{pmatrix} J_{p_1}(\lambda) & & & \mathrm{O} \\ & J_{p_2}(\lambda) & & \\ & & \ddots & \\ \mathrm{O} & & & J_{p_r}(\lambda) \end{pmatrix} \in \mathrm{M}_p(K), \tag{3.52}$$

where $p_1 + p_2 + \cdots + p_r = p$ is called a **Jordan block** of the order $p$.

**Example 3.69**. The matrices

$$B_5(\lambda) = \begin{pmatrix} J_2(\lambda) & \mathrm{O} \\ \mathrm{O} & J_3(\lambda) \end{pmatrix} = \begin{pmatrix} \lambda & 1 & 0 & 0 & 0 \\ 0 & \lambda & 1 & 0 & 0 \\ 0 & 0 & \lambda & 1 & 0 \\ 0 & 0 & 0 & \lambda & 1 \\ 0 & 0 & 0 & 0 & \lambda \end{pmatrix},$$

$$B_8(\lambda) = \begin{pmatrix} J_4(\lambda) & O \\ O & J_4(\lambda) \end{pmatrix} = \begin{pmatrix} \lambda & 1 & 0 & 0 & 0 & 0 & 0 & 0 \\ 0 & \lambda & 1 & 0 & 0 & 0 & 0 & 0 \\ 0 & 0 & \lambda & 1 & 0 & 0 & 0 & 0 \\ 0 & 0 & 0 & \lambda & 1 & 0 & 0 & 0 \\ 0 & 0 & 0 & 0 & \lambda & 1 & 0 & 0 \\ 0 & 0 & 0 & 0 & 0 & \lambda & 1 & 0 \\ 0 & 0 & 0 & 0 & 0 & 0 & \lambda & 1 \\ 0 & 0 & 0 & 0 & 0 & 0 & 0 & \lambda \end{pmatrix}$$

are some Jordan blocks of five and respectively eight order.

We shall use Sage to generate $B_5(\lambda)$ :

```
sage: la=var('la')
sage: A=jordan_block(la,2);B=jordan_block(la,3)
sage: block_matrix([[A,0],[0,B]])
[la  1| 0  0  0]
[ 0 la| 0  0  0]
[-----+--------]
[ 0  0|la  1  0]
[ 0  0| 0 la  1]
[ 0  0| 0  0 la]
```

We can also obtain $B_8(\lambda)$ in Sage, too:

```
sage: la=var('la')
sage: A=jordan_block(la,4);O=matrix(SR,4)
sage: block_matrix([[A,O],[O,A]])
[la  1  0  0| 0  0  0  0]
[ 0 la  1  0| 0  0  0  0]
[ 0  0 la  1| 0  0  0  0]
[ 0  0  0 la| 0  0  0  0]
[-----------+-----------]
[ 0  0  0  0|la  1  0  0]
[ 0  0  0  0| 0 la  1  0]
[ 0  0  0  0| 0  0 la  1]
[ 0  0  0  0| 0  0  0 la]
```

**Definition 3.70** (see [2], p. 47). A square matrix of order $n$, which has Jordan blocks on the main diagonal, i.e. of the form

$$\mathrm{J} = \begin{pmatrix} B_{n_1}(\lambda_1) & & & O \\ & B_{n_2}(\lambda_2) & & \\ & & \ddots & \\ O & & & B_{n_r}(\lambda_r) \end{pmatrix} \in \mathrm{M}_n(K), \tag{3.53}$$

where $n_1 + n_2 + \cdots + n_r = n$ is called a **matrix in the Jordan canonical form**.

**Example 3.71**. The matrix

$$J = \begin{pmatrix} B_3(\lambda_1) & O \\ O & B_3(\lambda_2) \end{pmatrix} = \begin{pmatrix} \lambda_1 & 1 & 0 & 0 & 0 & 0 \\ 0 & \lambda_1 & 1 & 0 & 0 & 0 \\ 0 & 0 & \lambda_1 & 1 & 0 & 0 \\ 0 & 0 & 0 & \lambda_2 & 1 & 0 \\ 0 & 0 & 0 & 0 & \lambda_2 & 1 \\ 0 & 0 & 0 & 0 & 0 & \lambda_2 \end{pmatrix}$$

is a matrix in the Jordan canonical form.

The matrix J can be generated in Sage:

```
sage: la1,la2=var('la1, la2')
sage: A=jordan_block(la1,2);B=jordan_block(la2,1)
sage: J=block_matrix([[A,0],[0,B]]);J
[la1   1|  0]
[  0 la1|  0]
[-------+---]
[  0   0|la2]
```

**Definition 3.72** (see [2], p. 47). The endomorphism $T\in\text{End}(V)$ is **jordanizable** if there is a basis of $V$ relative to which the associated matrix of $T$ has a Jordan canonical form.

**Definition 3.73** (see [2], p. 47). A matrix $A\in\text{M}_n(K)$ is called **jordanizable** if there is a nonsingular matrix $C\in\text{M}_n(K)$ such that $C^{-1}AC$ be a matrix in the Jordan canonical form.

**Theorem 3.74** (see [2], p. 47). Let $V$ be a $n$ dimensional vector space over $K$, $n\geq 1$ and $T\in\text{End}(V)$ be an andomorphism whose characteristic polynomial is

$$P(\lambda) = (-1)^n(\lambda-\lambda_1)^{m_1}\cdot\ldots\cdot(\lambda-\lambda_r)^{m_r}, \tag{3.54}$$

where the eigenvalues $\lambda_1, \lambda_2, \ldots, \lambda_r$ are distinct and $m_1+m_2+\cdots+m_r=n$.

Then there is a basis of $V$ relative to which the associated matrix of $T$ has the Jordan canonical form.

#### 3.4.4.1 Jordanization Algorithm of an Endomorphism

Let $V$ be a finite $n$ dimensional vector space over the field $K$, $n\geq 1$ and let $T\in\text{End}(V)$ be an endomorphism.

The jordanization algorithm of an endomorphism $T$ consists [2] of the following steps:

1. Choose a basis $B$ of $V$ and write the associated matrix of $T$ relative to this basis, i.e. the matrix $\widetilde{T}_{(B)}$.
2. Determine the characteristic polynomial $P(\lambda)$ using (3.37); there are two cases:

(a) the characteristic equation $P(\lambda) = 0$ hasn't $n$ roots in $K$, therefore $T$ is not jordanizable;
(b) the characteristic equation has the roots $\lambda_1, \lambda_2, \ldots, \lambda_r$ with the multiplicities $m_1, m_2, \ldots, m_r$, with $m_1 + m_2 + \cdots + m_r = n$.

3. For the eigenvalue $\lambda_i$ one computes $\widetilde{T}_{(B)} - \lambda_i \mathrm{I}_n$.
4. Determine the number of the Jordan cells for the eigenvalue $\lambda_i$,

$$n_i = \dim \operatorname{Ker}\left(T - \lambda_i \mathfrak{I}_V\right), \tag{3.55}$$

$\mathfrak{I}_V$ being the identity mapping.
Therefore, the number of Jordan cells for the eigenvalue $\lambda_i$ is equal to the maximum number of the corresponding linearly independent eigenvectors. There are two situations:

(a) if $n_i = m_i$ then a basis of the eigenspace $W_{\lambda_i}$ associated of the eigenvalue $\lambda_i$ will consist of $n_i$ linearly independent eigenvectors;
(b) if $n_i < m_i$ then go to the next step.

5. Find the smallest natural number $s_i \in \mathbb{N}^*$, $s_i \leq m_i$ such that

$$\dim \operatorname{Ker}\left(T - \lambda_i \mathfrak{I}_V\right)^{s_i} = m_i. \tag{3.56}$$

6. Determine the number of the Jordan cells of order $h \in \{1, 2, \ldots, s_i\}$ within the formula

$$d_h = \operatorname{rank}\left(T - \lambda_i \mathfrak{I}_V\right)^{h+1} + \operatorname{rank}\left(T - \lambda_i \mathfrak{I}_V\right)^{h-1} - 2 \cdot \operatorname{rank}\left(T - \lambda_i \mathfrak{I}_V\right)^{h}, \tag{3.57}$$

where

$$\begin{cases} \operatorname{rank}\left(T - \lambda_i \mathfrak{I}_V\right)^0 = n \\ \operatorname{rank}\left(T - \lambda_i \mathfrak{I}_V\right)^{s_i+1} = \operatorname{rank}\left(T - \lambda_i \mathfrak{I}_V\right)^{s_i} \\ \displaystyle\sum_{h=1}^{s_i} h \cdot d_h = m_i. \end{cases} \tag{3.58}$$

7. Repeat the steps 3–6 for each eigenvalue of $T$.
8. Write the matrix J (see (3.53)) of $T$ in the Jordan canonical form.
9. Achieve the basis $B'$ of $V$ relative to which $T$ has the matrix J.

**Example 3.75** (see [2], p. 88). Let $T : \mathbb{R}^3 \to \mathbb{R}^3$ be an endomorphism whose matrix relative to the canonical basis of $\mathbb{R}^3$ is

$$A = \begin{pmatrix} 3 & -2 & 1 \\ 2 & -2 & 2 \\ 3 & -6 & 5 \end{pmatrix}.$$

(a) Show that $T$ is jordanizable, write its matrix to the Jordan canonical form J and find the basis of $\mathbb{R}^3$ relative to which $T$ has the matrix J.
(b) Compute $A^n$, $n \in \mathbb{N}^*$.

**Solution**

(a) Using (3.54), the characteristic polynomial will be:

$$P(\lambda) = (2 - \lambda)^3$$

and the characteristic equation has the root $\lambda_1 = 2$, with $m_1 = 3 = n$; hence $T$ is jordanizable.

The associated matrix of the endomorphism $T - \lambda_i \Im_V$ is

$$A - 2I_3 = \begin{pmatrix} 1 & -2 & 1 \\ 2 & -4 & 2 \\ 3 & -6 & 3 \end{pmatrix}.$$

Since according to the Theorem 3.22, $\mathrm{Ker}(T - \lambda_i \Im_V)$ is a vector subspace of $\mathbb{R}^3$, from the Proposition 1.67 it results that

$$\dim \mathrm{Ker}\,(T - \lambda_i \Im_V) = 3 - r,$$

where

$$r = \mathrm{rank}\,(A - 2I_3) = 1;$$

therefore

$$n_1 = \dim \mathrm{Ker}\,(T - \lambda_i \Im_V) = 3 - 1 = 2.$$

As $n_1 < m_1$ we shall determine (with (3.56)) the smallest natural number $s_1 \in \mathbb{N}^*$, $s_1 \leq m_1$ such that

$$\dim \mathrm{Ker}\,(T - \lambda_1 \Im_V)^{s_1} = m_1.$$

We note that

$$(A - 2I_3)^2 = O_3 \text{ (null matrix of the third order)};$$

hence

$$\mathrm{rank}\,(A - 2I_3)^2 = 0 \Rightarrow \dim \mathrm{Ker}\,(T - \lambda_1 \Im_V)^2 = 3 - 0 = 3 = m_1,$$

i.e. $s_1 = 2$.

We have

$$d_1 = 0 + 3 - 2 = 1,$$

$$d_2 = 0 + 1 - 0 = 1.$$

Therefore, there will be two Jordan cells in the achieved matrix in the Jordan canonical form, namely a first order cell $J_1(2)$ and a second-order cell $J_2(2)$. The matrix J of $T$ will be the following in the Jordan canonical form:

$$\mathrm{J} = \begin{pmatrix} J_1(2) & \mathrm{O} \\ \mathrm{O} & J_2(2) \end{pmatrix} = \begin{pmatrix} 2 & 0 & 0 \\ 0 & 2 & 1 \\ 0 & 0 & 2 \end{pmatrix}$$

and the basis of $\mathbb{R}^3$ relative to which $T$ has this form is $B' = \{\overline{f}_1, \overline{f}_2, \overline{f}_3\}$.

Using the definition of the associated matrix to an endomorphism we shall have:

$$\begin{cases} T(\overline{f}_1) = 2\overline{f}_1 \\ T(\overline{f}_2) = 2\overline{f}_2 \\ T(\overline{f}_3) = \overline{f}_2 + 2\overline{f}_3. \end{cases}$$

As the number of Jordan cells for the eigenvalue $\lambda_1$ is equal to the maximum number of the corresponding linearly independent eigenvectors, it results that the vectors $\overline{f}_1, \overline{f}_2$ are some eigenvectors for $T$, i.e. their coordinates are the solution of system:

$$\begin{pmatrix} 3 & -2 & 1 \\ 2 & -2 & 2 \\ 3 & -6 & 5 \end{pmatrix} \begin{pmatrix} x \\ y \\ z \end{pmatrix} = 2 \begin{pmatrix} x \\ y \\ z \end{pmatrix} \Leftrightarrow$$

$$\begin{cases} 3x - 2y + z = 2x \\ 2x - 2y + 2z = 2y \\ 3x - 6y + 5z = 2z \end{cases} \Leftrightarrow \begin{cases} x - 2y + z = 0 \\ 2x - 4y + 2z = 0 \\ 3x - 6y + 3z = 0. \end{cases}$$

The rank of the associated matrix of the system is equal to 1, so we denote

$$y = t, z = u, (\forall)\, t, u \in \mathbb{R};$$

it results $x = 2t - u$.

We shall have

$$V_{\lambda_1} = \left\{ \overline{f} \in \mathbb{R}^3 | \overline{f} = (2t - u, t, u), t, u \in \mathbb{R} \right\}.$$

For $t = 1, u = 0$ we consider $\overline{f}_1 = (2, 1, 0)$. The eigenvector $\overline{f}_2$ belongs (as $\overline{f}_1$) to the family of vectors

$$\overline{f} = (2t - u, t, u), t, u \in \mathbb{R}, |t| + |u| \neq 0.$$

The vector $\overline{f}_3$ one determines from the relation

$$T\left(\overline{f}_3\right) = \overline{f}_2 + 2\overline{f}_3,$$

i.e. from the system

$$\begin{pmatrix} 3 & -2 & 1 \\ 2 & -2 & 2 \\ 3 & -6 & 5 \end{pmatrix} \begin{pmatrix} x \\ y \\ z \end{pmatrix} = \begin{pmatrix} 2t-u \\ t \\ u \end{pmatrix} + 2 \begin{pmatrix} x \\ y \\ z \end{pmatrix} \Leftrightarrow$$

$$\begin{cases} 3x - 2y + z = 2t - u + 2x \\ 2x - 2y + 2z = t + 2y \\ 3x - 6y + 5z = u + 2z \end{cases} \Leftrightarrow \begin{cases} x - 2y + z = 2t - u \\ 2x - 4y + 2z = t \\ 3x - 6y + 3z = u. \end{cases}$$

The previous system is compatible if and only if: $3t - 2u = 0$.

The rank of the associated matrix of the system is equal to 1, therefore we denote

$$y = v, z = w, (\forall)\, v, w \in \mathbb{R};$$

it results $x = 2t - u + 2v - w$.

For $t = 2, u = 3$ we consider $\overline{f}_2 = (1, 2, 3)$. In the case when $t = 2, u = 3$ for $v = w = 0$ we obtain $\overline{f}_3 = (1, 0, 0)$.

(b) The transition matrix from the canonical basis $B$ to the basis $B'$ will be

$$C = \mathrm{M}_{(B,B')} = \begin{pmatrix} 2 & 1 & 1 \\ 1 & 2 & 0 \\ 0 & 3 & 0 \end{pmatrix}.$$

We shall obtain

$$A = CJC^{-1} = \begin{pmatrix} 2 & 1 & 1 \\ 1 & 2 & 0 \\ 0 & 3 & 0 \end{pmatrix} \cdot \begin{pmatrix} 2 & 0 & 0 \\ 0 & 2 & 1 \\ 0 & 0 & 2 \end{pmatrix} \cdot \begin{pmatrix} 0 & 1 & -2/3 \\ 0 & 0 & 1/3 \\ -1 & 2 & 1 \end{pmatrix}$$

and

$$A^n = CJ^nC^{-1} = \begin{pmatrix} 2 & 1 & 1 \\ 1 & 2 & 0 \\ 0 & 3 & 0 \end{pmatrix} \cdot \begin{pmatrix} 2^n & 0 & 0 \\ 0 & 2^n & n \cdot 2^{n-1} \\ 0 & 0 & 2^n \end{pmatrix} \cdot \begin{pmatrix} 0 & 1 & -2/3 \\ 0 & 0 & 1/3 \\ -1 & 2 & 1 \end{pmatrix}$$

$$\Rightarrow A^n = \begin{pmatrix} 2^n + n \cdot 2^{n-1} & -n \cdot 2^n & n \cdot 2^{n-1} \\ n \cdot 2^n & (1-2n)\, 2^n & n \cdot 2^n \\ 3n \cdot 2^{n-1} & -3n \cdot 2^n & (2+3n)\, 2^{n-1} \end{pmatrix}.$$

We shall present the solution in Sage, too:

```
sage: Aa=matrix([[3,-2,1],[2,-2,2],[3,-6,5]])
sage: J=Aa.jordan_form(subdivide=False);J
[2 1 0]
[0 2 0]
[0 0 2]
sage: J.rank()
3
sage: var("x y z u t v w n")
(x, y, z, u, t, v, w, n)
sage: solve([3*x-2*y+z==2*x,2*x-2*y+2*z==2*y,3*x-6*y+5*z==2*z],x,y,z)
[[x == -r1 + 2*r2, y == r2, z == r1]]
sage: f1=vector([-u+2*t,t,u]).substitute(u=0,t=1);f1
(2, 1, 0)
sage: f2=vector([-u+2*t,t,u]).substitute(u=3,t=2);f2
(1, 2, 3)
sage: t=2;u=3;solve([x-2*y+z==2*t-u,2*x-4*y+2*z==t,3*x-6*y+3*z==u],x,y,z)
[[x == -r3 + 2*r4 + 1, y == r4, z == r3]]
sage: f3=vector([-v+2*w+1,w,v]).substitute(w=0,v=0);f3
(1, 0, 0)
sage: C=column_matrix([f1,f2,f3]);C
[2 1 1]
[1 2 0]
[0 3 0]
sage: A=C*J*C^(-1);C.inverse()
[   0    1 -2/3]
[   0    0  1/3]
[   1   -2    1]
sage: Jn=matrix([[2^n,0,0],[0,2^n,n*2^(n-1)],[0,0,2^n]])
sage: An=simplify(C*Jn*C^(-1));An
[  n*2^(n - 1) + 2^n                 -2^n*n          n*2^(n - 1)]
[               2^n*n  -n*2^(n + 1) + 2^n                 2^n*n]
[      3*n*2^(n - 1)              -3*2^n*n 3*n*2^(n - 1) + 2^n]
```

## 3.5 Problems

1. Let be the linear mapping

$$T : \mathbb{R}^2 \to \mathbb{R}^2, T(\overline{x}) = T(x_1, x_2) = (x_1, 3x_1 - x_2).$$

Show that $T^2 = \Im$ (the identity mapping).

**Solution**

Using Sage, we have:

```
sage: x1,x2 = var('x1,x2')
sage: f_symbolic(x1,x2)=[x1,3*x1-x2]
sage: f=linear_transformation(RR^2, RR^2, f_symbolic)
sage: f*f
Vector space morphism represented by the matrix:
[ 1.00000000000000 0.000000000000000]
[0.000000000000000  1.00000000000000]
Domain: Vector space of dimension 2 over Real Field with 53 bits of precision
Codomain: Vector space of dimension 2 over Real Field with 53 bits of precision
```

2. Find the matrix $A$ in each of the following cases:

   (a) $\left(3A^t + 2\begin{pmatrix} 1 & 0 \\ 0 & 2 \end{pmatrix}\right)^t = \begin{pmatrix} 8 & 0 \\ 3 & 1 \end{pmatrix}$;

   (b) $\left(2A - 3 \cdot \begin{pmatrix} 1 & 2 & 0 \end{pmatrix}\right)^t = 3A^t + \begin{pmatrix} 2 & 1 & -1 \end{pmatrix}^t$.

**Solution**

We shall use Sage to find the matrix $A$ :

```
sage: var("a b c d")
(a, b, c, d)
sage: A=matrix([[a,b],[c,d]])
sage: A1=matrix([[1,0],[0,2]]);A2=matrix([[8,0],[3,1]])
sage: B=(3*A.transpose()+2*A1).transpose()
sage: solve([B[0][0]==A2[0][0],B[0][1]==A2[0][1],B[1][0]==A2[1][0],B[1][1]==A2[1][1]],a,b,c,d)
[[a == 2, b == 0, c == 1, d == -1]]
```

3. One considers the mapping

$$T : \mathrm{R}_2[X] \to \mathrm{R}_2[X], T(P) = (4X + 1) P'.$$

   (a) Show that $T$ is a linear transformation.
   (b) Write the associated matrix of $T$ relative to the canonical basis $B = \{1, X, X^2\}$ of the vector space $\mathrm{R}_2[X]$.
   (c) Determine Ker$T$ and Im$T$.

4. In the case of the previous problem:

   (a) Find the eigenvalues and the eigenspace of the corresponding eigenvectors.
   (b) Decide if $T$ is diagonalizable or not and if so write the diagonal form of the matrix and specify the form relative of which the matrix is diagonal.

5. Using the Hamilton-Cayley theorem compute the inverse of the matrix

$$\begin{pmatrix} 1 & 3 & -1 \\ 1 & 0 & 2 \\ 2 & 2 & 1 \end{pmatrix}.$$

**Solution**

The solution of the problem in Sage is:

```
sage: A=matrix([[1,3,-1],[1,0,2],[2,2,1]])
sage: var("la")
la
sage: I3=identity_matrix(3)
sage: P=(A-la*I3).determinant();expand(P)
-la^3 + 2*la^2 + 4*la + 3
sage: -A^3+2*A^2+4*A+3*I3
```

```
[0 0 0]
[0 0 0]
[0 0 0]
sage: Ainv=1/3*A^2-2/3*A-4/3*I3;Ainv
[-4/3 -5/3    2]
[   1    1   -1]
[ 2/3  4/3   -1]
sage: Ainv==A.inverse()
True
```

6. Let be the endomorphism $T : \mathbb{R}^3 \to \mathbb{R}^3$, given through the matrix

$$A = \begin{pmatrix} 1 & -1 & 2 \\ 1 & 0 & 1 \\ 1 & 0 & -1 \end{pmatrix}$$

relative to the canonical basis of $\mathbb{R}^3$. Determine the matrix of $T$ relative to the basis

$$B_1 = \{\overline{f}_1 = (1, 2, 3), \overline{f}_2 = (3, 1, 2), \overline{f}_3 = (2, 3, 1)\}.$$

**Solution**

This matrix can be determined in Sage:

```
sage: F=matrix(RR,[[1,-2,1],[1,0,1],[1,0,-1]]);
sage: V=RR^3;v1=V([1,2,3]);v2=V([3,1,2]);v3=V([2,3,1])
sage: var("x1 x2 x3")
(x1, x2, x3)
sage: w=vector(SR,[x1,x2,x3])
sage: F*w.transpose()
[x1 - 2.00000000000000*x2 + x3]
[                      x1 + x3]
[                      x1 - x3]
sage: t_symbolic(x1,x2,x3)=[x1-2*x2+x3,x1+x3,x1-x3]
sage: T=linear_transformation(RR^3, RR^3, t_symbolic)
sage: W=V.subspace_with_basis([v1,v2,v3])
sage: w1=W.coordinate_vector(T(v1));w2=W.coordinate_vector(T(v2));w3=W.coordinate_vector(T(v3))
sage: column_matrix([w1,w2,w3]).n(digits=3)
[-0.556 -0.167   1.39]
[ -1.22 -0.167  -1.94]
[  2.11   1.83  0.722]
```

7. Let $V$ be a finite dimensional vector space, $\dim V = 3$, $T$ be an endomorphism of $V$ and $B = \{\overline{e}_1, \overline{e}_2, \overline{e}_3\}$ be a basis of $V$ such that

$$\widetilde{T}_{(B)} = \begin{pmatrix} 1 & 1 & 1 \\ 1 & 1 & 1 \\ 1 & 1 & 1 \end{pmatrix}.$$

(a) Determine the eigenvalues and the eigenvectors associated to the endomorphism $T$.
(b) Check if $T$ is diagonalizable and then determine a space basis relative to which the associated matrix of $T$ has the canonical diagonal form.

8. Let the mapping $T : \mathrm{R}_3[X] \to \mathbb{R}^4$, $T(P) = \left(P(-3)\ P(-1)\ P(1)\ P(3)\right)$.

(a) Show that $T$ is a linear transformation.
(b) Compute $T(Q)$, for $Q(t) = 5 + 2t - t^2$.
(c) Write the associated matrix of $T$ relative to the canonical basis $B_1 = \{1, X, X^2, X^3\}$ from $R_3[X]$ and respectively to the canonical basis from $\mathbb{R}^4$.

9. Let be the mapping

$$T : R_3[X] \to R_3[X], T(P) = \sum_{k=1}^{2} X^k \int_{-1}^{1} t^k P(t)\,dt.$$

(a) Write the associated matrix of $T$ relative to the canonical basis of the vector space $R_3[X]$.
(b) Determine $\mathrm{Ker}T$ and $\mathrm{Im}T$.

**Solution**

Using Sage to solve the problem we achieve:

```
sage: F=SR[x];var("a0 a1 a2 a3 t k")
(a0, a1, a2, a3, t, k)
sage: R.<x>=RR['x'];K=R^4;P=F([a0,a1,a2,a3]);
sage: f(t,k)=expand(t^k*P(t))
sage: u=integral(f(t,1),t,-1,1);u1=integral(f(t,2),t,-1,1)
sage: l=u*x+u1*x^2;l
2/15*(5*a0 + 3*a2)*x^2 + 2/15*(5*a1 + 3*a3)*x
sage: M=K.span([[1,x,x^2,x^3]])
sage: l1=l.substitute(a0=1,a1=0,a2=0,a3=0);l1
2/3*x^2
sage: l2=l.substitute(a0=0,a1=1,a2=0,a3=0);l2
2/3*x
sage: l3=l.substitute(a0=0,a1=0,a2=1,a3=0);l3
2/5*x^2
sage: l4=l.substitute(a0=0,a1=0,a2=0,a3=1);l4
2/5*x
sage: U=M([l1,l2,l3,l4])
sage: v0=vector([U[0][0],U[0][1],U[0][2],U[0][3]])
sage: v1=vector([U[1][0],U[1][1],U[1][2],U[1][3]])
sage: v2=vector([U[2][0],U[2][1],U[2][2],U[2][3]])
sage: v3=vector([U[3][0],U[3][1],U[3][2],U[3][3]])
sage: fB=column_matrix([v0,v1,v2,v3]).n(digits=3);fB
[0.000 0.000 0.000 0.000]
[0.000 0.667 0.000 0.400]
[0.667 0.000 0.400 0.000]
[0.000 0.000 0.000 0.000]
sage: eq=[l.coefficient(x^2),l.coefficient(x)]
sage: s=solve(eq,a0,a1,a2,a3);s
[[a0 == -3/5*r6, a1 == -3/5*r5, a2 == r6, a3 == r5]]
sage: Q=P.substitute(a0=-3/5*a2,a1=-3/5*a3);Q
a3*x^3 + a2*x^2 - 3/5*a3*x - 3/5*a2
sage: Q1=-3/5+x^2;Q2=-3/5*x+x^3
sage: expand(a2*Q1+a3*Q2)
a3*x^3 + a2*x^2 - 3/5*a3*x - 3/5*a2
```

Hence:

$$\mathrm{Ker} f = \{P \in R_3[X] \,|\, P = a_2 Q_1 + a_3 Q_2\}$$

and

$$\mathrm{Im} f = \{f(P) \,|\, P \in R_3[X]\} = \left\{\left(\frac{2}{3}a_1 + \frac{2}{3}a_3\right) X + \left(\frac{2}{3}a_0 + \frac{2}{5}a_2\right) X^2\right\}.$$

10. Let $T : \mathbb{R}^3 \to \mathbb{R}^3$ be an endomorphism whose matrix relative to the canonical basis of $\mathbb{R}^3$ is

$$A = \begin{pmatrix} 2 & 1 & 0 \\ -1 & 1 & -1 \\ 1 & 0 & 3 \end{pmatrix}.$$

(a) Show that $T$ is jordanizable, write its matrix in the Jordan canonical form J and determine the basis of $\mathbb{R}^3$ relative to which J is the matrix of $T$.
(b) Compute $A^n, n \in \mathbb{N}^*$.

## References

1. V. Postelnicu, S. Coatu, *Mică enciclopedie matematică, ed* (Tehnică, Bucureşti, 1980)
2. I. Vladimirescu, M. Popescu, *Algebră liniară şi geometrie analitică, ed* (Universitaria, Craiova, 1993)
3. I. Vladimirescu, M. Popescu, *Algebră liniară şi geometrie n- dimensională, ed* (Radical, Craiova, 1996)
4. C. Udrişte, *Algebră liniară, geometrie analitică* (Geometry Balkan Press, Bucureşti, 2005)
5. V. Balan, *Algebră liniară, geometrie analitică, ed* (Fair Partners, Bucureşti, 1999)
6. S. Chiriţă, *Probleme de matematici superioare, ed* (Didactică şi Pedagogică, Bucureşti, 1989)
7. Gh. Atanasiu, Gh. Munteanu, M. Postolache, *Algebr ă liniară, geometrie analitică şi diferenţială, ecua ţii diferenţiale, ed* (ALL, Bucureşti, 1998)
8. P. Matei, *Algebră liniară. Gometrie analitică şi diferenţială, ed* (Agir, Bucureşti, 2002)
9. I. Vladimirescu, M. Popescu, M. Sterpu, *Algebră liniară şi geometrie analitică* (Universitatea din Craiova, Note de curs şi aplicaţii, 1993)

# Chapter 4
# Euclidean Vector Spaces

## 4.1 Euclidean Vector Spaces

The study of the Euclidean vector space is required to obtain the orthonormal bases, whereas relative to these bases, the calculations are considerably simplified. In a Euclidean vector space, scalar product can be used to define the length of vectors and the angle between them.

Let E be a real vector space.

**Definition 4.1** (see [6], p. 101). The mapping $<,>$: E$\rightarrow$E is called a **scalar product** (or an **Euclidean structure**) on E if the following conditions are satisfied:

(a) $<\overline{x},\overline{y}>=<\overline{y},\overline{x}>$, $(\forall)\ \overline{x},\overline{y}\in$E

(b) $<\overline{x}+\overline{y},\overline{z}>=<\overline{x},\overline{z}>+<\overline{y},\overline{z}>$, $(\forall)\ \overline{x},\overline{y},\overline{z}\in$E

(c) $\alpha<\overline{x},\overline{y}>=<\alpha\overline{x},\overline{y}>$, $(\forall)\ \overline{x},\overline{y}\in$E, $(\forall)\ \alpha\in\mathbb{R}$

(d) $<\overline{x},\overline{x}>\geq 0$, $(\forall)\ \overline{x}\in$E; $<\overline{x},\overline{x}>=0\Leftrightarrow\overline{x}=\overline{0}$.

The scalar $<\overline{x},\overline{y}>\in\mathbb{R}$ is called the scalar product of vectors $\overline{x},\overline{y}\in$E.

**Definition 4.2** (see [6], p. 101). A real vector space on which a scalar product is defined, is called an **Euclidean real vector space** and it should be denoted by (E, $<,>$).

**Proposition 4.3** (see [6], p. 101). A scalar product on E has the following properties:

(i) $<\overline{0},\overline{x}>=<\overline{x},\overline{0}>=0$, $(\forall)\ \overline{x}\in$E

(ii) $<\overline{x},\overline{y}+\overline{z}>=<\overline{x},\overline{y}>+<\overline{x},\overline{z}>$, $(\forall)\ \overline{x},\overline{y},\overline{z}\in$E

(iii) $<\overline{x},\alpha\overline{y}>=\alpha<\overline{x},\overline{y}>$, $(\forall)\ \overline{x},\overline{y}\in$E, $(\forall)\ \alpha\in\mathbb{R}$

(iv) $<\sum_{i=1}^{n}\alpha^{(i)}\overline{x}_i,\sum_{j=1}^{m}\beta^{(j)}\overline{y}_j>=\sum_{i=1}^{n}\sum_{j=1}^{m}\alpha^{(i)}\beta^{(j)}<\overline{x}_i,\overline{y}_j>$, $(\forall)\ \overline{x}_i,\overline{y}_j\in$ E, $(\forall)\ \alpha^{(i)},\beta^{(j)}\in\mathbb{R}, i=\overline{1,n}, j=\overline{1,m}$.

G. A. Anastassiou and I. F. Iatan, *Intelligent Routines II*,
Intelligent Systems Reference Library 58, DOI: 10.1007/978-3-319-01967-3_4,

**Examples of Euclidean vector spaces**

1. $(\mathbb{R}^n, <, >)$ the canonical Euclidean real space. The mapping $<, >: \mathbb{R}^n \to \mathbb{R}^n$ defined by

$$< \overline{x}, \overline{y} >= \sum_{i=1}^{n} x^{(i)} y^{(i)}, (\forall) \overline{x}, \overline{y} \in \mathbb{R}^n, \ \overline{x} = \left(x^{(1)}, x^{(2)}, \ldots, x^{(n)}\right), \ \overline{y} = \left(y^{(1)}, y^{(2)}, \ldots, y^{(n)}\right)$$

is a scalar product on $\mathbb{R}^n$.

2. $\left(\mathrm{M}_{n,n}(\mathbb{R}), <, >\right)$ the real Euclidean space of the square matrices, with the scalar product

$$< A, B >= \mathrm{trace}\left(A^t B\right), (\forall) \ A, B \in \mathrm{M}_{n,n}(\mathbb{R}).$$

3. $(V_3, <, >)$ where the mapping $<, >: V_3 \times V_3 \to \mathbb{R}$ defined by

$$< \overline{x}, \overline{y} >= \begin{cases} \|\overline{x}\| \, \|\overline{y}\| \cos \sphericalangle (\overline{x}, \overline{y}), \ \overline{x} \neq \overline{0}, \overline{y} \neq \overline{0} \\ 0, \ \overline{x} = \overline{0} \text{ or } \overline{y} = \overline{0} \end{cases}$$

is a scalar product on $V_3$.

This concrete scalar product was the model from which, by abstraction has reached the concept of the scalar product.

**Example 4.4** (see [8]). Let $V_3$ be the vector space of geometric vectors of the usual physical space. Indicate if the next transformation is a scalar product on $V_3$:

$$<, >: V_3 \times V_3 \to \mathbb{R}, \quad < \overline{x}, \overline{y} >= \|\overline{x}\| \, \|\overline{y}\|, \quad (\forall) \overline{x}, \overline{y} \in V_3.$$

**Solution**

If the transformation would be a scalar product, then according to the Definition 4.1 it should be satisfied the condition:

$$< \alpha \overline{x}, \overline{y} >= \alpha < \overline{x}, \overline{y} >, (\forall) \ \overline{x}, \overline{y} \in V_3, (\forall) \ \alpha \in \mathbb{R},$$

i.e.

$$\|\alpha \overline{x}\| \, \|\overline{y}\| = \alpha \, \|\overline{x}\| \, \|\overline{y}\|, (\forall) \ \overline{x}, \overline{y} \in V_3, (\forall) \ \alpha \in \mathbb{R}.$$

Since the previous relationship can not be held for $\alpha < 0$ and $\overline{x}, \overline{y} \neq \overline{0}$ it results that the given transformation is not a scalar product.

This fact can also be checked using Sage:

```
sage: var("x1 x2 x3 y1 y2 y3")
(x1, x2, x3, y1, y2, y3)
sage: x=vector(SR,[x1,x2,x3]);y=vector(SR,[y1,y2,y3])
sage: def f(x,y): return (x.norm()*y.norm()).simplify_exp()
sage: al1=-2;al2=2
sage: f(al1*x,y)-al1*f(x,y)
4*sqrt(y1^2 + y2^2 + y3^2)*sqrt(x1^2 + x2^2 + x3^2)
sage: f(al2*x,y)-al2*f(x,y)
0
```

**Example 4.5** Let $B = \left\{1, 1+X, 1+X+X^2\right\}$ be a basis of the space $\mathrm{R}_2[X]$. We define the scalar product

$$< \overline{x}, \overline{y} >_B = x_1y_1 + x_2y_2 + x_3y_3,$$

where $x_i, y_i, i = \overline{1,3}$ are the coordinates of the vectors $\overline{x}$ and $\overline{y}$ in the basis $B$. Compute the scalar products:

(a) $< 1+X, 1+X >_B$
(b) $< 3-X^2, 2+4X+6X^2 >_B$.

**Solution**
(a) Noting that

$$1+X = 0\cdot 1 + 1\cdot(1+X) + 0\cdot\left(1+X+X^2\right)$$

it results

$$(1+X)_B = \begin{pmatrix} 0 \\ 1 \\ 0 \end{pmatrix},$$

i.e. the coordinates of $1+X$ in the basis $B$ are: 0, 1, 0; therefore

$$< 1+X, 1+X >_B = 0\cdot 1 + 1\cdot 1 + 0\cdot 0 = 1.$$

(b) Noting that

$$3-X^2 = 3+1+X-1-X-X^2 = 3\cdot 1 + 1\cdot(1+X) + (-1)\cdot\left(1+X+X^2\right)$$

and

$$2+4X+6X^2 = -2-2-2X+6+6X+6X^2$$
$$= (-2)\cdot 1 + (-2)\cdot(1+X) + 6\cdot\left(1+X+X^2\right)$$

it results

$$\left(3-X^2\right)_B = \begin{pmatrix} 3 \\ 1 \\ -1 \end{pmatrix},$$

$$\left(2+4X+6X^2\right)_B = \begin{pmatrix} -2 \\ -2 \\ 6 \end{pmatrix},$$

i.e. the coordinates of

- $3-X^2$ in the basis $B$ are: $3, 1, -1$;
- $2+4X+6X^2$ in the basis $B$ are: $-2, -2, 6$;

hence

$$< 3-X^2, 2+4X+6X^2 >_B = 3\cdot(-2) + 1\cdot(-2) + (-1)\cdot 6 = -14.$$

We shall give a solution in Sage, too:

```
sage: F=SR[x];a,b,c=var('a, b, c')
sage: U=F([a+b+c,b+c,c]);U
c*x^2 + (b + c)*x + a + b + c
sage: solve([U[0]==1,U[1]==1,U[2]==0],a,b,c)
[[a == 0, b == 1, c == 0]]
sage: v0=vector([0,1,0]);p1=v0.dot_product(v0);p1
1
sage: solve([U[0]==3,U[1]==0,U[2]==-1],a,b,c)
[[a == 3, b == 1, c == -1]]
sage: solve([U[0]==2,U[1]==4,U[2]==6],a,b,c)
[[a == -2, b == -2, c == 6]]
sage: v1=vector([3,1,-1]);v2=vector([-2,-2,6])
sage: p2=v1.dot_product(v2);p2
-14
```

**Definition 4.6** (see [7], p. 155). Let (E, $<,>$) be a real Euclidean vector space. It's called a **norm** on E, a mapping $\|\,\|$ :E$\to \mathbb{R}_+$ which satisfies the properties:

(i) $\|\overline{x}\| > 0$, $(\forall)\ \overline{x} \in$E and $\|\overline{x}\| = 0 \Leftrightarrow \overline{x} = \overline{0}$ (positivity)
(ii) $\|\alpha\overline{x}\| = |\alpha|\,\|\overline{x}\|$, $(\forall)\ \overline{x} \in$E, $(\forall)\ \alpha \in \mathbb{R}$ (homogeneity)
(iii) $\|\overline{x}+\overline{y}\| \leq \|\overline{x}\| + \|\overline{y}\|$, $(\forall)\ \overline{x}, \overline{y} \in$E (Minkowski's inequality or the triangle inequality).

**Theorem 4.7** (see [7], p. 154). Let (E, $<,>$) be a real Euclidean vector space. The function $\|\,\|$ :E$\rightarrow \mathbb{R}_+$, defined by

$$\|\overline{x}\| = \sqrt{<\overline{x},\overline{x}>}, (\forall)\ \overline{x} \in \mathrm{E} \tag{4.1}$$

is a norm on E. The norm defined in the Theorem 4.7 is called the **Euclidean norm.**
**Remark 4.8** (see [7], p. 154). If $\overline{x} \in V_3$, then its norm (length), in the sense of Theorem 4.7 coincides with the geometric meaning of its length.
**Proposition 4.9** (see [7], p. 154). Let (E, $<,>$) be a real Euclidean vector space.

(a) For all $\overline{x}, \overline{y}$ $\in$E, the *Cauchy- Schwartz- Buniakowski inequality* occurs

$$<\overline{x},\overline{y}>\leq \|\overline{x}\|\ \|\overline{y}\|; \tag{4.2}$$

(b) For all $\overline{x}, \overline{y}$ $\in$E, the *parallelogram identity* occurs

$$\|\overline{x}+\overline{y}\|^2 + \|\overline{x}-\overline{y}\|^2 = 2\left(\|\overline{x}\|^2 + \|\overline{y}\|^2\right). \tag{4.3}$$

**Definition 4.10** (see [7], p. 155). Let (E, $<,>$) be a real Euclidean vector space. It's called the **angle** of the non-zero vectors $\overline{x}, \overline{y}$ $\in$E, the unique number $\varphi \in [0,\pi]$ for which

$$\cos\varphi = \frac{<\overline{x},\overline{y}>}{\|\overline{x}\|\ \|\overline{y}\|}. \tag{4.4}$$

**Definition 4.11** (see [7], p. 156). Let (E, $<,>$) be a real Euclidean vector space. We shall say that the vectors $\overline{x}, \overline{y}$ $\in$E are **orthogonal** and we shall denote $\overline{x}\perp\overline{y}$ if $<\overline{x},\overline{y}>=0$.
**Remark 4.12** (see [7], p. 156). The null vector is orthogonal on any vector $\overline{x}$ $\in$E.
**Example 4.13** (see [8]). In the vector space $\mathbb{R}^2$ one considers $B=\{\overline{e}_1,\overline{e}_2\}$, $\|\overline{e}_1\|=2$, $\|\overline{e}_2\|=4$, $\prec(\overline{e}_1,\overline{e}_2)=\frac{\pi}{3}$. The vectors $\overline{a}=2\overline{e}_1-3\overline{e}_2$, $\overline{b}=-\overline{e}_1+\overline{e}_2$ are given. Compute $\prec\left(\overline{a},\overline{b}\right)$.
**Solution**

We have:

$$<\overline{e}_1,\overline{e}_1>=\|\overline{e}_1\|^2=4,$$
$$<\overline{e}_2,\overline{e}_2>=\|\overline{e}_2\|^2=16.$$

From the relation (4.4) we deduce

$$<\overline{x},\overline{y}>=\|\overline{x}\|\ \|\overline{y}\|\cos\varphi;$$

hence, we obtain:

$$<\overline{e}_1,\overline{e}_2>=\|\overline{e}_1\|\ \|\overline{e}_2\|\cos\prec(\overline{e}_1,\overline{e}_2)=2\cdot 4\cdot\frac{1}{2}=4.$$

It results that

$$< \overline{a}, \overline{a} >=< 2\overline{e}_1 - 3\overline{e}_2, 2\overline{e}_1 - 3\overline{e}_2 >= 4 < \overline{e}_1, \overline{e}_1 > -12 < \overline{e}_1, \overline{e}_2 > +9 < \overline{e}_2, \overline{e}_2 >= 112$$
$$< \overline{a}, \overline{b} >=< 2\overline{e}_1 - 3\overline{e}_2, -\overline{e}_1 + \overline{e}_2 >= -2 < \overline{e}_1, \overline{e}_1 > +5 < \overline{e}_1, \overline{e}_2 > -3 < \overline{e}_2, \overline{e}_2 >= -36$$
$$< \overline{b}, \overline{b} >=< -\overline{e}_1 + \overline{e}_2, -\overline{e}_1 + \overline{e}_2 >=< \overline{e}_1, \overline{e}_1 > -2 < \overline{e}_1, \overline{e}_2 > + < \overline{e}_2, \overline{e}_2 >= 12.$$

Using (4.1), we have:

$$\|\overline{a}\| = \sqrt{< \overline{a}, \overline{a} >} = 4\sqrt{7}$$
$$\|\overline{b}\| = \sqrt{< \overline{b}, \overline{b} >} = 2\sqrt{3}.$$

Therefore,

$$\cos \prec \left(\overline{a}, \overline{b}\right) \overset{(4.4)}{=} \frac{< \overline{a}, \overline{b} >}{\|\overline{a}\| \, \|\overline{b}\|} = -\frac{36}{4\sqrt{7} \cdot 2\sqrt{3}}.$$

We shall check this result in Sage, too:

```
sage: var("e11 e12 e13 e21 e22 e23")
(e11, e12, e13, e21, e22, e23)
sage: e1=vector([e11,e12,e13]);e2=vector([e21,e22,e23])
sage: a=2*e1-3*e2;b=-e1+e2;p1=4;p2=4;p3=16
sage: u1=expand(a.dot_product(a));u1
4*e11^2 - 12*e11*e21 + 4*e12^2 - 12*e12*e22 + 4*e13^2 - 12*e13*e23 + 9*e21^2 + 9*e22^2 + 9*e23^2
sage: v1=4*4-12*4+9*16;v1
112
sage: u2=expand(a.dot_product(b));u2
-2*e11^2 + 5*e11*e21 - 2*e12^2 + 5*e12*e22 - 2*e13^2 + 5*e13*e23 - 3*e21^2 - 3*e22^2 - 3*e23^2
sage: v2=-2*4+5*4-3*16;v2
-36
sage: u3=expand(b.dot_product(b));u3
e11^2 - 2*e11*e21 + e12^2 - 2*e12*e22 + e13^2 - 2*e13*e23 + e21^2 + e22^2 + e23^2
sage: v3=4-2*4+16;v3
12
sage: v2/(sqrt(v1)*sqrt(v3))
-3/14*sqrt(3)*sqrt(7)
```

**Proposition 4.14** (see [7], p. 156). Let (E, <, >) be a real Euclidean vector space and $\overline{a}_1, \overline{a}_2, \ldots, \overline{a}_p \in$E be some nonnull vectors which are pairwise orthogonal. Then, the vectors $\overline{a}_1, \overline{a}_2, \ldots, \overline{a}_p$ are linearly independent.

**Definition 4.15** (see [7], p. 156). If $\overline{x}$ is a nonnull vector of the real Euclidean vector space (E, <, >) then the vector

$$\overline{x}^0 = \frac{\overline{x}}{\|\overline{x}\|}$$

is called the versor of $\overline{x}$.

**Remark 4.16** (see [7], p. 156). The length of the vector $\overline{x}^0$ is equal to 1.

**Definition 4.17** (see [7], p. 156). The **dimesion** of the real Euclidean vector space (E, <, >) constitutes the dimension of the associated vector space E.

**Definition 4.18** (see [7], p. 156). Let (E, $<,>$) be a real Euclidean vector space and $S \subset$ E.

(a) The system $S$ is **orthogonal** if its vectors are nonnull and pairwise orthogonal.
(b) The system $S$ is **orthonormal** (or **orthonormat**) if it is orthogonal and each of its vectors has the length equal to 1.
(c) If $\dim \mathrm{E} = n < \infty$ then the basis $B = \{\overline{e}_1, \overline{e}_2, \ldots, \overline{e}_n\}$ of E is called **orthonormal** if $< \overline{e}_i, \overline{e}_j >= \delta_{ij}, (\forall)\, i, j = \overline{1,n}$,

where

$$\delta_{ij} = \begin{cases} 1, & i = j \\ 0, & i \neq j \end{cases}, i, j = \overline{1,n}$$

is called the *symbol of Kronecher*.

The canonical basis $\mathbb{R}^n$ is an orthonormal basis for canonical Euclidean real vector space $(\mathbb{R}^n, <,>)$.

**Definition 4.19** (see [7], p. 157). Let (E, $<,>$) be a $n$ finite dimensional real Euclidean vector space and $B = \{\overline{e}_1, \overline{e}_2, \ldots, \overline{e}_n\}$ is a basis of E. Let be $\overline{x}, \overline{y} \in$ E; it results

$$\begin{cases} \overline{x} = \sum\limits_{i=1}^{n} x^{(i)} \overline{e}_i \\ \overline{y} = \sum\limits_{j=1}^{n} y^{(j)} \overline{e}_j. \end{cases}$$

We have:

$$< \overline{x}, \overline{y} >=< \sum_{i=1}^{n} x^{(i)} \overline{e}_i, \sum_{j=1}^{n} y^{(j)} \overline{e}_j >= \sum_{i=1}^{n} \sum_{j=1}^{n} x^{(i)} y^{(j)} < \overline{e}_i, \overline{e}_j > . \quad (4.5)$$

We denote

$$< \overline{e}_i, \overline{e}_j >= g_{ij}, (\forall)\, i, j = \overline{1,n};$$

we note that $g_{ij} = g_{ji}, (\forall)\, i, j = \overline{1,n}$, i.e.

$$G = \begin{pmatrix} g_{11} & g_{12} & \cdots & g_{1n} \\ g_{12} & g_{22} & \cdots & g_{2n} \\ \cdots & \cdots & \cdots & \cdots \\ g_{1n} & g_{2n} & \cdots & g_{nn} \end{pmatrix};$$

the matrix $G$ signifies the **matrix of the scalar product** $<,>$ relative to the basis $B$.

**Remark 4.20** (see [7], p. 157). The matrix $G$ is symmetric ($G^t = G$) and positive definite ($\overline{x}^t G \overline{x} > 0, (\forall)\, \overline{x} \in \mathbb{R}^n \setminus \{\overline{0}\}$).

**Definition 4.21** (see [7], p. 157). From (4.5) we obtain

$$< \overline{x}, \overline{y} >= \sum_{i=1}^{n} \sum_{j=1}^{n} x^{(i)} y^{(j)} g_{ij}, \tag{4.6}$$

relation that constitutes the **analytical expression of the scalar product** $<,>$ relative to the basis $B$.

**Definition 4.22** (see [7], p. 157). The equality (4.6) is equivalent to the equality

$$< \overline{x}, \overline{y} >= \overline{x}^t_B G \overline{y}_B, \tag{4.7}$$

called the **matrix representation of the scalar product** $<,>$ relative to the basis $B$.

**Example 4.23**. One considers the mapping $<,>: \mathbb{R}^3 \times \mathbb{R}^3 \to \mathbb{R}$ which, relative to the canonical basis $B = \{\overline{e}_1, \overline{e}_2, \overline{e}_3\}$ of $\mathbb{R}^3$ has the analytical expression:

$$< \overline{x}, \overline{y} >= (x_1 - 2x_2)(y_1 - 2y_2) + x_2 y_2 + (x_2 + x_3)(y_2 + y_3),$$

$(\forall)\ \overline{x} = (x_1, x_2, x_3), \overline{y} = (y_1, y_2, y_3) \in \mathbb{R}^3$.

(a) Show that $\left(\mathbb{R}^3, <,>\right)$ is a real Euclidean vector space.
(b) Prove that the vectors $\overline{a} = \overline{e}_1 - \overline{e}_2 + 2\overline{e}_3$, $\overline{b} = -\overline{e}_1 + \overline{e}_2 + 9\overline{e}_3$ are orthogonal.
(c) Compute $\|\overline{x}\|$, where $\overline{x} = \overline{e}_1 - \overline{e}_2 + 2\overline{e}_3$.
(d) Compute the angle between the vectors $\overline{x} = \overline{e}_1 - \overline{e}_2 + 2\overline{e}_3$ and $\overline{y} = \overline{e}_2 + 2\overline{e}_3$.
(e) Write the matrix of the scalar product $<,>$ relative to the canonical basis of $\mathbb{R}^3$.

**Solution**

(a) $\left(\mathbb{R}^3, <,>\right)$ is a real Euclidean vector space if the mapping $<,>$ is a scalar product. We will check that the conditions of the Definition 4.1 of a scalar product are satisfied.

We note that

$$\begin{aligned}
&< \overline{x}, \overline{y} >= (x_1 - 2x_2)(y_1 - 2y_2) + x_2 y_2 + (x_2 + x_3)(y_2 + y_3) =< \overline{y}, \overline{x} >,\ (\forall)\ \overline{x}, \overline{y} \in \mathbb{R}^3 \\
&< \overline{x} + \overline{z}, \overline{y} >= (x_1 - 2x_2)(y_1 - 2y_2) + (z_1 - 2z_2)(y_1 - 2y_2) + x_2 y_2 + z_2 y_2 + \\
&\quad (x_2 + x_3)(y_2 + y_3) + (z_2 + z_3)(y_2 + y_3) =< \overline{x}, \overline{y} > + < \overline{z}, \overline{y} >,\ (\forall)\ \overline{x}, \overline{y}, \overline{z} \in \mathbb{R}^3 \\
&< \alpha \overline{x}, \overline{y} >= \alpha \left[(x_1 - 2x_2)(y_1 - 2y_2) + x_2 y_2 + (x_2 + x_3)(y_2 + y_3)\right] \\
&= \alpha < \overline{x}, \overline{y} >,\ (\forall)\ \overline{x} \in \mathbb{R}^3,\ (\forall)\ \alpha \in \mathbb{R} \\
&< \overline{x}, \overline{x} >= (x_1 - 2x_2)^2 + x_2^2 + (x_2 + x_3)^2 > 0,\ (\forall)\ \overline{x} \in \mathbb{R}^3 \setminus \{\overline{0}\}
\end{aligned}$$

$$< \overline{x}, \overline{x} >= 0 \Leftrightarrow \begin{cases} x_1 - 2x_2 = 0 \\ x_2 = 0 \\ x_2 + x_3 \end{cases} \Leftrightarrow x_1 = x_2 = x_3 = 0,$$

i.e. the mapping is a scalar product.

We shall prove that in Sage, too:

```
sage: var("x1 x2 x3 y1 y2 y3 z1 z2 z3 al")
(x1, x2, x3, y1, y2, y3, z1, z2, z3, al)
sage: x=vector(SR,[x1,x2,x3]);y=vector(SR,[y1,y2,y3]);z=vector(SR,[z1,z2,z3]);
sage: T(x,y)=[(x1-2*x2)*(y1-2*y2),x2*y2,(x2+x3)*(y2+y3)];T(x,y)
((y1 - 2*y2)*(x1 - 2*x2), x2*y2, (y2 + y3)*(x2 + x3))
sage: T+T.substitute(x1=z1,x2=z2,x3=z3)==T.substitute(x1=x1+z1,x2=x2+z2,x3=x3+z3)
True
sage: T.substitute(x1=al*x1,x2=al*x2,x3=al*x3)==al*T
True
sage: T.substitute(y1=x1,y2=x2,y3=x3)>0
True
sage: solve([T[0]==T[0].substitute(y1=x1),T[1]==T[1].substitute(y2=x2),T[2]==T[2].substitute(y3=x3)],x1,x2,x3)
[[x1 == y1, x2 == 0, x3 == 0], [x1 == 0, x2 == 0, x3 == 0], [x1 == y1, x2 == y2, x3 == -y2],
[x1 == 2*y2, x2 == y2, x3 == -y2], [x1 == y1, x2 == 0, x3 == y3], [x1 == 0, x2 == 0, x3 == y3],
[x1 == y1, x2 == y2, x3 == y3], [x1 == 2*y2, x2 == y2, x3 == y3]]
```

(b) As

- $\overline{a} = \overline{e}_1 - \overline{e}_2 + 2\overline{e}_3 \Rightarrow \overline{a} = (1, -1, 2)$,
- $\overline{b} = -\overline{e}_1 + \overline{e}_2 + 9\overline{e}_3 \Rightarrow \overline{b} = (-1, 1, 9)$

we shall obtain $< \overline{a}, \overline{b} >= 0$; hence $\overline{a}$ and $\overline{b}$ are orthogonal.

(c) As

$$\overline{x} = \overline{e}_1 - \overline{e}_2 + 2\overline{e}_3 = (1, 0, 0) - (0, 1, 0) + 2 \cdot (0, 0, 1) = (1, -1, 2)$$

it results that $< \overline{x}, \overline{x} >= 6$ and using (4.1): $\|\overline{x}\| = \sqrt{6}$.

(d) Taking into account that: $< \overline{x}, \overline{y} >= -4$, $\|\overline{y}\| = \sqrt{5}$, $\|\overline{x}\| = \sqrt{6}$, we can deduce:

$$\cos \prec (\overline{x}, \overline{y}) \overset{(4.4)}{=} -\frac{4}{\sqrt{6} \cdot \sqrt{5}}.$$

(e) We have

$$G = \begin{pmatrix} 1 & -2 & 0 \\ -2 & 6 & 1 \\ 0 & 1 & 1 \end{pmatrix},$$

where $g_{ij} =< \overline{e}_i, \overline{e}_j >, (\forall)\, i, j = \overline{1, 3}$.

Solving in Sage the points (b)–(e) of this problem, we achieve:

```
sage: V=RR^3;B=V.basis()
sage: e1=B[0];e2=B[1];e3=B[2]
sage: a=e1-e2+2*e3;b=-e1+e2+9*e3
sage: sum(T.substitute(x1=a[0],x2=a[1],x3=a[2],y1=b[0],y2=b[1],y3=b[2]))
(x, y) |--> 0.000000000000000
sage: x=a;nx=x.norm()
sage: y=e2+2*e3;ny=y.norm()
```

```
sage: s=sum(T.substitute(x1=x[0],x2=x[1],x3=x[2],y1=y[0],y2=y[1],y3=y[2]))
sage: u=s/(nx*ny);u
(x, y) |--> -0.730296743340221
sage: g0=sum(T.substitute(x1=e1[0],x2=e1[1],x3=e1[2],y1=e1[0],y2=e1[1],y3=e1[2]))
sage: g1=sum(T.substitute(x1=e1[0],x2=e1[1],x3=e1[2],y1=e2[0],y2=e2[1],y3=e2[2]))
sage: g2=sum(T.substitute(x1=e1[0],x2=e1[1],x3=e1[2],y1=e3[0],y2=e3[1],y3=e3[2]))
sage: g3=sum(T.substitute(x1=e2[0],x2=e2[1],x3=e2[2],y1=e2[0],y2=e2[1],y3=e2[2]))
sage: g4=sum(T.substitute(x1=e2[0],x2=e2[1],x3=e2[2],y1=e3[0],y2=e3[1],y3=e3[2]))
sage: g5=sum(T.substitute(x1=e3[0],x2=e3[1],x3=e3[2],y1=e3[0],y2=e3[1],y3=e3[2]))
sage: g=matrix([[g0,g1,g2],[g1,g3,g4],[g2,g4,g5]]);g.n(digits=3)
[ 1.00 -2.00 0.000]
[-2.00  6.00  1.00]
[0.000  1.00  1.00]
```

If $B' = \{\overline{e}_1, \overline{e}_2, \ldots, \overline{e}_n\}$ is an orthonormal basis of E, then the matrix of the scalar product relative to this basis is the unit matrix $\mathrm{I}_n$. In this case, from (4.6) we achieve:

$$< \overline{x}, \overline{y} >= \sum_{i=1}^{n}\sum_{j=1}^{n} x^{(i)} y^{(j)} \tag{4.8}$$

and from (4.7) we deduce

$$< \overline{x}, \overline{y} >= \overline{x}^t_{B'} \overline{y}_{B'}. \tag{4.9}$$

The equalities (4.8) and (4.9) justifies the importance of considering the orthonormal basis which consists in the fact that relative to these bases, the computations are more simplified.

**Definition 4.24** (see [7], p. 159). The matrix $A \in \mathrm{M}_n(\mathbb{R})$ is orthogonal if $A^t A = AA^t = \mathrm{I}_n$, $\mathrm{I}_n$ being unit matrix of order $n$.

**Theorem 4.25** (**theorem of change the orthonormal bases**, see [7], p. 160). Let (E, $<,>$) be a real Euclidean vector space of finite dimension $n$ and $B_1 = \{\overline{e}_1, \overline{e}_2, \ldots, \overline{e}_n\}$, $B_2 = \{\overline{u}_1, \overline{u}_2, \ldots, \overline{u}_n\}$ are two orthonormal bases of E. Then the transition matrix from the basis $B_1$ to the basis $B_2$ is orthogonal.

**Example 4.26** (see [6], p. 112). In the real Euclidean vector space E one considers the bases $B_1 = \{\overline{e}_1, \overline{e}_2, \overline{e}_3\}$, $B_2 = \{\overline{u}_1, \overline{u}_2, \overline{u}_3\}$. If $\overline{x}$ is an arbitrary vector from E, $x^{(i)}, y^{(i)}, i = \overline{1,3}$ are the coordinates of the vector relative to the bases $B_1$, $B_2$, $B_1$ is an orthonormal basis and

$$\begin{cases} x^{(1)} = \frac{2}{7}y^{(1)} + \frac{3}{7}y^{(2)} + \frac{6}{7}y^{(3)} \\ x^{(2)} = \frac{6}{7}y^{(1)} + \frac{2}{7}y^{(2)} + \alpha y^{(3)} \\ x^{(3)} = -\alpha y^{(1)} - \frac{6}{7}y^{(2)} + \frac{2}{7}y^{(3)} \end{cases}$$

determine $\alpha \in \mathbb{R}$ such that $B_2$ be an orthonormal basis, too.

**Solution**

In order that $B_2$ be an orthonormal basis it is necessary (within the Theorem 4.25) that the transition matrix from the basis $B_1$ to the basis $B_2$, i.e.

$$M_{(B_1,B_2)} = \begin{pmatrix} \frac{2}{7} & \frac{3}{7} & \frac{6}{7} \\ \frac{6}{7} & \frac{2}{7} & \alpha \\ -\alpha & -\frac{6}{7} & \frac{2}{7} \end{pmatrix}$$

to be an orthogonal one.

From the condition

$$M_{(B_1,B_2)} \cdot M^t_{(B_1,B_2)} = I_3$$

it results that $\alpha = -\frac{3}{7}$.

The solution is Sage is:

```
sage: al=var('al')
sage: M=matrix([[2/7,3/7,6/7],[6/7,2/7,al],[-al,-6/7,2/7]])
sage: U=M*M.transpose();U[1][1]
al^2 + 40/49
sage: solve([U[0][0]==1,U[0][1],U[0][2],U[1][0],U[1][1]==1,U[1][2],U[2][0],U[2][1],U[2][2]==1],al)
[[al == (-3/7)]]
```

**Theorem 4.27 (Gram-Schmidt orthogonalization**, see [1], p. 26). If (E, $<,>$) is a real Euclidean vector space of finite dimension $n$ and $B = \{\overline{a}_1, \overline{a}_2, \ldots, \overline{a}_n\}$ is a basis of E then there is a basis $B' = \{\overline{e}_1, \overline{e}_2, \ldots, \overline{e}_n\}$ of E which has the following properties:

(i) the basis $B'$ is orthonormal;
(ii) the sets $\{\overline{a}_1, \ldots, \overline{a}_k\}$ and $\{\overline{e}_1, \ldots, \overline{e}_k\}$ generate the same vector subspace $W_k \subset$ E, for each $k = \overline{1,n}$.

The Gram- Schmidt orthogonalization procedure (described in detail in [2], p. 150) can be summarized as:

(1) build an orthogonal set $B_1 = \{\overline{b}_1, \overline{b}_2, \ldots, \overline{b}_n\}$ which satisfies the property ii) of the Theorem 4.27, where:

$$\begin{cases} \overline{b}_1 = \overline{a}_1 \\ \overline{b}_2 = \alpha_2^{(1)}\overline{b}_1 + \overline{a}_2 \\ \overline{b}_3 = \alpha_3^{(1)}\overline{b}_1 + \alpha_3^{(2)}\overline{b}_2 + \overline{a}_3 \\ \vdots \\ \overline{b}_i = \alpha_i^{(1)}\overline{b}_1 + \ldots + \alpha_i^{(i-1)}\overline{b}_{i-1} + \overline{a}_i \\ \vdots \\ \overline{b}_n = \alpha_n^{(1)}\overline{b}_1 + \ldots + \alpha_n^{(n-1)}\overline{b}_{n-1} + \overline{a}_n, \end{cases} \tag{4.10}$$

where the scalars $\alpha_i^{(j)} \in \mathbb{R}$, $i = \overline{2,n}$, $j = \overline{1,i-1}$ are determined from the condition that $\overline{b}_i \perp \overline{b}_j, i, j = \overline{1,n}, i \neq j$.

(2) determine the orthonormal basis $B' = \{\overline{e}_1, \overline{e}_2, \ldots, \overline{e}_n\}$ of E, such that:

$$\overline{e}_i = \frac{\overline{b}_i}{\left\|\overline{b}_i\right\|}, i = \overline{1, n}. \tag{4.11}$$

**Example 4.28** (see [8]). In the space $R_2[X]$ we define

$$< P, Q >= \int_{-1}^{1} P(t)\, Q(t)\, \mathrm{d}t.$$

Orthonormate the canonical basis of the space $R_2[X]$ (i.e. $B = \{1, X, X^2\}$), with respect to this scalar product.

**Solution**

*Stage I.* We build the orthogonal basis $B' = \{\overline{f}_1, \overline{f}_2, \overline{f}_3\}$, with

$$\begin{cases} \overline{f}_1 = 1 \\ \overline{f}_2 = X + \alpha \overline{f}_1 \\ \overline{f}_3 = X^2 + \alpha_1 \overline{f}_1 + \alpha_2 \overline{f}_2. \end{cases}$$

From the orthogonality condition of $\overline{f}_1$ and $\overline{f}_2$ we deduce:

$$< \overline{f}_1, \overline{f}_2 >= 0 \Leftrightarrow < \overline{f}_1, X + \alpha \overline{f}_1 >= 0 \Leftrightarrow$$

$$< \overline{f}_1, X > + \alpha < \overline{f}_1, \overline{f}_1 >= 0 \Leftrightarrow \alpha = -\frac{< \overline{f}_1, X >}{< \overline{f}_1, \overline{f}_1 >}.$$

As

$$< \overline{f}_1, X >=< 1, X >= \int_{-1}^{1} 1 \cdot t \mathrm{d}t = 0$$

it results that $\alpha = 0$ and $\overline{f}_2 = X$.

From the orthogonality condition $\overline{f}_1$ and $\overline{f}_3$ we deduce:

$$< \overline{f}_1, \overline{f}_3 >= 0 \Leftrightarrow < \overline{f}_1, X^2 + \alpha_1 \overline{f}_1 + \alpha_2 \overline{f}_2 >= 0 \Leftrightarrow$$

$$< \overline{f}_1, X^2 > + \alpha_1 < \overline{f}_1, \overline{f}_1 > + \alpha_2 \underbrace{< \overline{f}_1, \overline{f}_2 >}_{=0} = 0 \Leftrightarrow \alpha_1 = -\frac{< \overline{f}_1, X^2 >}{< \overline{f}_1, \overline{f}_1 >}.$$

We obtain

$$< \overline{f}_1, X^2 >=< 1, X^2 >= \int_{-1}^{1} 1 \cdot t^2 \mathrm{d}t = \frac{2}{3}$$

and

$$<\overline{f}_1, \overline{f}_1>=<1,1>=\int_{-1}^{1} 1\cdot 1\,\mathrm{d}t = 2;$$

hence $\alpha_1 = -1/3$.

From the orthogonality condition $\overline{f}_2$ and $\overline{f}_3$ we deduce:

$$<\overline{f}_2, \overline{f}_3>=0 \Leftrightarrow <\overline{f}_2, X^2+\alpha_1\overline{f}_1+\alpha_2\overline{f}_2>=0 \Leftrightarrow$$

$$<\overline{f}_2, X^2> +\alpha_1 \underbrace{\overline{f}_2, \overline{f}_1>}_{=0} +\alpha_2 <\overline{f}_2, \overline{f}_2>=0 \Leftrightarrow \alpha_2 = -\frac{<\overline{f}_2, X^2>}{<\overline{f}_2, \overline{f}_2>}.$$

As

$$<\overline{f}_2, X^2>=<X, X^2>=\int_{-1}^{1} t\cdot t^2\mathrm{d}t = 0$$

it results $\alpha_2 = 0$. Therefore

$$\overline{f}_3 = X^2 - \frac{1}{3}.$$

*Stage II.* We build the orthonormal basis $B'' = \{\overline{g}_1, \overline{g}_2, \overline{g}_3\}$ :

$$\overline{g}_i \overset{(4.11)}{=} \frac{\overline{f}_i}{\|\overline{f}_i\|},\ (\forall)\ i=\overline{1,3}.$$

Whereas

$$\|\overline{f}_1\| = \sqrt{<\overline{f}_1, \overline{f}_1>} = \sqrt{2}$$

it results that

$$\overline{g}_1 = \frac{\overline{f}_1}{\|\overline{f}_1\|} = \frac{1}{\sqrt{2}}.$$

We compute

$$<\overline{f}_2, \overline{f}_2>=<X, X>=\int_{-1}^{1} t\cdot t\,\mathrm{d}t = \frac{2}{3};$$

we shall have

$$\overline{g}_2 = \frac{\overline{f}_2}{\|\overline{f}_2\|} = \frac{\sqrt{3}}{\sqrt{2}}X.$$

We achieve that:

$$< \overline{f}_3, \overline{f}_3 >=< X^2 - \frac{1}{3}, X^2 - \frac{1}{3} >= \int_{-1}^{1} \left(t^2 - \frac{1}{3}\right)^2 \mathrm{d}t = \frac{8}{45}$$

and

$$\overline{g}_3 = \frac{\overline{f}_3}{\|\overline{f}_3\|} = \frac{3\sqrt{5}}{2\sqrt{2}} \left(X^2 - \frac{1}{3}\right).$$

We can also solve this problem in Sage:

```
sage: var("t x al al1 al2 F G")
(t, x, al, al1, al2, F, G)
sage: f1=1; f2=x+al*f1;f3=x^2+al1*f1+al2*f2;g=F*G
sage: sp1=integral(g.substitute(F=f1,G=f1).substitute(x=t),t,-1,1);sp1
2
sage: sp2=integral(g.substitute(F=f1,G=x).substitute(x=t),t,-1,1);sp2
0
sage: all=-sp2/sp1;all
0
sage: f2=f2.substitute(al=all);f2
x
sage: sp3=integral(g.substitute(F=f1,G=x^2).substitute(x=t),t,-1,1);sp3
2/3
sage: all1=-sp3/sp1;all1
-1/3
sage: sp4=integral(g.substitute(F=f2,G=x^2).substitute(x=t),t,-1,1);sp4
0
sage: sp5=integral(g.substitute(F=f2,G=f2).substitute(x=t),t,-1,1);sp5
2/3
sage: all2=-sp4/sp5;all2
0
sage: f3=f3.substitute(al1=all1,al2=all2);f3
x^2 - 1/3
sage: sp6=integral(g.substitute(F=f3,G=f3).substitute(x=t),t,-1,1);sp6
8/45
sage: g1=f1/sqrt(sp1);g1
1/2*sqrt(2)
sage: g2=f2/sqrt(sp5);g2
3/2*sqrt(2/3)*x
sage: g3=f3/sqrt(sp6);g3
5/4*(3*x^2 - 1)*sqrt(2/5)
```

**Example 4.29** We consider the real vector space of the symmetric matrices, of the order $n$, with real elements, $\mathbf{M}_n^s(\mathbb{R})$ and

$$<,>: \mathbf{M}_n^s(\mathbb{R}) \times \mathbf{M}_n^s(\mathbb{R}) \to \mathbb{R}, \quad < A, B >= \text{trace}\left(A^t B\right).$$

Orthonormate the system of matrices:

$$\left\{A_1 = \begin{pmatrix} 1 & 1 \\ 1 & 0 \end{pmatrix}, A_2 = \begin{pmatrix} 0 & 1 \\ 1 & 2 \end{pmatrix}, A_3 = \begin{pmatrix} -1 & 0 \\ 0 & 1 \end{pmatrix}\right\}.$$

**Solution**
We consider the orthogonal system $\{B_1, B_2, B_3\}$, trace $\left(B_i^t B_j\right) = 0$, $(\forall)\ i \neq j$, as follows:

$$\begin{cases} B_1 = A_1 \\ B_2 = A_2 + \alpha B_1 \\ B_3 = A_3 + \alpha_1 B_1 + \alpha_2 B_2. \end{cases}$$

From the condition $< B_2, B_1 >= 0$, we have

$$0 =< A_2 + \alpha B_1, B_1 >=< A_2, B_1 > +\alpha < B_1, B_1 >,$$

i.e.

$$0 = \text{trace}\,(A_2 B_1) + \alpha \text{trace}\,(B_1 B_1)\,;$$

therefore

$$\alpha = -\frac{\text{trace}\,(A_2 B_1)}{\text{trace}\,(B_1 B_1)} = -\frac{2}{3}.$$

We obtain

$$B_2 = \begin{pmatrix} 0 & 1 \\ 1 & 2 \end{pmatrix} - \frac{2}{3}\begin{pmatrix} 1 & 1 \\ 1 & 0 \end{pmatrix} = \begin{pmatrix} -2/3 & 1/3 \\ 1/3 & 2 \end{pmatrix}.$$

The condition $< B_3, B_1 >= 0$ involves

$$0 = < A_3 + \alpha_1 B_1 + \alpha_2 B_2, B_1 >=< A_3, B_1 > +\alpha_1 < B_1, B_1 > +\alpha_2 < B_2, B_1 >$$
$$= \text{trace}\,(A_3 B_1) + \alpha_1 \text{trace}\,(B_1 B_1) \Rightarrow \alpha_1 = -\frac{\text{trace}\,(A_3 B_1)}{\text{trace}\,(B_1 B_1)} = \frac{1}{3}.$$

The condition $< B_3, B_2 >= 0$ involves

$$0 = < A_3 + \alpha_1 B_1 + \alpha_2 B_2, B_2 >=< A_3, B_2 > +\alpha_1 < B_1, B_2 > +\alpha_2 < B_2, B_2 >$$
$$= \text{trace}\,(A_3 B_2) + \alpha_2 \text{trace}\,(B_2 B_2) \Rightarrow \alpha_2 = -\frac{\text{trace}\,(A_3 B_2)}{\text{trace}\,(B_2 B_2)} = -\frac{4}{7}.$$

We shall obtain

$$B_3 = A_3 + \frac{1}{3}B_1 - \frac{4}{7}B_2.$$

The orthonotormal system $\{C_1, C_2, C_3\}$ will be:

$$C_i = \frac{B_i}{\| B_i \|}, i = \overline{1, 3}.$$

We shall also determine these matrices in Sage:

```
sage: al,al1,al2=var("al al1 al2");F=matrix(SR,2);G=matrix(SR,2)
sage: A1=matrix([[1,1],[1,0]]);A2=matrix([[0,1],[1,2]]);A3=matrix([[-1,0],[0,1]])
sage: B1=A1;B2=A2+al*B1;B3=A3+al1*B1+al2*B2
sage: sp1=(B1.transpose()*B1).trace()
sage: sp2=(A2.transpose()*B1).trace()
sage: all=-sp2/sp1;all
-2/3
sage: B2=B2.substitute(al=all);B2
[-2/3  1/3]
[ 1/3    2]
sage: sp3=(A3.transpose()*B1).trace()
sage: all1=-sp3/sp1;all1
1/3
sage: sp4=(A3.transpose()*B2).trace()
sage: sp5=(B2.transpose()*B2).trace()
sage: all2=-sp4/sp5;all2
-4/7
sage: B3=B3.substitute(al1=all1,al2=all2,al=all);B3
[-2/7  1/7]
[ 1/7 -1/7]
sage: sp6=(B3.transpose()*B3).trace()
sage: C1=B1/sqrt(sp1);C1
[1/3*sqrt(3) 1/3*sqrt(3)]
[1/3*sqrt(3)           0]
sage: C2=B2/sqrt(sp5);C2
[-1/7*sqrt(14/3) 1/14*sqrt(14/3)]
[1/14*sqrt(14/3)  3/7*sqrt(14/3)]
sage: C3=B3/sqrt(sp6);B3
[-2/7  1/7]
[ 1/7 -1/7]
```

## 4.2 Linear Operators in Euclidean Vector Spaces

In investigating the Euclidean vector spaces are very useful the linear transformations compatible with the scalar product, i.e. the *orthogonal transformations.*

**Definition 4.30** (see [7], p. 199). Let (E, <, >) be a finite dimensional real Euclidean space. The endomorphism $T \in \mathrm{End}(V)$ is called **orthogonal operator** or **orthogonal transformation** if $T$ transforms the orthonormal basis into some orthonormal basis, i.e. if $B = \{\overline{e}_1, \overline{e}_2, \ldots, \overline{e}_n\}$ is an orthonormal basis of E then $B' = \{T(\overline{e}_1), T(\overline{e}_2), \ldots, T(\overline{e}_n)\}$ is an orthonormal basis of E, too.

**Theorem 4.31** (see [7], p. 199). For an operator $T \in \mathrm{End}(V)$ the following statements are equivalent:

1. $T$ is orthogonal,
2. $T$ is bijective and $T^{-1}$ is orthogonal,
3. $T$ preserves the scalar product, i.e., $< T(\overline{x}), T(\overline{y}) >=< \overline{x}, \overline{y} >, (\forall)\, \overline{x}, \overline{y} \in \mathrm{E}$,
4. $T$ stores the length of vectors, i.e., $\|T(\overline{x})\| = \|\overline{x}\|, (\forall)\, \overline{x} \in \mathrm{E}$,
5. the operator matrix $T$ relative to an orthonormal basis of E is orthogonal.

**Corollary 4.32** (see [7], p. 200). If $T \in \mathrm{End}(V)$ is orthogonal then $T$ preserves the vector angles, i.e.

$$\cos \prec (\overline{x}, \overline{y}) = \frac{< \overline{x}, \overline{y} >}{\|\overline{x}\| \, \|\overline{y}\|} = \frac{< T(\overline{x}), T(\overline{y}) >}{\|T(\overline{x})\| \, \|T(\overline{y})\|} = \cos \prec (T(\overline{x}), T(\overline{y})). \quad (4.12)$$

**Proposition 4.33** (see [7], p. 200). Let $T, S \in \mathrm{End}(V)$ be two orthogonal operators and be $\alpha \in \mathbb{R}$. Then:

1. $T \circ S$ is an orthogonal operator,
2. $\alpha T$ is orthogonal$\Leftrightarrow \alpha = \pm 1$.

We denote by

$$\mathrm{O}(\mathrm{E}) = \{T \in \mathrm{End}(V) \mid T \text{ orthogonal}\}.$$

**Proposition 4.34** (see [7], p. 200). If $T \in \mathrm{O}(\mathrm{E})$ and $A$ is the associated matrix of $T$ relative to an orthonormal basis B of E then $\det A = \pm 1$.

**Definition 4.35** (see [7], p. 200). It's called an **orthogonal operator of the first kind** or the **rotation operator**, an orthogonal operator for which the determinant of the associated matrix in an orthonormal basis of E is equal to $-1$.

**Definition 4.36** (see [7], p. 200). It's called an **orthogonal operator of the second kind**, an orthogonal operator for which the determinant of the associated matrix in an orthonormal basis of E is equal to $-1$.

We denote by:

- $\mathrm{O}^{+}(\mathrm{E})$ = the set of the orthogonal operators of the first kind,
- $\mathrm{O}^{-}(\mathrm{E})$ = the set of the orthogonal operators of the second kind.

**Proposition 4.37** (see [7], p. 200). The roots of the characteristic equation of an orthogonal operator have their absolute values equal to 1. In particular, the eigenvalues of an orthogonal operator are equal to $\pm 1$.

**Proposition 4.38** (see [7], p. 202). For an orthogonal operator, the eigenvectors that correspond to different eigenvalues are orthogonal.

**Theorem 4.39** (see [3], p. 95). The orthogonal matrices of $\mathrm{M}_2(\mathbb{R})$ are of the form:

$$\begin{pmatrix} \cos\varphi & \sin\varphi \\ \sin\varphi & -\cos\varphi \end{pmatrix}, \begin{pmatrix} -\cos\varphi & \sin\varphi \\ -\sin\varphi & -\cos\varphi \end{pmatrix}, \quad (4.13)$$
$$\begin{pmatrix} -\cos\varphi & \sin\varphi \\ \sin\varphi & \cos\varphi \end{pmatrix}, \begin{pmatrix} -\cos\varphi & -\sin\varphi \\ -\sin\varphi & \cos\varphi \end{pmatrix}, \begin{pmatrix} \cos\varphi & -\sin\varphi \\ \sin\varphi & \cos\varphi \end{pmatrix},$$
$$\begin{pmatrix} \cos\varphi & \sin\varphi \\ -\sin\varphi & \cos\varphi \end{pmatrix}, \begin{pmatrix} -\cos\varphi & -\sin\varphi \\ \sin\varphi & -\cos\varphi \end{pmatrix},$$

$(\forall)\ \varphi \in [0, 2\pi]$.

**Definition 4.40** (see [7], p. 203 and [3], p. 95). An orthogonal matrix with $\det A = 1$ is called a **rotation matrix** in $\mathbb{R}^n$.

### 4.2.1 Orthogonal Transformations in the Euclidean Plane

**Theorem 4.41** (see [8]). The orthogonal transformations in the Euclidean plane are: the rotations, the reflections or the compositions of rotations with reflections.

**Proposition 4.42** (see [8]). The rotation of the plane vectors around the origin, in the counterclockwise, with the angle $\varphi$, $r_\varphi : \mathbb{R}^2 \to \mathbb{R}^2$,

$$r_\varphi\left(\overline{x}\right) = r_\varphi\left(x^{(1)}, x^{(2)}\right) = \left(x^{(1)} \cos\varphi - x^{(2)} \sin\varphi,\ x^{(1)} \sin\varphi + x^{(2)} \cos\varphi\right) \tag{4.14}$$

is an orthogonal transformation.

**Remark 4.43** (see [8]). If O is the center of rotation then each point $M$ has associated the point $M'$, such that (see Fig. 4.1):

$$\begin{cases} \left\|\overline{OM}\right\| = \left\|\overline{OM'}\right\| = a \\ \prec MOM' = \text{the rotation angle in the counterclockwise.} \end{cases}$$

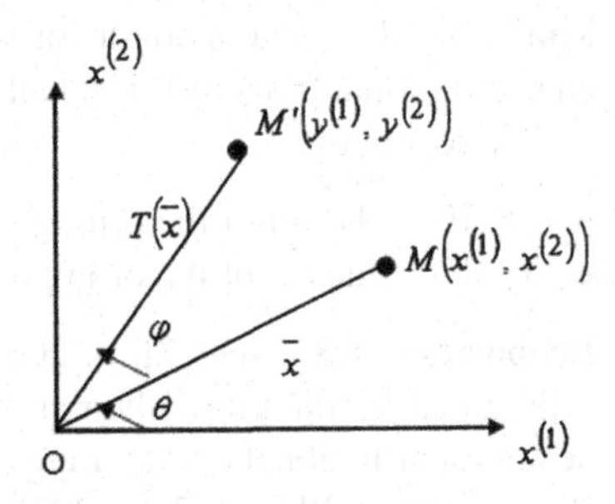


**Fig. 4.1** Rotation through angle $\varphi$

We have:

$$T\left(\overline{x}\right) = r_\varphi\left(\overline{x}\right) = A\overline{x},$$

where

$$A = \begin{pmatrix} \cos\varphi & -\sin\varphi \\ \sin\varphi & \cos\varphi \end{pmatrix}$$

determines a rotation through the angle $\varphi$ in the plane.

**Proposition 4.44** (see [8]). The rotation through the angle $\pi$ around the origin,

$$T(\overline{x}) = s_O(\overline{x}) = \left(-x^{(1)}, -x^{(2)}\right) \tag{4.15}$$

coincides with the reflection with respect to the origin (see Fig. 4.2).

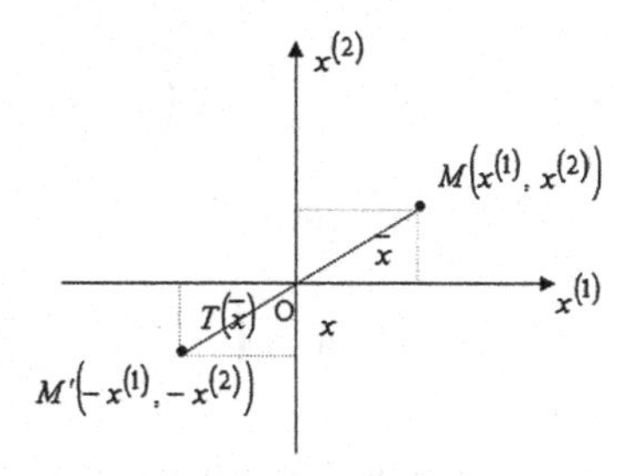


**Fig. 4.2** Rotation through angle $\pi$ around the origin

**Proposition 4.45** (see [8]). The reflection across the O$x$ axis (see Fig. 4.3),

$$T(\overline{x}) = s_d(\overline{x}) = \left(x^{(1)}, -x^{(2)}\right) \tag{4.16}$$

is an orthogonal transformation.

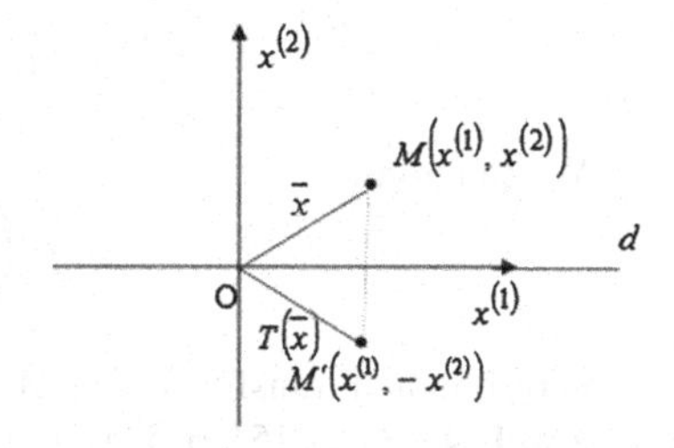


**Fig. 4.3** Reflection across the axis O$x$

**Proposition 4.46** (see [8]). The reflection across the O$y$ axis (see Fig. 4.4),

$$T(\overline{x}) = s_d'(\overline{x}) = \left(-x^{(1)}, x^{(2)}\right) \tag{4.17}$$

is an orthogonal transformation.

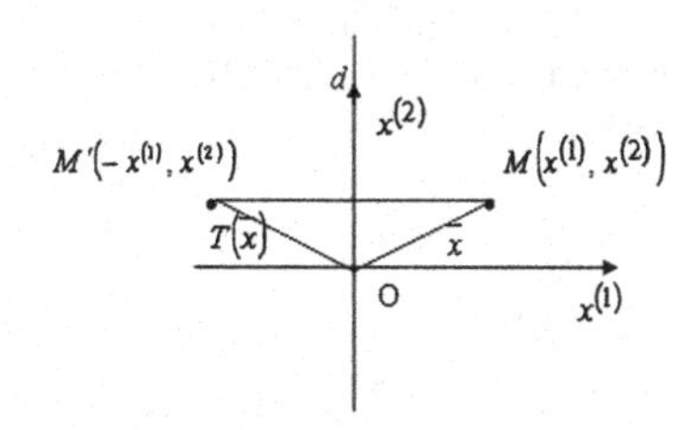


**Fig. 4.4** Reflection across the axis O$y$

**Proposition 4.47** (see [8]). The composition of the rotation $r_\varphi$ with the reflection $s_d$ is an orthogonal transformation.

**Proof**

We shall have

$$T(\overline{x}) = (r_\varphi \circ s_d)(\overline{x}) = r_\varphi(s_d(\overline{x})) \overset{(4.15)}{=} r_\varphi\left(x^{(1)}, -x^{(2)}\right)$$
$$\overset{(4.13)}{=} \left(x^{(1)}\cos\varphi + x^{(2)}\sin\varphi,\ x^{(1)}\sin\varphi - x^{(2)}\cos\varphi\right).$$

We can note that

$$A = \begin{pmatrix} \cos\varphi & \sin\varphi \\ \sin\varphi & -\cos\varphi \end{pmatrix} = \begin{pmatrix} \cos\varphi & -\sin\varphi \\ \sin\varphi & \cos\varphi \end{pmatrix}\begin{pmatrix} 1 & 0 \\ 0 & -1 \end{pmatrix}$$

is an orthogonal transformation.

**Proposition 4.48** (see [8]). The composition of the rotation $r_\varphi$ with the reflection $s_O$ is an orthogonal transformation.

**Proof**

We shall achieve

$$T(\overline{x}) = (r_\varphi \circ s_O)(\overline{x}) = r_\varphi(s_O(\overline{x})) \overset{(4.14)}{=} r_\varphi\left(-x^{(1)}, -x^{(2)}\right)$$
$$\overset{(4.13)}{=} \left(-x^{(1)}\cos\varphi + x^{(2)}\sin\varphi,\ -x^{(1)}\sin\varphi - x^{(2)}\cos\varphi\right).$$

We can note

$$A = \begin{pmatrix} -\cos\varphi & \sin\varphi \\ -\sin\varphi & -\cos\varphi \end{pmatrix}$$

is an orthogonal transformation, $A$ being a matrix from (4.12).

**Example 4.49** (see [5], p. 51)**.** The coordinate axes O$x$ and O$y$ one rotates with the angle $\varphi = \frac{\pi}{3}$ and one considers the new system is oriented opposite to the original system. Knowing that a point $A$ has the coordinates $\left(\sqrt{3}, -2\sqrt{3}\right)$ in the new system, find its coordinates in the old coordinate system.

**Solution**

*Case 1*. We have a rotation, followed by a reflection across the O$y'$axis (Fig. 4.5).

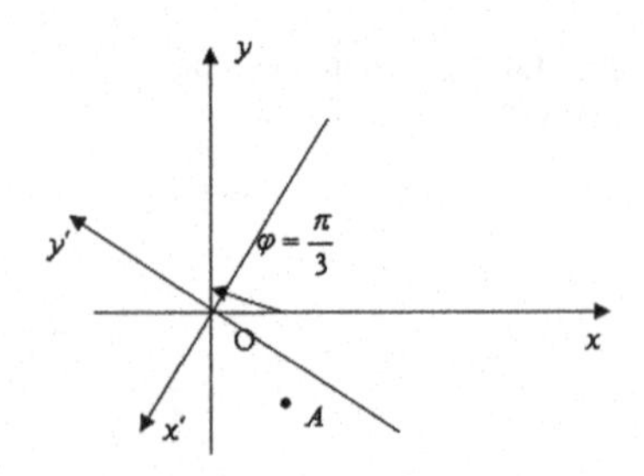


**Fig. 4.5** Rotation, followed by a reflection across the O$y'$ axis

We deduce

$$T(\overline{x}) = (x', y') = (r_\varphi \circ s'_d)(\overline{x}) = r_\varphi(s'_d(\overline{x})) \overset{(4.16)}{=} r_\varphi(-x, y)$$
$$\overset{(4.13)}{=} (-x\cos\varphi - y\sin\varphi, \quad -x\sin\varphi + y\cos\varphi).$$

The transformation $T$ has the equations:

$$\begin{cases} x' = -x\cos\varphi - y\sin\varphi \\ y' = -x\sin\varphi + y\cos\varphi. \end{cases}$$

As $x' = \sqrt{3}$, $y' = -2\sqrt{3}$, then solving the above system it results

$$\begin{cases} x = 3 - \frac{\sqrt{3}}{2} = 2.134 \\ y = -\frac{3}{2} - \sqrt{3} = -3.232. \end{cases}$$

Using Sage, we achieve:

```
sage: x,y=var('x,y');phi=pi/3
sage: f_symbolic(x,y)=[x*cos(phi)-y*sin(phi),x*sin(phi)+y*cos(phi)]
sage: rphi=linear_transformation(RR^2,RR^2,f_symbolic)
sage: T_symbolic(x,y)=[-x,y]
sage: spd=linear_transformation(RR^2,RR^2,T_symbolic)
sage: D=(rphi*spd).matrix(side='right');D.n(digits=3)
[-0.500 -0.866]
[-0.866  0.500]
sage: v=vector([x,y])
sage: D1=D*v.transpose();D1
[-0.500000000000000*x - 0.866025403784439*y]
[-0.866025403784439*x + 0.500000000000000*y]
sage: s=solve([D1[0][0]==sqrt(3),D1[1][0]==-2*sqrt(3)],x,y)
sage: s[0][0].right().n(digits=4)
2.134
sage: s[0][1].right().n(digits=4)
-3.232
sage: po2=point2d((3-sqrt(3)/2,-3/2-sqrt(3)),size=20,color='red')
sage: l1=arrow((2,3.4641),(-2,-3.4641),color="blue");l2=arrow((-4,0),(4,0),color="purple")
sage: l3=arrow((3.4641,-2),(-3.4641,2),color="blue");l4=arrow((0,-4),(0,4),color="purple")
sage: (l1+l2+l3+l4+po2).show(aspect ratio=1.5,axes=false)
```

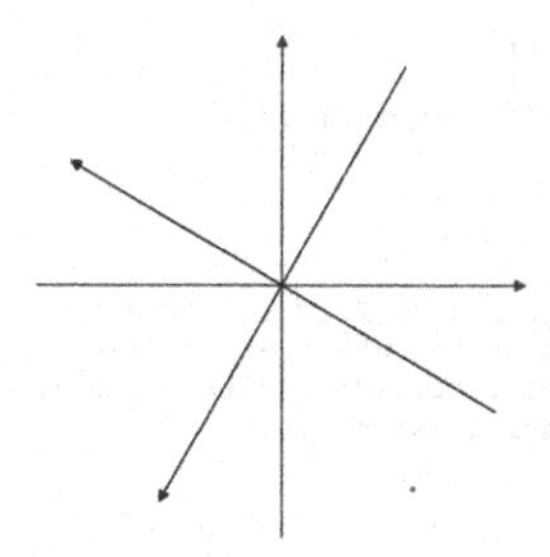

*Case* 2. We have a rotation, followed by a reflection across the $Ox'$ axis (Fig. 4.6). We obtain

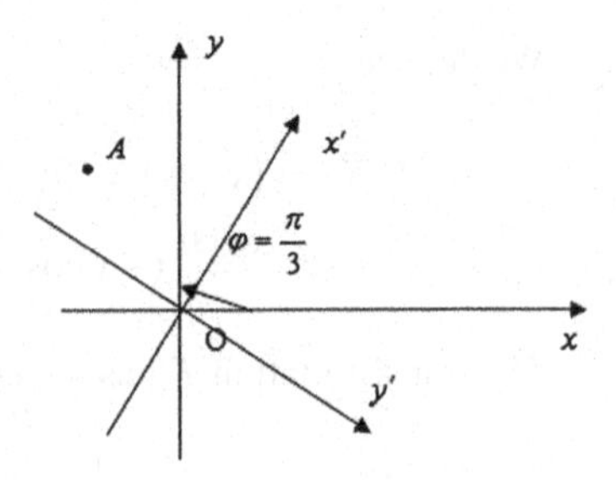


**Fig. 4.6** Rotation, followed by a reflection across the $Ox'$ axis

$$T(\overline{x}) = (x', y') = (r_\varphi \circ s_d)(\overline{x}) = r_\varphi(s_d(\overline{x})) \overset{(4.15)}{=} r_\varphi(x, -y)$$
$$\overset{(4.13)}{=} (x\cos\varphi + y\sin\varphi,\ x\sin\varphi - y\cos\varphi).$$

The transformation $T$ has the equations:

$$\begin{cases} x' = x\cos\varphi + y\sin\varphi \\ y' = x\sin\varphi - y\cos\varphi. \end{cases}$$

As $x' = \sqrt{3}$, $y' = -2\sqrt{3}$, then solving the above system it results

$$\begin{cases} x = \frac{\sqrt{3}}{2} - 3 = -2.134 \\ y = \sqrt{3} + \frac{3}{2} = 3.232. \end{cases}$$

We can obtain these coordinates with Sage:

```
sage: x,y=var('x,y');phi=pi/3
sage: f_symbolic(x,y)=[x*cos(phi)-y*sin(phi),x*sin(phi)+y*cos(phi)]
sage: rphi=linear_transformation(RR^2,RR^2,f_symbolic)
sage: T_symbolic(x,y)=[x,-y]
sage: spd=linear_transformation(RR^2,RR^2,T_symbolic)
sage: D=(rphi*spd).matrix(side='right');D.n(digits=3)
[ 0.500  0.866]
[ 0.866 -0.500]
sage: v=vector([x,y])
sage: D1=D*v.transpose();D1
[0.500000000000000*x + 0.866025403784439*y]
[0.866025403784439*x - 0.500000000000000*y]
sage: s=solve([D1[0][0]==sqrt(3),D1[1][0]==-2*sqrt(3)],x,y)
sage: s[0][0].right().n(digits=4)
-2.134
sage: s[0][1].right().n(digits=4)
3.232
sage: po2=point2d((-3+sqrt(3)/2,3/2+sqrt(3)),size=20,color='red')
sage: l1=arrow((-2,-3.4641),(2,3.4641),color="blue");l2=arrow((-4,0),(4,0),color="purple")
sage: l3=arrow((-3.4641,2),(3.4641,-2),color="blue");l4=arrow((0,-4),(0,4),color="purple")
sage: (l1+l2+l3+l4+po2).show(aspect_ratio=1.5,axes=false)
```

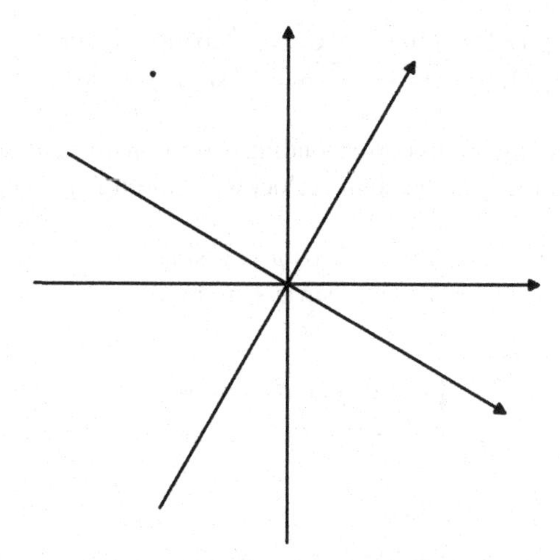

**Proposition 4.50** (see [4], p. 463). The rotation of a rectangular coordinate system around the origin, in the counterclockwise, through the angle $\varphi$, $r_{\varphi} : \mathbb{R}^2 \to \mathbb{R}^2$ is an orthogonal transformation.

**Proof**

By rotating the rectangular coordinate system $xOy$ around the origin, in the counterclockwise, through the angle $\varphi$ one gets the system $x'Oy'$. A point $M$ which has the coordinates $(x, y)$ in the old system will have the coordinates $\left(x', y'\right)$ in the new system.

We choose in the plane an orthonormal reference with the origin in the center of rotation (Fig. 4.7).

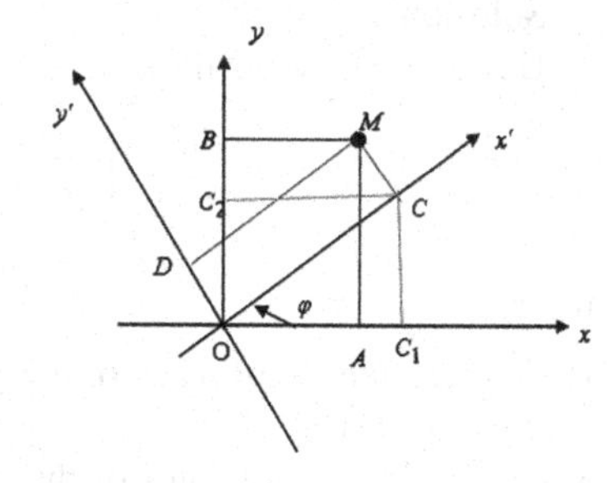


**Fig. 4.7** Rotation of a coordonate system through the angle $\varphi$

We note that:

$$\begin{cases} \overline{OC_1} = x' \cos\varphi \\ \overline{AC_1} = y' \sin\varphi \\ \overline{OC_2} = x' \sin\varphi \\ \overline{C_2B} = y' \cos\varphi \end{cases}$$

and

$$\begin{cases} \overline{OA} = \overline{OC_1} - \overline{AC_1} = x' \cos\varphi - y' \sin\varphi \\ \overline{OB} = \overline{OC_2} + \overline{C_2B} = x' \sin\varphi + y' \cos\varphi. \end{cases}$$

It turns out that the equations corresponding to the transformation of the coordinate system $xOy$ by rotating it in the counterclockwise, through the angle $\varphi$ will be:

$$\begin{cases} x = x' \cos\varphi - y' \sin\varphi \\ y = x' \sin\varphi + y' \cos\varphi \end{cases}$$

i.e.

$$\begin{cases} x' = x \cos\varphi + y \sin\varphi \\ y' = -x \sin\varphi + y \cos\varphi. \end{cases} \tag{4.18}$$

We obtain:

$$R\varphi\left(\overline{x}\right) = R\varphi\left(x, y\right) = \left(x' \cos\varphi - y' \sin\varphi,\ x' \sin\varphi + y' \cos\varphi\right),$$

$$T\left(\overline{x}\right) = R\varphi\left(\overline{x}\right) = A\overline{x},$$

where

$$A = \begin{pmatrix} \cos\varphi & -\sin\varphi \\ \sin\varphi & \cos\varphi \end{pmatrix};$$

$R\varphi$ is an orthogonal transformation as $A$ is a matrix from (4.12) and $A^t \cdot A = I_2$.

**Example 4.51**. One gives the point $M\,(1, 1)$ in the plane reported to the rectangular axes O$x$, O$y$. Determine that angle with which the axes should be rotated so that the point $M$ belongs to the O$x'$ axis. Find the new coordinates of $M$ in these conditions.

**Solution**

Using (4.17) we achieve:

$$\begin{cases} x' = x \cos\theta + y \sin\theta \\ y' = -x \sin\theta + y \cos\theta, \end{cases}$$

where:

- $x'$, $y'$ are the coordinates of the point $M$ in the plane reported to the rectangular axes O$x'$, O$y'$;
- $x$, $y$, are the coordinates of the point $M$ in the plane reported to the rectangular axes O$x$, O$y$;
- $\theta$ is the angle to be rotated the axes.

Multiplying the first equation with $\cos\varphi$ and the second with $-\sin\varphi$ and adding the obtained equations, we deduce:

$$x' \cos\theta - y' \sin\theta = x,$$

while multiplying the first equation with $\sin\theta$ and the second with $\cos\theta$ and adding the obtained equations, we deduce:

$$x' \sin\theta + y' \cos\theta = y.$$

By emphasizing the condition that the point $M$ belongs to the $\mathrm{O}x'$ axis (i.e. $y' = 0$) we have

$$\begin{cases} x' \cos\theta = x \\ x' \sin\theta = y \end{cases} \Leftrightarrow \begin{cases} x' \cos\theta = 1 \\ x' \sin\theta = 1 \end{cases} \Leftrightarrow x'^2 = 2 \Longleftrightarrow x' = \pm\sqrt{2}.$$

In the case when $x' = \sqrt{2}$ it results

$$\begin{cases} \cos\theta = \frac{1}{\sqrt{2}} = \frac{\sqrt{2}}{2} \\ \sin\theta = \frac{1}{\sqrt{2}} = \frac{\sqrt{2}}{2} \end{cases}$$

i.e. $\theta = \pi/4$.

In the case when $x' = -\sqrt{2}$ it results

$$\begin{cases} \cos\theta = -\frac{1}{\sqrt{2}} = -\frac{\sqrt{2}}{2} \\ \sin\theta = \frac{1}{\sqrt{2}} = \frac{\sqrt{2}}{2} \end{cases}$$

i.e.

$$\theta = \pi + \frac{\pi}{4} = \frac{5\pi}{4}.$$

So, the new coordinates of $M$ if:

- the axes one rotate with the angle $\theta = \frac{\pi}{4}$ are $M\left(\sqrt{2}, 0\right)$,
- axes one rotate with the angle $\theta = \frac{5\pi}{4}$ are $M\left(-\sqrt{2}, 0\right)$.

We shall given a solution in Sage, too:

```
sage: xp,th=var('xp,th');solve([xp^2==2],xp)
[xp == -sqrt(2), xp == sqrt(2)]
sage: solve([cos(th)==1/sqrt(2),sin(th)==1/sqrt(2)],th)
[[th == 1/4*pi + 2*pi*z164]]
sage: l1=arrow((-1,-1),(1.2,1.2),color="purple")
sage: l2=arrow((-sqrt(2),0),(1.2*sqrt(2),0),color="blue")
sage: l3=arrow((1,-1),(-1,1),color="purple");
sage: l4=arrow((0,-sqrt(2)),(0,sqrt(2)),color="blue")
sage: po2=point2d((1,1),size=45,color='red');
sage: (l1+l2+l3+l4+po2).show(aspect_ratio=1.4,axes=False)
```

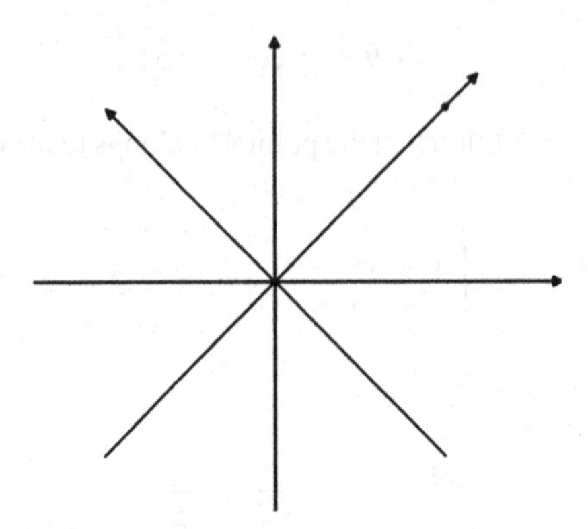

## 4.3 Problems

1. Show that the function $<,>: \mathbb{R}^2 \times \mathbb{R}^2 \to \mathbb{R}^2$, defined by

$$< \overline{x}, \overline{y} >= 5x_1y_1 - 2x_1y_2 - 2x_2y_1 + 3x_2y_2, \ (\forall)\ \overline{x} = (x_1, x_2), \\ \overline{y} = (y_1, y_2) \in \mathbb{R}^2$$

is a scalar product.

2. In the real Euclidean vector space (E, <, >) having a basis $B = \{\overline{e}_1, \overline{e}_2\}$ such that: $\|\overline{e}_1\| = 1$, $\|\overline{e}_2\| = 4$, $\prec (\overline{e}_1, \overline{e}_2) = \pi/4$ one considers the vector $\overline{x} = \overline{e}_1 + 5\overline{e}_2$. Compute $\|\overline{x}\|$.
3. In the real Euclidean vector space $\mathbb{R}^3$ one assumes the vectors: $\overline{a}_1 = (1, 0, 3)$, $\overline{a}_2 = (1, 1, 0)$, $\overline{a}_3 = (1, 1, 1)$.

   (a) Determine if $B_1 = \{\overline{a}_1, \overline{a}_2, \overline{a}_3\}$ constitutes a basis of $\mathbb{R}^3$.
   (b) Map the system of vectors $B_1$ into one orthonormal.

**Solution**

Using Sage we shall have:

```
sage: A=matrix(QQ,[[1,0,3],[1,1,0],[1,1,1]])
sage: V=VectorSpace(RR,3);a1=V([1,0,3]);a2=V([1,1,0]);a3=V([1,1,1])
sage: V.linear dependence([a1,a2,a3])==[]
True
sage: A.gram schmidt()
(
[ 1  0  3] [ 1  0  0]
[9/10  1 -3/10] [1/10  1  0]
[-3/19 3/19 1/19], [ 2/5 16/19  1]
),
```

4. Let $\mathbb{R}^3$ be the arithmetic vector space and the scalar products $<,>_1, <,>_2$: $\mathbb{R}^3 \times \mathbb{R}^3 \to \mathbb{R}^3$,defined by:

$$<\overline{x}, \overline{y}>_1 = x_1y_1 + 2x_1y_2 + 2x_2y_1 + 5x_2y_2 + x_3y_3,$$
$$<\overline{x}, \overline{y}>_2 = 2x_1y_1 + 3x_1y_2 + 3x_2y_1 + 7x_2y_2 + 2x_3y_3,$$

$(\forall)\ \overline{x} = (x_1, x_2, x_3)\,,\ \overline{y} = (y_1, y_2, y_3) \in \mathbb{R}^3$.
Compute the length of the vector $\overline{x} = (1, 2, 3)$ in the Euclidean vector spaces $(\mathbb{R}^3, <,>_1)$ and $(\mathbb{R}^3, <,>_2)$.

**Solution**
Using Sage, we shall have:

```
sage: x1,x2,x3,y1,y2,y3=var('x1,x2,x3,y1,y2,y3')
sage: x=vector(SR,[x1,x2,x3]);y=vector(SR,[y1,y2,y3])
sage: T1=x1*y1+2*x1*y2+2*x2*y1+5*x2*y2+x3*y3
sage: u1=T1.substitute(x1=1,x2=2,x3=3,y1=1,y2=2,y3=3)
sage: n1=sqrt(u1).n(digits=4);n1
6.164
sage: T2=2*x1*y1+3*x1*y2+3*x2*y1+7*x2*y2+2*x3*y3
sage: u2=T2.substitute(x1=1,x2=2,x3=3,y1=1,y2=2,y3=3)
sage: n2=sqrt(u2).n(digits=4);n2
7.746
```

5. In the real vector space $M_2(\mathbb{R})$ one assumes the matrices:

$$A_1 = \begin{pmatrix} 1 & 2 \\ 1 & 1 \end{pmatrix},\ A_2 = \begin{pmatrix} 2 & 3 \\ 1 & 0 \end{pmatrix},\ A_3 = \begin{pmatrix} 3 & 1 \\ 1 & -2 \end{pmatrix},\ A_4 = \begin{pmatrix} 4 & 2 \\ -1 & -6 \end{pmatrix}$$

(a) Check if $B_1 = \{A_1, A_2, A_3, A_4\}$ determine a basis of $M_2(\mathbb{R})$.
(b) Transform the basis $B_1$ into an orthonormal one.

6. Let be the vectors $\overline{v}_1 = \left(\frac{1}{2}, \frac{1}{2}, \frac{1}{2}, \frac{1}{2}\right)$ and $\overline{v}_2 = \left(\frac{1}{6}, \frac{1}{6}, \frac{1}{2}, -\frac{5}{6}\right)$ in the Euclidean space $\mathbb{R}^4$. Check that these vectors have their norm equal to 1 and they are orthogonal. Then, bulid an ortonormate basis of this space, that contains the vectors $\overline{v}_1$ and $\overline{v}_2$.

**Solution**
With Sage, it will result:

```
sage: V=VectorSpace(RR,4)
sage: v1=V([1/2,1/2,1/2,1/2]);v2=V([1/6,1/6,1/2,-5/6])
sage: v1.norm().n(digits=3);v2.norm().n(digits=3)
1.00
1.00
sage: (v1.dot_product(v2)).n(digits=3)
-5.55e-17
sage: A=matrix(QQ,[[1/2,1/2,1/2,1/2],[1/6,1/6,1/2,-5/6]])
sage: A.gram_schmidt()
(
[ 1/2  1/2  1/2  1/2] [1 0]
[ 1/6  1/6  1/2 -5/6], [0 1]
)
```

7. Prove that the transformation $T : \mathbb{R}^3 \to \mathbb{R}^3$, where

$$T\left(\overline{x}\right) = \left(\frac{2}{3}x_1 + \frac{2}{3}x_2 - \frac{1}{3}x_3,\ \frac{2}{3}x_1 - \frac{1}{3}x_2 + \frac{2}{3}x_3,\ -\frac{1}{3}x_1 + \frac{2}{3}x_2 + \frac{2}{3}x_3\right)$$

$(\forall)\ \overline{x} = (x_1, x_2, x_3) \in \mathbb{R}^3$ is orthogonal in $\mathbb{R}^3$ with the usually scalar product.

**Solution**

We shall use Sage:

```
sage: M=matrix([[2/3,2/3,-1/3],[2/3,-1/3,2/3],[-1/3,2/3,2/3]])
sage: M*M.transpose()
[1 0 0]
[0 1 0]
[0 0 1]
```

8. One gives the point $M\,(1, 1)$ in the plane reported to the rectangular axes $Ox$, $Oy$. Determine the angle that the axes should be rotated so that the point $M$ belongs to the $Oy'$ axis. Find the new coordinates of $M$ in these conditions.
9. Let be a triangle, having the vertices $A\,(3, 1)$, $B\,(7, 1)$, $C\,(7, 4)$. Find its image through the rotation with the center $O$ and the angle $\frac{\pi}{3}$.

**Solution**

Solving this problem in Sage, we achieve:

```
sage: phi=pi/3;var("x,y")
(x, y)
sage: xp=x*cos(phi)+y*sin(phi);yp=-x*sin(phi)+y*cos(phi)
sage: Ap=(xp.subs(x=3,y=1),yp.subs(x=3,y=1));(Ap[0].n(digits=3),Ap[1].n(digits=3))
(2.37, -2.10)
sage: Bp=(xp.subs(x=7,y=1),yp.subs(x=7,y=1));(Bp[0].n(digits=3),Bp[1].n(digits=3))
(4.37, -5.56)
sage: Cp=(xp.subs(x=7,y=4),yp.subs(x=7,y=4));(Cp[0].n(digits=3),Cp[1].n(digits=3))
(6.96, -4.06)
```

```
sage: l1=line([(3,1), (7,1)], color='blue');l2=line([(7,1), (7,4)], color='blue');
sage: l3=line([(7,4), (3,1)], color='blue');
sage: t1=text("A",(2.98,0.9));t2=text("B",(6.98,0.9));t3=text("C",(7,4.09))
sage: Ap=(xp.subs(x=3,y=1),yp.subs(x=3,y=1));(Ap[0].n(digits=3),Ap[1].n(digits=3))
(2.37, -2.10)
sage: Bp=(xp.subs(x=7,y=1),yp.subs(x=7,y=1));(Bp[0].n(digits=3),Bp[1].n(digits=3))
(4.37, -5.56)
sage: Cp=(xp.subs(x=7,y=4),yp.subs(x=7,y=4));(Cp[0].n(digits=3),Cp[1].n(digits=3))
(6.96, -4.06)
```

```
sage: tt1=text("Ap",(Ap[0],Ap[1]+0.1));tt2=text("Bp",(Bp[0]-0.02,Bp[1]-0.1));
sage: tt3=text("Cp",(Cp[0],Cp[1]+0.09))
sage: ll1=line([(Ap[0],Ap[1]), (Bp[0],Bp[1])], color='orange')
sage: ll2=line([(Bp[0],Bp[1]), (Cp[0],Cp[1])], color='orange')
sage: ll3=line([(Cp[0],Cp[1]), (Ap[0],Ap[1])], color='orange')
sage: (l1+l2+l3+t1+t2+t3+ll1+ll2+ll3+tt1+tt2+tt3).show(aspect_ratio=1)
```

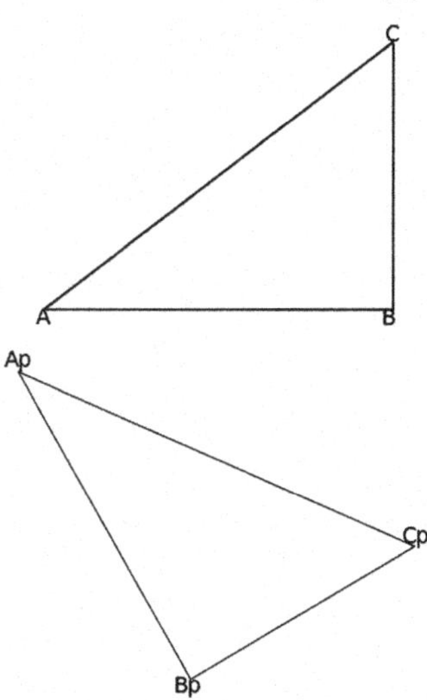


10. One considers the rotation through the angle $\varphi$ in the counterclockwise, $r_\varphi : \mathbb{R}^2 \to \mathbb{R}^2$.

    (a) Justify the linearity of the transformation $r_\varphi$.
    (b) Build the associated matrix in the canonical basis from $\mathbb{R}^2$ and in the basis $B_1 = \{\overline{e}_1 + \overline{e}_2, \overline{e}_1 - \overline{e}_2\}$.

(c) What is the relationship between the two matrices?
(d) Compute the kernel and the image of this linear mapping.
(e) Is $r_{\varphi}$ injective? But is it surjective?

## References


1. V. Balan, *Algebră liniară, geometrie analitică, ed* (Fair Partners, Bucureşti, 1999)
2. I. Iatan, *Advances Lectures on Linear Algebra with Applications* (Lambert Academic Publishing, 2011)
3. P. Matei, Algebră liniară. Gometrie analitică şi diferenţială, ed. (Agir, Bucureşti, 2002)
4. V. Postelnicu, S. Coatu, *Mică enciclopedie matematică, ed* (Tehnică, Bucureşti, 1980)
5. C. Udrişte, *Aplicaţii de algebră, geometrie şi ecuaţii diferenţiale, ed* (Didactică şi Pedagogică R.A, Bucureşti, 1993)
6. I. Vladimirescu, M. Popescu, *Algebră liniară şi geometrie analitică, ed.* (Universitaria, Craiova, 1993)
7. I. Vladimirescu, M. Popescu, M. Sterpu, *Algebră liniară şi geometrie n- dimensională, ed.* (Radical, Craiova, 1996)
8. I. Vladimirescu, M. Popescu, M. Sterpu, *Algebră liniară şi geometrie analitică, Note de curs şi aplicaţii*, (Universitatea din Craiova, 1993)

# Chapter 5
# Bilinear and Quadratic Forms

## 5.1 Bilinear and Quadratic Forms

The theory of bilinear form and quadratic form is used [5] in the analytic geometry for getting the classification of the conics and of the quadrics.

It is also used in physics, in particular to describe physical systems subject to small vibrations. The coefficients of a bilinear form one behave to certain transformations like the tensors coordinates. Tensors are useful in theory of elasticity (the deformation of an elastic medium is described through the deformation tensor).

**Definition 5.1** (see [1], p. 150). A mapping $b : V \times V \to K$ is called a **bilinear form** on $V$ if it satisfies the conditions:

1. $b(\alpha\overline{x} + \beta\overline{y}, \overline{z}) = \alpha b(\overline{x}, \overline{z}) + \beta b(\overline{y}, \overline{z}),\ (\forall)\,\alpha, \beta \in K,\ (\forall)\,\overline{x}, \overline{y}, \overline{z} \in V,$
2. $b(\overline{x}, \alpha\overline{y} + \beta\overline{z}) = \alpha b(\overline{x}, \overline{y}) + \beta b(\overline{x}, \overline{z}),\ (\forall)\,\alpha, \beta \in K,\ (\forall)\,\overline{x}, \overline{y}, \overline{z} \in V.$

**Definition 5.2** (see [1], p. 150). We say that the bilinear form $b : V \times V \to K$ is **symmetric (antisymmetric)** if $b(\overline{x}, \overline{y}) = b(\overline{y}, \overline{x})$ (respectively, $b(\overline{x}, \overline{y}) = -b(\overline{y}, \overline{x})$).

**Consequences 5.3** (see [2], p. 116). If the mapping $b : V \times V \to K$ is a bilinear form then:

(1) $b(\overline{0}, \overline{x}) = b(\overline{x}, \overline{0}) = 0,\ (\forall)\,\overline{x} \in V$

(2) (a) $b\left(\sum_{i=1}^{n} \alpha^{(i)}\overline{x}_i, \overline{y}\right) = \sum_{i=1}^{n} \alpha^{(i)} b(\overline{x}_i, \overline{y}),$

$(\forall)\,\alpha^{(1)}, \ldots, \alpha^{(n)} \in K,\ (\forall)\,\overline{x}_1, \ldots, \overline{x}_n, \overline{y} \in V$

(b) $b\left(\overline{x}, \sum_{i=1}^{n} \beta^{(i)}\overline{y}_i\right) = \sum_{i=1}^{n} b(\overline{x}, \overline{y}_i),$

$(\forall)\,\beta^{(1)}, \ldots, \beta^{(n)} \in K,\ (\forall)\,\overline{x}, \overline{y}_1, \ldots, \overline{y}_n \in V.$

G. A. Anastassiou and I. F. Iatan, *Intelligent Routines II*,
Intelligent Systems Reference Library 58, DOI: 10.1007/978-3-319-01967-3_5,

**Definition 5.4** (see [1], p. 150). If $b : V \times V \to K$ is a symmetric bilinear form, the mapping $f : V \to K$, defined by $f(\overline{x}) = b(\overline{x}, \overline{x})$, for all $\overline{x} \in V$ is called the **quadratic form** associated to $b$.

**Remark 5.5** (see [1], p. 150). Knowing the quadratic form $f$, allows to get the symmetric bilinear form, associated to $f$ as:

$$b(\overline{x}, \overline{y}) = \frac{1}{2}[f(\overline{x}+\overline{y}) - f(\overline{x}) - f(\overline{y})], (\forall)\ \overline{x}, \overline{y} \in V \tag{5.1}$$

**Definition 5.6** (see [1], p. 150). The symmetric bilinear form $b$ associated to the quadratic form $f$ is called the **polar form** of the quadratic form $f$.

**Example 5.7** (see [3], p. 93). The quadratic form corresponding to the real scalar product (which is a symmetric bilinear form) is the square of the Euclidean norm:

$$f(\overline{x}) = <\overline{x}, \overline{x}> = \|\overline{x}\|^2, (\forall)\ \overline{x} \in V.$$

Let $V$ be an $n$ finite dimensional vector space over $K, n \geq 1$ and $B = \{\overline{a}_1, \ldots, \overline{a}_n\}$ one of its basis. If $b : V \times V \to K$ is a bilinear form then $(\forall)\ \overline{x}, \overline{y} \in V$ it results:

$$\overline{x} = \sum_{i=1}^{n} x^{(i)} \overline{a}_i,\ \overline{y} = \sum_{j=1}^{n} y^{(j)} \overline{a}_j;$$

therefore

$$b(\overline{x}, \overline{y}) = \sum_{i=1}^{n}\sum_{j=1}^{n} a_{ij} x^{(i)} y^{(j)}, \tag{5.2}$$

where

$$a_{ij} = b(\overline{a}_i, \overline{a}_j),\ (\forall)\ i, j = \overline{1, n}.$$

The expression (5.2) constitutes [1] the *analytic expression of the bilinear form b relative to the basis B*, and $A \in M_n(K)$, $A = (a_{ij})_{1 \leq i,j \leq n}$ represents the *associated matrix of the bilinear form b* relative to the basis $B$.

From (5.2) one obtains [1] *the analytic expression of the bilinear form* $f : V \to K$ relative to the basis $B$ of $V$:

$$f(\overline{x}) = \sum_{i=1}^{n}\sum_{j=1}^{n} a_{ij} x^{(i)} y^{(j)},\ (\forall)\ \overline{x} = \sum_{i=1}^{n} x^{(i)} \overline{a}_i \in V. \tag{5.3}$$

**Definition 5.8** (see [1], p. 151) We call the **associated matrix of a quadratic form** $f : V \to K$ relative to a basis of $V$, the matrix of the bilinear mapping $b : V \times V \to K$ from which derives $f$ relative to the considered basis.

**Example 5.9** (see [4]) Let be $b : \mathbb{R}^4 \times \mathbb{R}^4 \to \mathbb{R}$,

$$b(\overline{x}, \overline{y}) = x_1 y_1 + 2x_2 y_1 + 2x_2 y_2 - 4x_2 y_3 + x_3 y_3 + x_4 y_1 - x_4 y_4.$$

(a) Prove that $b$ is a bilinear functional.
(b) Find the associated matrix of $b$ relative to the basis

$$B' = \{\overline{f}_1 = (1,1,0,0),\ \overline{f}_2 = (0,1,0,0),\ \overline{f}_3 = (0,1,0,1),\ \overline{f}_4 = (1,0,0,1)\}$$

and relative to the canonical basis and highlight the link between them.
(c) Determine the expression of the quadratic form $f$ associated to $b$.

**Solution**

(a) According to the Definition 5.1, $b$ is a bilinear functional if those two conditions are accomplished. We shall check the first condition as for the others one proceeds similarly. Let be $\alpha, \beta \in K$ and $\overline{x}, \overline{y}, \overline{z} \in V$ ; we have:

$$\begin{aligned} b(\alpha\overline{x} + \beta\overline{y}, \overline{z}) &= (\alpha x_1 + \beta y_1) z_1 + 2(\alpha x_2 + \beta y_2) z_1 + 2(\alpha x_2 + \beta y_2) z_2 \\ &\quad - 4(\alpha x_2 + \beta y_2) z_3 + (\alpha x_3 + \beta y_3) z_3 \\ &\quad + (\alpha x_4 + \beta y_4) z_1 - (\alpha x_4 + \beta y_4) z_4 \\ &= \alpha x_1 z_1 + \beta y_1 z_1 + 2\alpha x_2 z_1 + 2\beta y_2 z_1 + 2\alpha x_2 z_2 + 2\beta y_2 z_2 \\ &\quad - 4\alpha x_2 z_3 - 4\beta y_2 z_3 + \alpha x_3 z_3 + \beta y_3 z_3 + \alpha x_4 z_1 + \beta y_4 z_1 \\ &\quad - \alpha x_4 z_4 - \beta y_4 z_4 \\ &= \alpha (x_1 z_1 + 2x_2 z_1 + 2x_2 z_2 - 4x_2 z_3 + x_3 z_3 + x_4 z_1 - x_4 z_4) \\ &\quad + \beta (y_1 z_1 + 2y_2 z_1 + 2y_2 z_2 - 4y_2 z_3 + y_3 z_3 + y_4 z_1 - y_4 z_4) \\ &= \alpha b(\overline{x}, \overline{z}) + \beta b(\overline{y}, \overline{z}). \end{aligned}$$

(b) To determine the associated matrix of $b$ relative to the basis $B'$ we have to compute:

$$\begin{cases} b(\overline{f}_1, \overline{f}_1) = 5 \\ b(\overline{f}_1, \overline{f}_2) = 2 \\ b(\overline{f}_1, \overline{f}_3) = 2 \\ \quad\vdots \end{cases}$$

We achieve:

$$A' = \begin{pmatrix} 5 & 2 & 2 & 3 \\ 4 & 2 & 2 & 2 \\ 5 & 2 & 1 & 2 \\ 2 & 0 & -1 & 2 \end{pmatrix}.$$

The associated matrix of $b$ relative to the canonical basis is

$$A = \begin{pmatrix} 1 & 0 & 0 & 0 \\ 2 & 2 & -4 & 0 \\ 0 & 0 & 1 & 0 \\ 1 & 0 & 0 & -1 \end{pmatrix}.$$

We can note that:

$$\mathrm{M}_{(B,B')} = \begin{pmatrix} 1\,0\,0\,1 \\ 1\,1\,1\,0 \\ 0\,0\,0\,0 \\ 0\,0\,1\,1 \end{pmatrix}.$$

We have:

$$b(\overline{x}, \overline{y}) = \overline{x}_B^t A \overline{y}_B$$

and

$$A' = \mathrm{M}_{(B,B')}^t A \mathrm{M}_{(B,B')}.$$

(c) The expression of the quadratic form $f$ associated to $b$ is

$$f(\overline{x}) = x_1^2 + 2x_2x_1 + 2x_2^2 - 4x_2x_3 + x_3^2 + x_4x_1 - x_4^2.$$

The Solution in Sage will be presented, too:

```
sage: var("x1 x2 x3 x4 y1 y2 y3 y4 z1 z2 z3 z4 al be")
(x1, x2, x3, x4, y1, y2, y3, y4, z1, z2, z3, z4, al, be)
sage: b(x1,x2,x3,x4,y1,y2,y3,y4)=x1*y1+2*x2*y1+2*x2*y2-4*x2*y3+x3*y3+x4*y1-x4*y4
sage: b1=b(al*x1+be*y1,al*x2+be*y2,al*x3+be*y3,al*x4+be*y4,z1,z2,z3,z4)
sage: b2=al*b(x1,x2,x3,x4,z1,z2,z3,z4);b3=be*b(y1,y2,y3,y4,z1,z2,z3,z4);
sage: expand(b1-(b2+b3))
0
sage: a00=b(1,1,0,0,1,1,0,0);a01=b(1,1,0,0,0,1,0,0);a02=b(1,1,0,0,0,1,0,1);a03=b(1,1,0,0,1,0,0,1)
sage: a10=b(0,1,0,0,1,1,0,0);a11=b(0,1,0,0,0,1,0,0);a12=b(0,1,0,0,0,1,0,1);a13=b(0,1,0,0,1,0,0,1)
sage: a20=b(0,1,0,1,1,1,0,0);a21=b(0,1,0,1,0,1,0,0);a22=b(0,1,0,1,0,1,0,1);a23=b(0,1,0,1,1,0,0,1)
sage: a30=b(1,0,0,1,1,1,0,0);a31=b(1,0,0,1,0,1,0,0);a32=b(1,0,0,1,0,1,0,1);a33=b(1,0,0,1,1,0,0,1)
sage: A=matrix([[a00,a01,a02,a03],[a10,a11,a12,a13],[a20,a21,a22,a23],[a30,a31,a32,a33]])
sage: A
[5 2 2 3]
[4 2 2 2]
[5 2 1 2]
[2 0 -1 1]
sage: c00=b(1,0,0,0,1,0,0,0);c01=b(1,0,0,0,0,1,0,0);c02=b(1,0,0,0,0,0,1,0);c03=b(1,0,0,0,0,0,0,1)
sage: c10=b(0,1,0,0,1,0,0,0);c11=b(0,1,0,0,0,1,0,0);c12=b(0,1,0,0,0,0,1,0);c13=b(0,1,0,0,0,0,0,1)
sage: c20=b(0,0,1,0,1,0,0,0);c21=b(0,0,1,0,0,1,0,0);c22=b(0,0,1,0,0,0,1,0);c23=b(0,0,1,0,0,0,0,1)
sage: c30=b(0,0,0,1,1,0,0,0);c31=b(0,0,0,1,0,1,0,0);c32=b(0,0,0,1,0,0,1,0);c33=b(0,0,0,1,0,0,0,1)
sage: C=matrix([[c00,c01,c02,c03],[c10,c11,c12,c13],[c20,c21,c22,c23],[c30,c31,c32,c33]])
sage: C
[1 0 0 0]
[2 2 -4 0]
[0 0 1 0]
[1 0 0 -1]
sage: V=VectorSpace(RR,4)
sage: u1=V([1,1,0,0]);u2=V([0,1,0,0]);u3=V([0,1,0,1]);u4=V([1,0,0,1])
sage: w1=V.coordinate_vector(u1);w2=V.coordinate_vector(u2)
sage: w3=V.coordinate_vector(u3);w4=V.coordinate_vector(u4)
sage: M=column_matrix([w1,w2,w3,w4]).n(digits=2);M
[ 1.0 0.00 0.00  1.0]
[ 1.0  1.0  1.0 0.00]
[0.00 0.00 0.00 0.00]
[0.00 0.00  1.0  1.0]
sage: A==M.transpose()*C*M
True
sage: f=b(x1,x2,x3,x4,x1,x2,x3,x4);f
x1^2 + 2*x1*x2 + x1*x4 + 2*x2^2 - 4*x2*x3 + x3^2 - x4^2
```

**Definition 5.10** (see [1], p. 152) The **rank** of the quadratic form $f$ is the rank of its matrix relative to a basis of $V$ and one denotes with $rank\ f$.
**Remark 5.11** (see [2], p. 122) Because of the symmetry of the associated matrix of a quadratic form, relative to a basis $B$ of $V$, the relation (5.3) is written

$$f(\overline{x}) = \sum_{i=1}^{n} a_{ii} \left(x^{(i)}\right)^2 + 2 \sum_{\substack{i,\, j=1 \\ i<j}}^{n} a_{ij} x^{(i)} x^{(j)}. \tag{5.4}$$

**Definition 5.12** (see [2], p. 122) If the associated matrix of the quadratic form $f : V \to K$ relative to the basis $B = \{\overline{e}_1, \overline{e}_2, \ldots, \overline{e}_n\}$ of $V$ is diagonal, i.e. $A = \text{diag}\,(\alpha_1, \ldots, \alpha_n)$ ; we shall say that:

- the basis $B$ is a **canonical basis** for $f$,
- the analytical expression of $f$ relative to the basis $B$, i.e.

$$f(\overline{x}) = \sum_{i=1}^{n} \alpha_i \left(x^{(i)}\right)^2, \ (\forall)\ \overline{x} = \sum_{i=1}^{n} x^{(i)} \overline{e}_i \in V \tag{5.5}$$

is a **canonical expression** for $f$.

We shall present three methods for obtaining a canonical expression for a quadratic form.

## 5.2 Gauss-Lagrange Method for Reducing a Quadratic Form to a Canonical Expression

**Theorem 5.13 (Gauss-Lagrange**, see [1], p. 152) Let $V$ be an $n$ finite dimensional vector space over $K$ and $f : V \to K$ a quadratic form. Then there is a basis $B' = \left\{\overline{e}'_1, \overline{e}'_2, \ldots, \overline{e}'_n\right\}$ of $V$, relative to which $f$ has a canonical expression.
**Proof**

We present the proof this theorem because it provides a method for getting a canonical expression for a quadratic form.

Let $B = \{\overline{e}_1, \overline{e}_2, \ldots, \overline{e}_n\}$ be a basis of $V$ relative to which $f$ has the analytical expression:

$$f(\overline{x}) = \sum_{i=1}^{n} \sum_{j=1}^{n} a_{ij} x^{(i)} y^{(j)}, \ (\forall)\ \overline{x} = \sum_{i=1}^{n} x^{(i)} \overline{e}_i.$$

If $f$ is the quadratic null form, then $f$ has the canonical expression in any basis of $V$. Hence, we can assume that isn't null.

We can also assume that $(\exists)\ i = \overline{1, n}$ such that $a_{ii} \neq 0$. Otherwise, if $a_{rp} \neq 0$, for $r \neq p$ then we make the change of coordinates:

$$\begin{cases} x^{(r)} = t^{(r)} + t^{(p)} \\ x^{(p)} = t^{(r)} - t^{(p)} \\ x^{(i)} = t^{(i)},\ i \in \{1, \ldots, n\} \setminus \{r, p\} \end{cases} \tag{5.6}$$

and we get an analytical expression, having non null its coefficients.

We suppose that $a_{11} \neq 0$. By grouping the terms that contain the variable $x^{(1)}$, from (5.4) we obtain:

$$f(\overline{x}) = a_{11} \left(x^{(1)}\right)^2 + 2 \sum_{k=2}^{n} a_{1k} x^{(1)} x^{(k)} + \sum_{i,\, j \neq 1}^{n} a_{ij} x^{(i)} x^{(j)}. \tag{5.7}$$

We shall add and subtract the necessary terms in (5.7) to write it in the form:

$$f(\overline{x}) = \frac{1}{a_{11}} \left(a_{11} x^{(1)} + a_{12} x^{(2)} + \ldots + a_{1n} x^{(n)}\right)^2 + \sum_{i,\, j=2}^{n} a'_{ij} x^{(i)} x^{(j)}, \tag{5.8}$$

where $\sum_{i,\, j=2}^{n} a'_{ij} x^{(i)} x^{(j)}$ doesn't contain $x^{(1)}$.

We make the change of coordinates:

$$\begin{cases} z^{(1)} = a_{11} x^{(1)} + a_{12} x^{(2)} + \ldots + a_{1n} x^{(n)} \\ z^{(2)} = x^{(2)} \\ \vdots \\ z^{(n)} = x^{(n)} \end{cases}$$

hence

$$\begin{cases} x^{(1)} = \frac{1}{a_{11}} z^{(1)} - \frac{a_{12}}{a_{11}} z^{(2)} - \ldots - \frac{a_{1n}}{a_{11}} z^{(n)} \\ x^{(2)} = z^{(2)} \\ \vdots \\ x^{(n)} = z^{(n)}. \end{cases}$$

The transition to the new coordinates $\left(z^{(1)}, z^{(2)}, \ldots, z^{(n)}\right)$ is achieved through the relation:

$$\overline{x}_B = \mathbf{M}_{(B, B_1)} \cdot \overline{x}_{B_1}, \tag{5.9}$$

with the transition matrix

$$\mathbf{M}_{(B, B_1)} = \begin{pmatrix} \frac{1}{a_{11}} & -\frac{a_{12}}{a_{11}} & \cdots & -\frac{a_{1n}}{a_{11}} \\ 0 & 1 & \cdots & 0 \\ \vdots & \vdots & \ddots & \vdots \\ 0 & 0 & 0 & 1 \end{pmatrix}.$$

The new coordinates correspond to the new basis $B_1 = \left\{\overline{f}_1, \overline{f}_2, \ldots, \overline{f}_n\right\}$, where:

$$\begin{cases} \overline{f}_1 = \frac{1}{a_{11}}\overline{e}_1 \\ \overline{f}_2 = -\frac{a_{12}}{a_{11}}\overline{e}_1 + \overline{e}_2 \\ \quad\vdots \\ \overline{f}_n = -\frac{a_{1n}}{a_{11}}\overline{e}_1 + \overline{e}_n. \end{cases}$$

The form $Q$ has the following analytical expression relative to the basis $B_1$:

$$f(\overline{x}) = \frac{1}{a_{11}}\left(z^{(1)}\right)^2 + \sum_{i,\, j=2}^{n} a'_{ij} x^{(i)} x^{(j)}. \tag{5.10}$$

The sum

$$Q_1 = \sum_{i,\, j=2}^{n} a'_{ij} x^{(i)} x^{(j)}$$

from the right member of the relation (5.10) is a quadratic form in $n-1$ variables, therefore can be treated by the process described above, as well as the form Q.

Finally, after at most $n-1$ steps we obtain a basis $B' = \left\{\overline{e}'_1, \overline{e}'_2, \ldots, \overline{e}'_n\right\}$ of $V$, relative to which the quadratic form $Q$ is reduced to the canonical expression.

**Example 5.14** (see [4]). Let be the quadratic form

$$Q : \mathbb{R}^4 \to \mathbb{R},\ Q(\overline{x}) = \left(x^{(1)}\right)^2 + 2x^{(1)}x^{(2)} + 2\left(x^{(2)}\right)^2 - 4x^{(2)}x^{(3)} + \left(x^{(3)}\right)^2 + x^{(1)}x^{(4)} - \left(x^{(4)}\right)^2.$$

Using the Gauss-Lagrange method, we can bring $Q$ to the canonical expression and highlight the transition matrix from the initial basis to that basis, relative to which $Q$ has the canonical expression.

**Solution**

The associated matrix of $Q$ relative to the canonical basis of the space $\mathbb{R}^4$, i.e.

$$B = \{\overline{e}_1 = (1, 0, 0, 0),\ \overline{e}_2 = (0, 1, 0, 0),\ \overline{e}_3 = (0, 0, 1, 0),\ \overline{e}_4 = (0, 0, 0, 1)\}$$

is

$$A = \begin{pmatrix} 1 & 1 & 0 & 1/2 \\ 1 & 2 & -2 & 0 \\ 0 & -2 & 1 & 0 \\ 1/2 & 0 & 0 & -1 \end{pmatrix}.$$

We note that $a_{11} \neq 0$. We can write $Q$ in the form

$$Q(\overline{x}) = \left(x^{(1)} + x^{(2)} + \frac{x^{(4)}}{2}\right)^2 + \left(x^{(2)}\right)^2 - 4x^{(2)}x^{(3)}$$
$$- \frac{5}{4}\left(x^{(4)}\right)^2 - x^{(2)}x^{(4)} + \left(x^{(3)}\right)^2.$$

By making the change of coordinates:

$$\begin{cases} y^{(1)} = x^{(1)} + x^{(2)} + \frac{x^{(4)}}{2} \\ y^{(2)} = x^{(2)} \\ y^{(3)} = x^{(3)} \\ y^{(4)} = x^{(4)} \end{cases}$$

it results

$$\begin{cases} x^{(1)} = y^{(1)} - y^{(2)} - \frac{y^{(4)}}{2} \\ x^{(2)} = y^{(2)} \\ x^{(3)} = y^{(3)} \\ x^{(4)} = y^{(4)}. \end{cases}$$

The transition matrix associated with this change of coordinates will be:

$$\mathrm{M}_{(B,B_1)} = \begin{pmatrix} 1 & -1 & 0 & -1/2 \\ 0 & 1 & 0 & 0 \\ 0 & 0 & 1 & 0 \\ 0 & 0 & 0 & 1 \end{pmatrix},$$

the new basis being

$$B_1 = \left\{\overline{f}_1 = \overline{e}_1,\ \overline{f}_2 = -\overline{e}_1 + \overline{e}_2,\ \overline{f}_3 = \overline{e}_3,\ \overline{f}_4 = -\frac{1}{2}\overline{e}_1 + \overline{e}_4\right\}.$$

The expression of the quadratic form $Q$ relative to the basis $B_1$ is

$$Q = \left(y^{(1)}\right)^2 + \left(y^{(2)}\right)^2 - 4y^{(2)}y^{(3)} - \frac{5}{4}\left(y^{(4)}\right)^2 - y^{(2)}y^{(4)} + \left(y^{(3)}\right)^2.$$

The associated matrix of $Q$ relative to the basis $B_1$ is

$$A' = \begin{pmatrix} 1 & 0 & 0 & 0 \\ 0 & 1 & -2 & -1/2 \\ 0 & -2 & 1 & 0 \\ 0 & -1/2 & 0 & -5/4 \end{pmatrix}.$$

Noting that $a'_{22} \neq 0$, we can write $Q$ in the form:

$$Q = \left(y^{(1)}\right)^2 + \left(y^{(2)} - 2y^{(3)} - \frac{1}{2}y^{(4)}\right)^2 - 3\left(y^{(3)}\right)^2 - \frac{3}{2}\left(y^{(4)}\right)^2 - 2y^{(3)}y^{(4)}.$$

We make the change of coordinates:

$$\begin{cases} z^{(1)} = y^{(1)} \\ z^{(2)} = y^{(2)} - 2y^{(3)} - \frac{1}{2}y^{(4)} \\ z^{(3)} = y^{(3)} \\ z^{(4)} = y^{(4)}; \end{cases}$$

hence

$$\begin{cases} y^{(1)} = z^{(1)} \\ y^{(2)} = z^{(2)} + 2z^{(3)} + \frac{1}{2}z^{(4)} \\ y^{(3)} = z^{(3)} \\ y^{(4)} = z^{(4)}. \end{cases}$$

The transition to the new coordinates $\left(z^{(1)}, z^{(2)}, \ldots, z^{(n)}\right)$ is achieved through the relation

$$\overline{x}_{B_1} = \mathrm{M}_{(B_1,B_2)} \cdot \overline{x}_{B_2},$$

with the transition matrix

$$\mathrm{M}_{(B_1,B_2)} = \begin{pmatrix} 1 & 0 & 0 & 0 \\ 0 & 1 & 2 & 1/2 \\ 0 & 0 & 1 & 0 \\ 0 & 0 & 0 & 1 \end{pmatrix}.$$

The new coordinates correspond to the new basis

$$B_2 = \left\{\overline{g}_1 = \overline{f}_1, \ \overline{g}_2 = \overline{f}_2, \ \overline{g}_3 = 2\overline{f}_2 + \overline{f}_3, \ \overline{g}_4 = \frac{1}{2}\overline{f}_2 + \overline{f}_4\right\}.$$

The expression of the quadratic form $Q$ relative to the basis $B_2$ is

$$Q = \left(z^{(1)}\right)^2 + \left(z^{(2)}\right)^2 - 3\left(z^{(3)}\right)^2 - \frac{3}{2}\left(z^{(4)}\right)^2 - 2z^{(2)}z^{(4)}.$$

The associated matrix of $Q$ relative to the basis $B_2$ is

$$A'' = \begin{pmatrix} 1 & 0 & 0 & 0 \\ 0 & 1 & 0 & 0 \\ 0 & 0 & -3 & -1 \\ 0 & 0 & -1 & -3/2 \end{pmatrix}.$$

We note that $a''_{33} \neq 0$. We will form a perfect square in $Q$ for those terms that contain $z^{(3)}$; is follows

$$Q = \left(z^{(1)}\right)^2 + \left(z^{(2)}\right)^2 - \frac{1}{3}\left(3z^{(3)} + z^{(4)}\right) - \frac{7}{6}\left(z^{(4)}\right)^2.$$

We shall make the change of coordinates:

$$\begin{cases} t^{(1)} = z^{(1)} \\ t^{(2)} = z^{(2)} \\ t^{(3)} = 3z^{(3)} + z^{(4)} \\ t^{(4)} = z^{(4)}; \end{cases}$$

we have

$$\mathrm{M}_{(B_2,B_3)} = \begin{pmatrix} 1 & 0 & 0 & 0 \\ 0 & 1 & 0 & 0 \\ 0 & 0 & 1/3 & -1 \\ 0 & 0 & 0 & 1 \end{pmatrix}.$$

The expression of the quadratic form $Q$ relative to the basis

$$B_3 = \left\{ \overline{h}_1 = \overline{g}_1, \, \overline{h}_2 = \overline{g}_2, \, \overline{h}_3 = \frac{1}{3}\overline{g}_3, \, \overline{h}_4 = -\overline{g}_3 + \overline{g}_4 \right\}$$

is

$$Q = \left(t^{(1)}\right)^2 + \left(t^{(2)}\right)^2 - \frac{1}{3}\left(t^{(3)}\right)^2 - \frac{7}{6}\left(t^{(4)}\right)^2,$$

so we have obtained the canonical expression of $Q$.

We shall get:

$$\overline{x}_B = \mathrm{M}_{(B,B_1)}\overline{x}_{B_1} = \mathrm{M}_{(B,B_1)}\mathrm{M}_{(B_1,B_2)}\overline{x}_{B_2} = \mathrm{M}_{(B,B_1)}\mathrm{M}_{(B_1,B_2)}\mathrm{M}_{(B_2,B_3)}\overline{x}_{B_3}.$$

It follows that transition matrix from the initial basis $B$ of the space $\mathbb{R}^4$ to the basis $B_3$, relative to which $Q$ has the canonical expression is:

$$\mathrm{M}_{(B,B_3)} = \mathrm{M}_{(B,B_1)}\mathrm{M}_{(B_1,B_2)}\mathrm{M}_{(B_2,B_3)}.$$

A solution in Sage will be given, too:

```
sage: var("x1 x2 x3 x4 y1 y2 y3 y4 z1 z2 z3 z4 t1 t2 t3 t4")
(x1, x2, x3, x4, y1, y2, y3, y4, z1, z2, z3, z4, t1, t2, t3, t4)
sage: Qq(x1,x2,x3,x4)=x1^2+2*x1*x2+2*x2^2-4*x2*x3+x3^2+x1*x4-x4^2
sage: Q(x1,x2,x3,x4)=(x1+x2+x4/2)^2+expand(Qq-(x1+x2+x4/2)^2)
sage: Qq1(y1,y2,y3,y4)=Q(y1-y2-y4/2,y2,y3,y4);Qq1
(y1, y2, y3, y4) |--> y1^2 + y2^2 - 4*y2*y3 - y2*y4 + y3^2 - 5/4*y4^2
```

```
sage: Q1(y1,y2,y3,y4)=(y2-2*y3-y4/2)^2+expand(Qq1-(y2-2*y3-y4/2)^2)
sage: Qq2(z1,z2,z3,z4)=Q1(z1,z2+2*z3+z4/2,z3,z4);Qq2
(z1, z2, z3, z4) |--> z1^2 + z2^2 - 3*z3^2 - 2*z3*z4 - 3/2*z4^2
sage: Q2(z1,z2,z3,z4)=-1/3*(3*z3+z4)^2+expand(Qq2+1/3*(3*z3+z4)^2)
sage: Q3(t1,t2,t3,t4)=Q2(t1,t2,(t3-t4)/3,t4);Q3
(t1, t2, t3, t4) |--> t1^2 + t2^2 - 1/3*t3^2 - 7/6*t4^2
```

## 5.3 Reducing a Quadratic Form to a Canonical Expression by Jacobi Method

**Theorem 5.15** (**Jacobi**, see [3], p. 100). Let $V$ be an $n$ finite dimensional vector space over $K$, $f : V \to K$, a quadratic form and $A = \left(a_{ij}\right)_{1\le i,j\le n}$ its relative matrix to the basis $B = \{\overline{e}_1, \overline{e}_2, \ldots, \overline{e}_n\}$ of $V$.

If all the principal minors

$$\begin{cases} \Delta_1 = a_{11} \\ \Delta_2 = \begin{vmatrix} a_{11} & a_{12} \\ a_{21} & a_{22} \end{vmatrix} \\ \quad\vdots \\ \Delta_n = \det A \end{cases} \tag{5.11}$$

are all non- null, then there is a basis $B' = \left\{\overline{e}'_1, \overline{e}'_2, \ldots, \overline{e}'_n\right\}$ of $V$, relative to which the quadratic form $Q$ has the canonical expression

$$f\left(\overline{x}\right) = \sum_{i=1}^{n} \frac{\Delta_{i-1}}{\Delta_i}\left(y^{(i)}\right)^2, \tag{5.12}$$

where

- $y^{(i)}$, $i = \overline{1,n}$ are the coordinates of $\overline{x}$ in the basis $B'$,
- $\Delta_0 = 1$.

**Proof**

We are looking for the vectors $\overline{e}'_1, \overline{e}'_2, \ldots, \overline{e}'_n$ by the form

$$\begin{cases} \overline{e}'_1 = c_{11}\overline{e}_1 \\ \overline{e}'_2 = c_{21}\overline{e}_1 + c_{22}\overline{e}_2 \\ \quad\vdots \\ \overline{e}'_i = c_{i1}\overline{e}_1 + c_{i2}\overline{e}_2 + \ldots + c_{ii}\overline{e}_i \\ \quad\vdots \\ \overline{e}'_n = c_{n1}\overline{e}_1 + c_{n2}\overline{e}_2 + \ldots + c_{nn}\overline{e}_n, \end{cases} \tag{5.13}$$

where $c_{ij}$, $i, j = \overline{1,n}$ will be determined by imposing the conditions:

$$b\left(\overline{e}_i',\overline{e}_j\right)=\begin{cases}0, & 1\le j<i\le n\\ 1, & i=j\end{cases} \tag{5.14}$$

and $b : V\times V\to K$ is the bilinear form from which $f$ derives.

We compute

$$\begin{aligned} b\left(\overline{e}_i',\overline{e}_j\right) &= b\left(c_{i1}\overline{e}_1+c_{i2}\overline{e}_2+\ldots+c_{ii}\overline{e}_i,\overline{e}_j\right)\\ &= c_{i1}b\left(\overline{e}_1,\overline{e}_j\right)+c_{i2}b\left(\overline{e}_2,\overline{e}_j\right)+\ldots+c_{ii}b\left(\overline{e}_i,\overline{e}_j\right)\\ &= c_{i1}a_{1j}+c_{i2}a_{2j}+\ldots+c_{ii}a_{ij}. \end{aligned} \tag{5.15}$$

We obtain:

$$\begin{cases} j=1: b\left(\overline{e}_i',\overline{e}_1\right)=c_{i1}a_{11}+c_{i2}a_{12}+\ldots+c_{ii}a_{1i}=0\\ j=2: b\left(\overline{e}_i',\overline{e}_2\right)=c_{i1}a_{21}+c_{i2}a_{22}+\ldots+c_{ii}a_{2i}=0\\ \vdots\\ j=i-1: b\left(\overline{e}_i',\overline{e}_{i-1}\right)=c_{i1}a_{i-1,1}+c_{i2}a_{i-1,2}+\ldots+c_{ii}a_{i-1,i}=0\\ j=i: b\left(\overline{e}_i',\overline{e}_i\right)=c_{i1}a_{i1}+c_{i2}a_{i2}+\ldots+c_{ii}a_{ii}=1 \end{cases} \tag{5.16}$$

i.e. a compatible system that is determined, since its determinant is $\Delta_i\neq 0$ (hence the vector $\overline{e}_i'$ is uniquely determined).

Using the formulas of Crammer we get solutions of the system (5.16):

$$c_{ii}=\frac{\begin{vmatrix} a_{11} & \cdots & a_{1,i-1} & 0\\ \vdots & \ddots & \vdots & \vdots\\ a_{i-1,1} & \cdots & a_{i-1,i-1} & 0\\ a_{i1} & \cdots & a_{i,i-1} & 1\end{vmatrix}}{\Delta_i}=\frac{\Delta_{i-1}}{\Delta_i},\ (\forall)\ i=\overline{1,n}. \tag{5.17}$$

To determine the expression of the quadratic form in the basis $B'=\left\{\overline{e}_1',\overline{e}_2',\ldots,\overline{e}_n'\right\}$ we shall calculate the elements of the matrix $A'$, associated of $f$ relative to the basis $B'$.

We have

$$\begin{aligned} a_{ij}' &= b\left(\overline{e}_i',\overline{e}_j'\right)=b\left(\overline{e}_i',c_{j1}\overline{e}_1+\ldots+c_{jj}\overline{e}_j\right)\\ &= c_{j1}b\left(\overline{e}_i',\overline{e}_1\right)+c_{j2}b\left(\overline{e}_i',\overline{e}_2\right)+\ldots+c_{jj}b\left(\overline{e}_i',\overline{e}_j\right),\ (\forall)\ i,j=\overline{1,n}. \end{aligned}$$

But, from (5.14) we know that $b\left(\overline{e}_i',\overline{e}_j'\right)=0$ for $j<i$; hence $a_{ij}'=0$ for $j<i$. Because of the symmetry of the bilinear form $b$ it results $a_{ij}'=0$ for $j>i$. Therefore $a_{ij}'=0$ for $j\neq i$. For $j=i$ we have

$$\begin{aligned} a'_{ii} &= b\left(\overline{e}'_i, \overline{e}'_i\right) = b\left(\overline{e}'_i, c_{i1}\overline{e}_1 + \ldots + c_{ii}\overline{e}_i\right) \\ &= c_{i1} b\left(\overline{e}'_i, \overline{e}_1\right) + c_{i2} b\left(\overline{e}'_i, \overline{e}_2\right) + \ldots + c_{i,i-1} b\left(\overline{e}'_i, \overline{e}_{i-1}\right) + c_{ii} b\left(\overline{e}'_i, \overline{e}_i\right) \\ &= c_{ii} = \frac{\Delta_{i-1}}{\Delta_i}, (\forall)\ i, j = \overline{1, n}. \end{aligned}$$

We deduce that the quadratic form has the following canonical expression in the basis $B'$ :

$$f\left(\overline{x}\right) = \sum_{i,\, j=1}^{n} a'_{ij} y^{(i)} y^{(j)} = \frac{\Delta_{i-1}}{\Delta_i} \left(y^{(i)}\right)^2$$

and its associated matrix is

$$A' = \begin{pmatrix} \frac{\Delta_0}{\Delta_1} & & \mathrm{O} \\ & \ddots & \\ \mathrm{O} & & \frac{\Delta_{n-1}}{\Delta_n} \end{pmatrix}.$$

**Example 5.16** (see [3], p. 101). Using the Jacobi method find the canonical expression and the basis in which to do this for the quadratic form

$$Q : \mathbb{R}^3 \to \mathbb{R},\ Q\left(\overline{x}\right) = x_1^2 + 7x_2^2 + x_3^2 - 8x_1x_2 - 8x_2x_3 - 16x_1x_3, \\ \times (\forall)\ \overline{x} = (x_1, x_2, x_3) \in \mathbb{R}^3. \tag{5.18}$$

**Solution**

The matrix of the quadratic form relative to the canonical basis of the space $\mathbb{R}^3$ is

$$A = \begin{pmatrix} 1 & -4 & -8 \\ -4 & 7 & -4 \\ -8 & -4 & 1 \end{pmatrix}.$$

Its principal minors $\Delta_i$, $i = \overline{0, 3}$ are:

$$\begin{cases} \Delta_0 = 1 \\ \Delta_1 = a_{11} = 1 \\ \Delta_2 = \begin{vmatrix} 1 & -4 \\ -4 & 7 \end{vmatrix} = -9 \\ \Delta_3 = \det A = -729. \end{cases}$$

The quadratic form $Q$ will have the following canonical expression:

$$Q(\overline{x}) = \sum_{i=1}^{n} \frac{\Delta_{i-1}}{\Delta_i} y_i^2 = \frac{\Delta_0}{\Delta_1} y_1^2 + \frac{\Delta_1}{\Delta_2} y_2^2 + \frac{\Delta_2}{\Delta_3} y_3^2$$
$$= y_1^2 - \frac{1}{9} y_2^2 + \frac{1}{81} y_3^2.$$

We shall determine the new basis $B' = \{\overline{e}'_1, \overline{e}'_2, \ldots, \overline{e}'_n\}$, relative to which $Q$ has the canonical expression:

$$\begin{cases} \overline{e}'_1 = c_{11}\overline{e}_1 \\ \overline{e}'_2 = c_{21}\overline{e}_1 + c_{22}\overline{e}_2 \\ \overline{e}'_3 = c_{31}\overline{e}_1 + c_{32}\overline{e}_2 + c_{33}\overline{e}_3, \end{cases}$$

where $c_{ij}, i, j = \overline{1,3}$ will be determined by imposing the conditions (5.14), $b$ being the associated bilinear form $Q$ of the quadratic form in the basis $B'$, i.e.

$$b(\overline{x}, \overline{y}) = x_1 y_1 - \frac{1}{9} x_2 y_2 + \frac{1}{81} x_3 y_3.$$

We have:

$$\left.\begin{array}{c} b\left(\overline{e}'_i, \overline{e}_1\right) = b(c_{11}\overline{e}_1, \overline{e}_1) = c_{11} b(\overline{e}_1, \overline{e}_1) = c_{11} a_{11} = c_{11} \\ b\left(\overline{e}'_1, \overline{e}_1\right) = 1 \end{array}\right\} \Rightarrow c_{11} = 1;$$

therefore: $\overline{e}'_1 = \overline{e}_1$.

We shall compute:

$$\begin{aligned} b\left(\overline{e}'_2, \overline{e}_1\right) &= b(c_{21}\overline{e}_1 + c_{22}\overline{e}_2, \overline{e}_1) = c_{21} b(\overline{e}_1, \overline{e}_1) + c_{22} b(\overline{e}_2, \overline{e}_1) \\ &= c_{21} a_{11} + c_{22} a_{21} = c_{21} - 4c_{22} \\ b\left(\overline{e}'_2, \overline{e}_2\right) &= b(c_{21}\overline{e}_1 + c_{22}\overline{e}_2, \overline{e}_2) = c_{21} b(\overline{e}_1, \overline{e}_2) + c_{22} b(\overline{e}_2, \overline{e}_2) \\ &= c_{21} a_{12} + c_{22} a_{22} = -4c_{21} + 7c_{22}. \end{aligned}$$

Taking into account (5.14) we obtain the system:

$$\begin{cases} c_{21} - 4c_{22} = 0 \\ -4c_{21} + 7c_{22} = 1 \end{cases} \Rightarrow c_{21} = -\frac{4}{9}, \; c_{22} = -\frac{1}{9},$$

i.e.

$$\overline{e}'_2 = -\frac{4}{9}\overline{e}_1 - \frac{1}{9}\overline{e}_2.$$

We have also to calculate:

$$\begin{aligned}
b\left(\overline{e}_3', \overline{e}_1\right) &= b\left(c_{31}\overline{e}_1 + c_{32}\overline{e}_2 + c_{33}\overline{e}_3, \overline{e}_1\right) = c_{31}b\left(\overline{e}_1, \overline{e}_1\right) \\
&\quad + c_{32}b\left(\overline{e}_2, \overline{e}_1\right) + c_{33}b\left(\overline{e}_3, \overline{e}_1\right) \\
&= c_{31}a_{11} + c_{32}a_{12} + c_{33}a_{13} = c_{31} - 4c_{32} - 8c_{33} \\
b\left(\overline{e}_3', \overline{e}_2\right) &= b\left(c_{31}\overline{e}_1 + c_{32}\overline{e}_2 + c_{33}\overline{e}_3, \overline{e}_2\right) = c_{31}b\left(\overline{e}_1, \overline{e}_2\right) \\
&\quad + c_{32}b\left(\overline{e}_2, \overline{e}_2\right) + c_{33}b\left(\overline{e}_3, \overline{e}_2\right) \\
&= c_{31}a_{21} + c_{32}a_{22} + c_{33}a_{23} = -4c_{31} + 7c_{32} - 4c_{33} \\
b\left(\overline{e}_3', \overline{e}_3\right) &= b\left(c_{31}\overline{e}_1 + c_{32}\overline{e}_2 + c_{33}\overline{e}_3, \overline{e}_3\right) = c_{31}b\left(\overline{e}_1, \overline{e}_3\right) \\
&\quad + c_{32}b\left(\overline{e}_2, \overline{e}_3\right) + c_{33}b\left(\overline{e}_3, \overline{e}_3\right) \\
&= c_{31}a_{31} + c_{32}a_{32} + c_{33}a_{33} = -8c_{31} - 4c_{32} + c_{33}.
\end{aligned}$$

Taking into account (5.14) we obtain the system:

$$\begin{cases} c_{31} - 4c_{32} - 8c_{33} = 0 \\ -4c_{31} + 7c_{32} - 4c_{33} = 0 \\ -8c_{31} - 4c_{32} + c_{33} = 1 \end{cases} \Rightarrow c_{31} = -\frac{8}{81},\ c_{32} = -\frac{4}{81},\ c_{33} = \frac{1}{81};$$

it results:

$$\overline{e}_3' = -\frac{8}{81}\overline{e}_1 - \frac{4}{81}\overline{e}_2 + \frac{1}{81}\overline{e}_3.$$

The solution in Sage will be given, too:

```
sage: var("x x1 x2 x3 y1 y2 y3")
(x, x1, x2, x3, y1, y2, y3)
sage: A=matrix([[1,-4,-8],[-4,7,-4],[-8,-4,1]])
sage: B=A[0:2,0:2]
sage: d1=1;d2=A[0][0];d3=B.determinant();d4=A.determinant()
sage: f=d1/d2*y1^2+d2/d3*y2^2+d3/d4*y3^2;f
y1^2 - 1/9*y2^2 + 1/81*y3^2
sage: I=identity_matrix(RR,3)
sage: s=solve([x*A[0][0]==1],x);c11=s[0].right()
sage: v=vector(SR,[x1,x2]);vv=vector(SR,[x1,x2,x3])
sage: v1=B*v.transpose()
sage: ss=solve([v1[0][0],v1[1][0]==1],x1,x2)
sage: c21=ss[0][0].right();c22=ss[0][1].right()
```

```
sage: vv1=A*vv.transpose()
sage: ss1=solve([vv1[0][0],vv1[1][0],vv1[2][0]==1],x1,x2,x3)
sage: c31=ss1[0][0].right();c32=ss1[0][1].right();c33=ss1[0][2].right()
sage: ep1=c11*l[0];ep1.n(digits=2)
(1.0, 0.00, 0.00)
sage: ep2=c21*l[0]+c22*l[1];ep2.n(digits=2)
(-0.44, -0.11, 0.00)
sage: ep3=c31*l[0]+c32*l[1]+c33*l[2];ep3.n(digits=2)
(-0.099, -0.049, 0.012)
```

## 5.4 Eigenvalue Method for Reducing a Quadratic Form into Canonical Expression

**Theorem 5.17 (Eigenvalue method**, see [1], p. 153). Let $V$ be an Euclidean real vector space and let $f : V \to \mathbb{R}$ be a real quadratic form. Then there is an orthonormal basis $B' = \{\overline{e}'_1, \overline{e}'_2, \ldots, \overline{e}'_n\}$ of the vector space $V$ relative to which the canonical expression of the form is

$$f(\overline{x}) = \sum_{i=1}^{n} \lambda_i \left(y^{(i)}\right)^2, \tag{5.19}$$

where:

- $\lambda_1, \ldots, \lambda_n$ are the eigenvalues of the associated matrix of the quadratic form, relative to an orthonormal basis $B$ (each eigenvalue being included in sum such many times as its multiplicity),
- $y^{(1)}, \ldots, y^{(n)}$ are the coordinates of the vector $\overline{x}$ relative to the basis $B'$.

To apply the eigenvalue method for reducing a quadratic form to canonical expression one determines as follows:

1. choose an orthonormal basis $B = \{\overline{e}_1, \overline{e}_2, \ldots, \overline{e}_n\}$ of $V$ and write the matrix $A$, associated to $f$ relative to the basis $B$;
2. determine the eigenvalues: $\lambda_1, \ldots, \lambda_r \in \mathbb{R}$ of the matrix $A$, with the corresponding algebraic multiplicities $a_{\lambda_1}, \ldots, a_{\lambda_r}$, with $a_{\lambda_1} + \ldots + a_{\lambda_r} = n$ ;
3. for the eigensubspaces $W_{\lambda_1}, \ldots, W_{\lambda_r}$ associated to the eigenvalues $\lambda_1, \ldots, \lambda_r$ determine the orthonormal bases $B_1, \ldots, B_r$, using the *Gram-Schmidt orthogonalization procedure*;
4. one considers the orthonormal basis: $B' = B_1 \cup \ldots \cup B_r$ of $V$ and one writes the canonical expression of $f$ relative to the basis $B'$ with (5.19), where $\overline{x}_{B'} = \left(y^{(1)}, \ldots, y^{(n)}\right)^t$.

**Example 5.18** (see [4]). Use the eigenvalue method to determine the canonical expression and the basis relative to which can be made this, for the quadratic form:

$$f:\mathbb{R}^3\to\mathbb{R},\ f(\overline{x})=x_1^2+x_2^2+x_3^2+x_1x_2+x_2x_3+x_1x_3,\ (\forall)\ \overline{x}=(x_1,x_2,x_3)\in\mathbb{R}^3.$$

**Solution**

The associated matrix of $f$ relative to the canonical basis of the space $\mathbb{R}^3$ is

$$A=\begin{pmatrix}1 & 1/2 & 1/2\\ 1/2 & 1 & 1/2\\ 1/2 & 1/2 & 1\end{pmatrix}.$$

We have:

$$P(\lambda)=\begin{vmatrix}1-\lambda & 1/2 & 1/2\\ 1/2 & 1-\lambda & 1/2\\ 1/2 & 1/2 & 1-\lambda\end{vmatrix}=(2-\lambda)\left(\frac{1}{2}-\lambda\right)^2,$$

which has the roots

$$\begin{cases}\lambda_1=2,\ a_{\lambda_1}=1\\ \lambda_2=1/2,\ a_{\lambda_2}=2.\end{cases}$$

The associated eigenspace of the eigenvalue $\lambda_1$ is

$$W_{\lambda_1}=\left\{\overline{x}\in\mathbb{R}^3|\ A\overline{x}=\lambda_1\overline{x}\right\}.$$

We deduce:

$$\begin{cases}x_1+\frac{1}{2}x_2+\frac{1}{2}x_3=2x_1\\ \frac{1}{2}x_1+x_2+\frac{1}{2}x_3=2x_2\\ \frac{1}{2}x_1+\frac{1}{2}x_2+x_3=2x_3\end{cases}\Leftrightarrow\begin{cases}-x_1+\frac{1}{2}x_2+\frac{1}{2}x_3=0\\ \frac{1}{2}x_1-x_2+\frac{1}{2}x_3=0\\ \frac{1}{2}x_1+\frac{1}{2}x_2-x_3=0.\end{cases}$$

Denoting $x_3=t,\ t\in\mathbb{R}$ we achieve:

$$\begin{cases}-2x_1+x_2=-t\\ x_1-2x_2=-t\end{cases}\Rightarrow x_1=t,x_2=t.$$

Therefore

$$W_{\lambda_1}=\left\{\overline{x}\in\mathbb{R}^3|\ \overline{x}=(t,t,t)=t\cdot\underbrace{(1,1,1)}_{\overline{c}_1}=t\overline{c}_1\right\}.$$

The orthonormal basis $B_1$ will be $B_1=\{\overline{f}_1\}$, where

$$\overline{f}_1=\frac{\overline{c}_1}{\|\overline{c}_1\|}=\frac{1}{\sqrt{3}}\overline{c}_1=\left(\frac{1}{\sqrt{3}},\frac{1}{\sqrt{3}},\frac{1}{\sqrt{3}}\right).$$

The associated eigenspace of the eigenvalue $\lambda_2$ is

$$W_{\lambda_2} = \left\{ \overline{x} \in \mathbb{R}^3 \middle| A\overline{x} = \lambda_2 \overline{x} \right\}.$$

We achieve:

$$\begin{cases} x_1 + \frac{1}{2}x_2 + \frac{1}{2}x_3 = \frac{1}{2}x_1 \\ \frac{1}{2}x_1 + x_2 + \frac{1}{2}x_3 = \frac{1}{2}x_2 \\ \frac{1}{2}x_1 + \frac{1}{2}x_2 + x_3 = \frac{1}{2}x_3 \end{cases} \Leftrightarrow$$

$$\frac{1}{2}x_1 + \frac{1}{2}x_2 + \frac{1}{2}x_3 = 0 \Leftrightarrow x_1 + x_2 + x_3 = 0.$$

We denote $x_1 = t_1$, $x_2 = t_2$, $t_1, t_2 \in \mathbb{R}$; hence $x_3 = -t_1 - t_2$.
Therefore:

$$W_{\lambda_2} = \left\{ \overline{x} \in \mathbb{R}^3 \middle| \overline{x} = (t_1, t_2, -t_1 - t_2) = t_1 \underbrace{(1, 0, -1)}_{\overline{c}_2} + t_2 \underbrace{(0, 1, -1)}_{\overline{c}_3} = t_1 \overline{c}_2 + t_2 \overline{c}_3 \right\}. \tag{5.20}$$

We consider the orthogonal system $\left\{ \overline{f}'_2, \overline{f}'_3 \right\}$, where

$$\begin{cases} \overline{f}'_2 = \overline{c}_2 \\ \overline{f}'_3 = \overline{c}_3 + \alpha \overline{f}'_2, \end{cases}$$

where $\alpha$ is obtained from the condition that $\overline{f}'_3$ and $\overline{f}'_2$ to be orthogonal, i.e.

$$< \overline{f}'_3, \overline{f}'_2 >= 0.$$

From

$$< \overline{c}_3 + \alpha \overline{f}'_2, \overline{f}'_2 >= 0 \Rightarrow \alpha = -\frac{< \overline{c}_3, \overline{f}'_2 >}{< \overline{f}'_2, \overline{f}'_2 >} = -\frac{1}{2}.$$

It results

$$\overline{f}'_3 = \overline{c}_3 - \frac{1}{2}\overline{f}'_2 = (0, 1, -1) - \frac{1}{2}(1, 0, -1) = \left( -\frac{1}{2}, 1, -\frac{1}{2} \right).$$

The basis $B = \{\overline{f}_2, \overline{f}_3\}$ is orthonormal, where

$$\overline{f}_2 = \frac{\overline{f}_2'}{\left\|\overline{f}_2'\right\|} = \frac{1}{\sqrt{2}}\overline{f}_2' = \left(\frac{1}{\sqrt{2}}, 0, -\frac{1}{\sqrt{2}}\right),$$

$$\overline{f}_3 = \frac{\overline{f}_3'}{\left\|\overline{f}_3'\right\|} = \frac{\sqrt{2}}{\sqrt{3}}\overline{f}_3' = \left(-\frac{1}{\sqrt{6}}, \frac{\sqrt{2}}{\sqrt{3}}, -\frac{1}{\sqrt{6}}\right).$$

We achieve:

$$B' = B_1 \cup B_2 = \{\overline{f}_1, \overline{f}_2, \overline{f}_3\}.$$

The associated matrix of $f$ in the basis $B'$ will be

$$A' = \begin{pmatrix} 2 & 0 & 0 \\ 0 & 1/2 & 0 \\ 0 & 0 & 1/2 \end{pmatrix}$$

and the canonical expression of $f$ relative to the basis $B'$:

$$f(\overline{x}) = 2y_1^2 + \frac{1}{2}y_2^2 + \frac{1}{2}y_3^2.$$

We need the following Sage code to implement this method:

```
sage: A=matrix(RR,[[1,1/2,1/2],[1/2,1,1/2],[1/2,1/2,1]])
sage: I=identity_matrix(RR,3);var("la  x1 x2 x3 y1 y2 y3")
(la, x1, x2, x3, y1, y2, y3)
sage: v=vector(SR,[x1, x2, x3]);v1=vector(SR,[y1^2, y2^2, y3^2])
sage: s=solve([(A-la*I).determinant()],la,multiplicities=True);s
([la == 2, la == (1/2)], [1, 2])
sage: s1=s[0][0].right();s2=s[0][1].right()
sage: m=A-s1*I;m1=m*v.transpose()
sage: ss=solve([m1[0][0],m1[1][0],m1[2][0]],x1,x2,x3);ss
[[x1 == r2, x2 == r2, x3 == r2]]
sage: mm=A-s2*I;m2=mm*v.transpose()
sage: ss1=solve([m2[0][0],m2[1][0],m2[2][0]],x1,x2,x3)
sage: l=vector([1,1/2,1/2])
sage: f=v1*l.transpose();f[0]
y1^2 + 1/2*y2^2 + 1/2*y3^2
```

```
sage: c1=vector([ss[0][0].right().subs(r2=1),ss[0][1].right().subs(r2=1),ss[0][2].right().subs(r2=1)])
sage: f1=c1/c1.norm();f1
(1/3*sqrt(3), 1/3*sqrt(3), 1/3*sqrt(3))
sage: c2=vector([ss1[0][0].right().subs(r3=-1,r4=0),ss1[0][1].right().subs(r3=-
1,r4=0),ss1[0][2].right().subs(r3=-1,r4=0)])
sage: f2=c2/c2.norm();f2
(1/2*sqrt(2), 0, -1/2*sqrt(2))
sage: c3=vector([ss1[0][0].right().subs(r3=-1/2,r4=1),ss1[0][1].right().subs(r3=-
1/2,r4=1),ss1[0][2].right().subs(r3=-1/2,r4=1)])
sage: f3=c3/c3.norm();f3
(-1/3*sqrt(3/2), 2/3*sqrt(3/2), -1/3*sqrt(3/2))
```

## 5.5 Characterization Criteria for Positive (Negative) Definite Matrices

**Definition 5.19** (see [2], p. 125). Let $V$ be a real vector space.

(a) The quadratic form $f : V \to \mathbb{R}$ is called **positive definite (negative definite)** if $f(\overline{x}) > 0$ ( respectively, $f(\overline{x}) < 0$ ), $(\forall)\, \overline{x} \in V, \overline{x} \neq \overline{0}$;
(b) The quadratic form $f : V \to \mathbb{R}$ is called **positive semidefinite (negative semidefinite)** if $f(\overline{x}) \geq 0$ ( respectively, $f(\overline{x}) \leq 0$), $(\forall)\, \overline{x} \in V$ and $(\exists)\, \overline{a} \in V$, $\overline{a} \neq \overline{0}$, for which $f(\overline{a}) = 0$;
(c) The quadratic form $f : V \to \mathbb{R}$ is called **nondefinite** if $(\exists)\, \overline{a}, \overline{b} \in V$ such that $f(\overline{a}) > 0$ and $f\left(\overline{b}\right) < 0$.

**Definition 5.20** (see [3], p. 104). A symmetric matrix is **positive (negative) definite** if its associated quadratic form is positive (negative) defined.
**Proposition 5.21** (see [2], p. 125). Let $V$ be an $n$ finite dimensional real vector space and $A = \left(a_{ij}\right)_{1 \leq i,j \leq n}$, $A \in M_n(\mathbb{R})$ the associated symmetric matrix of the positive definite quadratic form $f : V \to \mathbb{R}$, relative to the basis $B = \{\overline{e}_1, \overline{e}_2, \ldots, \overline{e}_n\}$ of $V$. Then, the following statements take place:

(a) $a_{ii} > 0$, $(\forall)\, i = \overline{1,n}$
(b) $det\ A > 0$
(c) $rank\ f = n$.

**Theorem 5.22 (Sylvester's criterion, the law of inertia**, see [1], p. 154). Let $V$ be an $n$ finite dimensional real vector space and $f : V \to \mathbb{R}$ a quadratic form. Then the number of positive and respective negative coefficients from a canonical expression of $f$ doesn't depend on the choice of the canonical basis.
**Definition 5.23** (see [3], p. 104).

(i) The number $p$ of the positive coefficients from a canonical expression of the quadratic form $f$ is called the **positive index** of $f$.

(ii) The number $q$ of the negative coefficients from a canonical expression of the quadratic form $f$ is called the **negative index** of $f$.
(iii) The pair $(p, q, d)$ is called the **signature** of the quadratic form, where $d = n - (p + q)$ is the number of the null coefficients.

The following theorem allows us to decide if a quadratic form is positive or negative definite, without being obliged to determine one of its canonical expression.

**Theorem 5.24 (Sylvester's criterion, inertia theorem**, see [3], p. 104). Let $V$ be an $n$ finite dimensional real vector space and $A = \left(a_{ij}\right)_{1\leq i,j\leq n}$, $A \in M_n(\mathbb{R})$ be a symmetric matrix associated of the quadratic form $f : V \to \mathbb{R}$ relative to the basis $B = \{\overline{e}_1, \overline{e}_2, \ldots, \overline{e}_n\}$ of $V$. Then

1. $f$ is positive definite if and only if all the principal minors $\Delta_1, \Delta_2, \ldots, \Delta_n$ of the matrix $A$ are strictly positive,
2. $f$ is negative definite if and only if $(-1)^k \Delta_k > 0$, $(\forall)\ k = \overline{1, n}$.

**Remark 5.25** (see [3], p. 104).

(i) The quadratic form $f$ is positive (negative) definite if and only if $rank\ f = n = p$ (respectively $rank\ f = n = q$).
(ii) The law of inertia states that following any of the three methods to obtain the canonical expression of a quadratic form, the signature of the quadratic form (inferred from obtained canonical the expression) is always the same.
(iii) Given a quadratic form $f : V \to \mathbb{R}$, its associated matrix relative to a basis of the space $V$, $f$ is positive definite if and only if any of the following conditions are satisfied:

- the quadratic form $f$ has the signature $(n, 0, 0)$
- the determinants $\Delta_i > 0$, $(\forall)\ i = \overline{1, n}$
- the eigenvalues of the matrix $A$ are strictly positive.

**Example 5.26** (see [1], p. 156). Let $f : \mathbb{R}^4 \to \mathbb{R}$ be a quadratic form whose analytical expression form relative to the canonical basis of $\mathbb{R}^4$ is

$$f(\overline{x}) = x_1x_2 - x_2x_3 + x_3x_4 + x_4x_1,\ \ (\forall)\ \overline{x} = (x_1, x_2, x_3, x_4) \in \mathbb{R}^4.$$

(a) Write the matrix of $f$ relative to the canonical basis of $\mathbb{R}^4$ and the analytical expression of the polar of $f$, relative to the same basis.
(b) Use the Gauss method to determine a canonical expression for $f$ and a basis of $\mathbb{R}^4$, relative to which $f$ has this canonical expression.
(c) Indicate the signature of $f$.

**Solution**

(a) The matrix of the quadratic form relative to the canonical basis of $\mathbb{R}^4$ is

$$A=\begin{array}{c} \\ x_1 \\ x_2 \\ x_3 \\ x_4 \end{array}\begin{array}{c} \begin{array}{cccc} y_1 & y_2 & y_3 & y_4 \end{array} \\ \begin{pmatrix} 0 & 1/2 & 0 & 1/2 \\ 1/2 & 0 & -1/2 & 0 \\ 0 & -1/2 & 0 & 1/2 \\ 1/2 & 0 & 1/2 & 0 \end{pmatrix} \end{array}.$$

**Remark 5.27** (see [4]). In the writing of the matrix $A$ occurs both $x_1, x_2, x_3, x_4$ and $y_1, y_2, y_3, y_4$ to obtain the analytical expression of the polar of $f$: multiply each element of the matrix $A$ with the index cooresponding to the line denoted by $x_i$ respectively of the column, denoted by $y_j$ to intersection which is this element.

The analytical expression of the polar of $f$ relative to the canonical basis of $\mathbb{R}^4$ will be

$$\begin{aligned} b(\overline{x},\overline{y}) &= \frac{1}{2}x_1y_2+\frac{1}{2}x_1y_4+\frac{1}{2}x_2y_1-\frac{1}{2}x_2y_3-\frac{1}{2}x_3y_2+\frac{1}{2}x_3y_4 \\ &+\frac{1}{2}x_4y_1+\frac{1}{2}x_4y_3, \quad (\forall)\ \overline{x},\overline{y}\in\mathbb{R}^4. \end{aligned} \tag{5.21}$$

(b) As $a_{12}\neq 0$ we make the change of coordinates:

$$\begin{cases} x_1=y_1+y_2 \\ x_2=y_1-y_2 \\ x_3=y_3 \\ x_4=y_4 \end{cases} \Rightarrow \begin{cases} y_1=\frac{1}{2}x_1+\frac{1}{2}x_2 \\ y_2=\frac{1}{2}x_1-\frac{1}{2}x_2 \\ y_3=x_3 \\ y_4=x_4. \end{cases}$$

The transition associated matrix to this change of coordinates will be:

$$M_{(B,B_1)}=\begin{pmatrix} 1 & 1 & 0 & 0 \\ 1 & -1 & 0 & 0 \\ 0 & 0 & 1 & 0 \\ 0 & 0 & 0 & 1 \end{pmatrix},$$

the new basis being

$$B_1=\left\{\overline{f}_1=\overline{e}_1+\overline{e}_2,\ \overline{f}_2=\overline{e}_1-\overline{e}_2,\ \overline{f}_3=\overline{e}_3,\ \overline{f}_4=\overline{e}_4\right\}.$$

The expression of the quadratic form $f$ relative to the basis $B_1$ is

$$f(\overline{x})=(y_1+y_2)(y_1-y_2)-(y_1-y_2)y_3+y_3y_4+y_4(y_1+y_2),$$

i.e.

$$f(\overline{x})=y_1^2-y_2^2-y_1y_3+y_2y_3+y_3y_4+y_1y_4+y_2y_4.$$

The associated matrix of $f$ relative to the basis $B_1$ is

$$A' = \begin{pmatrix} 1 & 0 & -1/2 & 1/2 \\ 0 & -1 & 1/2 & 1/2 \\ -1/2 & 1/2 & 0 & 1/2 \\ 1/2 & 1/2 & 1/2 & 0 \end{pmatrix}.$$

We note that $a'_{11} \neq 0$. We can write $f$ in the form

$$f(\overline{x}) = \left(y_1 - \frac{1}{2}y_3 + \frac{1}{2}y_4\right)^2 - \frac{1}{4}y_3^2 - \frac{1}{4}y_4^2 - y_2^2 + \frac{3}{2}y_3y_4 + y_2y_3 + y_2y_4.$$

By making the change of coordinates:

$$\begin{cases} z_1 = y_1 - \frac{1}{2}y_3 + \frac{1}{2}y_4 \\ z_2 = y_2 \\ z_3 = y_3 \\ z_4 = y_4; \end{cases}$$

it results

$$\begin{cases} y_1 = z_1 + \frac{1}{2}z_3 - \frac{1}{2}z_4 \\ y_2 = z_2 \\ y_3 = z_3 \\ y_4 = z_4. \end{cases}$$

The transition matrix associated with this change of coordinates, through the relation

$$\overline{x}_{B_1} = \mathrm{M}_{(B_1,B_2)}\overline{x}_{B_2}$$

will be:

$$\mathrm{M}_{(B_1,B_2)} = \begin{pmatrix} 1 & 0 & 1/2 & -1/2 \\ 0 & 1 & 0 & 0 \\ 0 & 0 & 1 & 0 \\ 0 & 0 & 0 & 1 \end{pmatrix},$$

the new basis being

$$B_2 = \left\{\overline{g}_1 = \overline{f}_1,\ \overline{g}_2 = \overline{f}_2,\ \overline{g}_3 = \frac{1}{2}\overline{e}_1 + \overline{f}_3,\ \overline{f}_4 = -\frac{1}{2}\overline{f}_1 + \frac{1}{2}\overline{f}_4\right\}.$$

The expression of the quadratic form $f$ relative to the basis $B_2$ is

$$f(\overline{x}) = z_1^2 - z_2^2 - \frac{1}{4}z_3^2 - \frac{1}{4}z_4^2 + \frac{3}{2}z_3z_4 + z_2z_3 + z_2z_4.$$

The associated matrix of $f$ relative to the basis $B_2$ is

$$A'' = \begin{pmatrix} 1 & 0 & 0 & 0 \\ 0 & -1 & 1/2 & 1/2 \\ 0 & 1/2 & -1/4 & 3/4 \\ 0 & 1/2 & 3/4 & -1/4 \end{pmatrix}.$$

We note that $a''_{22} \neq 0$. We can write $f$ in the form

$$f(\overline{x}) = z_1^2 - \left(-z_2 + \frac{1}{2}z_3 + \frac{1}{2}z_4\right)^2 + 2z_3z_4.$$

By making the change of coordinates:

$$\begin{cases} t_1 = z_1 \\ t_2 = -z_2 + \frac{1}{2}z_3 + \frac{1}{2}z_4 \\ t_3 = z_3 \\ t_4 = z_4; \end{cases}$$

it results

$$\begin{cases} z_1 = t_1 \\ z_2 = -t_2 + \frac{1}{2}t_3 + \frac{1}{2}t_4 \\ z_3 = t_3 \\ z_4 = t_4. \end{cases}$$

The transition matrix associated with this change of coordinates, through the relation

$$\overline{x}_{B_2} = \mathrm{M}_{(B_2,B_3)}\overline{x}_{B_3}$$

will be:

$$\mathrm{M}_{(B_2,B_3)} = \begin{pmatrix} 1 & 0 & 0 & 0 \\ 0 & -1 & 1/2 & 1/2 \\ 0 & 0 & 1 & 0 \\ 0 & 0 & 0 & 1 \end{pmatrix},$$

the new basis being

$$B_3 = \left\{\overline{h}_1 = \overline{g}_1, \overline{h}_2 = -\overline{g}_2, \overline{h}_3 = \frac{1}{2}\overline{g}_2 + \overline{g}_3, \overline{g}_4 = \frac{1}{2}\overline{f}_2 + \overline{f}_4\right\}.$$

The expression of the quadratic form $f$ relative to the basis $B_3$ is

$$f(\overline{x}) = t_1^2 - t_2^2 + 2t_3t_4.$$

The associated matrix of $f$ relative to the basis $B_3$ is

$$A''' = \begin{pmatrix} 1 & 0 & 0 & 0 \\ 0 & -1 & 0 & 0 \\ 0 & 0 & 0 & 1 \\ 0 & 0 & 1 & 0 \end{pmatrix}.$$

We note that $a'''_{33} = 0,\ a'''_{34} \neq 0.$
Making the change of coordinates

$$\begin{cases} t_1 = u_1 \\ t_2 = u_2 \\ t_3 = u_3 + u_4 \\ t_4 = u_3 - u_4; \end{cases}$$

it results

$$\begin{cases} u_1 = t_1 \\ u_2 = t_2 \\ u_3 = \frac{1}{2}(t_3 + t_4) \\ u_4 = \frac{1}{2}(t_3 - t_4). \end{cases}$$

The associated transition matrix of this change of coordinates, through the relation

$$\overline{x}_{B_3} = \mathrm{M}_{(B_3, B_4)} \overline{x}_{B_4}$$

will be

$$\mathrm{M}_{(B_3, B_4)} = \begin{pmatrix} 1 & 0 & 0 & 0 \\ 0 & 1 & 0 & 0 \\ 0 & 0 & 1 & 1 \\ 0 & 0 & 1 & -1 \end{pmatrix},$$

the new basis being

$$B_4 = \left\{ \overline{v}_1 = \overline{h}_1,\ \overline{v}_2 = \overline{h}_2,\ \overline{v}_3 = \overline{h}_3 + \overline{h}_4,\ \overline{v}_4 = \overline{h}_3 - \overline{h}_4 \right\}.$$

The expression of the quadratic form $f$ relative to the basis $B_4$ is

$$f(\overline{x}) = u_1^2 - u_2^2 + 2u_3^2 - 2u_4^2.$$

The associated matrix of $f$ relative to the basis $B_4$ is

$$B = \begin{pmatrix} 1 & 0 & 0 & 0 \\ 0 & -1 & 0 & 0 \\ 0 & 0 & 2 & 0 \\ 0 & 0 & 0 & -2 \end{pmatrix}.$$

(c) We have

$$rank\ f = rank\ A = 4,\ p = 2, q = 2.$$

We obtain that the quadratic form $f$ has the signature $(2, 2, 0)$.
We can also solve this problem in Sage, too:

```
sage: var("x1 x2 x3 x4 y1 y2 y3 y4 z1 z2 z3 z4 t1 t2 t3 t4 u1 u2 u3 u4")
(x1, x2, x3, x4, y1, y2, y3, y4, z1, z2, z3, z4, t1, t2, t3, t4, u1, u2, u3, u4)
sage: A=matrix([[0,1/2,0,1/2],[1/2,0,-1/2,0],[0,-1/2,0,1/2],[1/2,0,1/2,0]])
sage: x=vector([x1,x2,x3,x4]);y=vector([y1,y2,y3,y4])
sage: b=A[0][0]*x[0]*y[0]+A[0][1]*x[0]*y[1]+A[0][2]*x[0]*y[2]+A[0][3]*x[0]*y[3]
sage: b+=A[1][0]*x[1]*y[0]+A[1][1]*x[1]*y[1]+A[1][2]*x[1]*y[2]+A[1][3]*x[1]*y[3]
sage: b+=A[2][0]*x[2]*y[0]+A[2][1]*x[2]*y[1]+A[2][2]*x[2]*y[2]+A[2][3]*x[2]*y[3]
sage: b+=A[3][0]*x[3]*y[0]+A[3][1]*x[3]*y[1]+A[3][2]*x[3]*y[2]+A[3][3]*x[3]*y[3]
sage: b
1/2*x1*y2 + 1/2*x1*y4 + 1/2*x2*y1 - 1/2*x2*y3 - 1/2*x3*y2 + 1/2*x3*y4 + 1/2*x4*y1 + 1/2*x4*y3
sage: f(x1,x2,x3,x4)=b.subs(y1=x1,y2=x2,y3=x3,y4=x4);f
(x1, x2, x3, x4) |--> x1*x2 + x1*x4 - x2*x3 + x3*x4
sage: ff(y1,y2,y3,y4)=expand(f(y1+y2,y1-y2,y3,y4));ff
(y1, y2, y3, y4) |--> y1^2 - y1*y3 + y1*y4 - y2^2 + y2*y3 + y2*y4 + y3*y4
sage: ff1(y1,y2,y3,y4)=(y1-1/2*y3+1/2*y4)^2+expand(ff-(y1-1/2*y3+1/2*y4)^2)
sage: ff2(z1,z2,z3,z4)=ff1(z1+1/2*z3-1/2*z4,z2,z3,z4);ff2
(z1, z2, z3, z4) |--> z1^2 - z2^2 + z2*z3 + z2*z4 - 1/4*z3^2 + 3/2*z3*z4 - 1/4*z4^2
sage: ff3(z1,z2,z3,z4)=(-z2+1/2*z3+1/2*z4)^2+expand(ff2-(-z2+1/2*z3+1/2*z4)^2)
sage: ff4(t1,t2,t3,t4)=ff3(t1,-t2+1/2*t3+1/2*t4,t3,t4);expand(ff4)
(t1, t2, t3, t4) |--> t1^2 - t2^2 + 2*t3*t4
sage: ff5(u1,u2,u3,u4)=expand(ff4(u1,u2,u3+u4,u3-u4));ff5
(u1, u2, u3, u4) |--> u1^2 - u2^2 + 2*u3^2 - 2*u4^2
sage: n=A.rank();p=2;q=2;d=n-(p+q);sig=(p,q,d);sig
(2, 2, 0)
```

## 5.6 Problems

1. Let $f : \mathbb{R}^3 \to \mathbb{R}$ be a quadratic form whose analytical expression form relative to the canonical basis of $\mathbb{R}^3$ is

$$f(\overline{x}) = x_1x_2 + x_2x_3.$$

   (a) Write the matrix of $f$ relative to the canonical basis of $\mathbb{R}^3$ and the analytical expression corresponding to the polar of $f$, relative to the same basis.

(b) Use the Gauss method to determine a canonical expression for $f$ and a basis of $\mathbb{R}^3$, relative to which $f$ has this canonical expression.

(c) Indicate the signature of $f$.

**Solution**

Solving this problem in Sage, we obtain:

```
sage: A=matrix([[0,1/2,0],[1/2,0,1/2],[0,1/2,0]])
sage: var("x1 x2 x3 y1 y2 y3 z1 z2 z3 t1 t2 t3 u1 u2 u3")
(x1, x2, x3, y1, y2, y3, z1, z2, z3, t1, t2, t3, u1, u2, u3)
sage: x=vector([x1,x2,x3]);y=vector([y1,y2,y3])
sage: b=A[0][0]*x[0]*y[0]+A[0][1]*x[0]*y[1]+A[0][2]*x[0]*y[2]
sage: b+=A[1][0]*x[1]*y[0]+A[1][1]*x[1]*y[1]+A[1][2]*x[1]*y[2]
sage: b+=A[2][0]*x[2]*y[0]+A[2][1]*x[2]*y[1]+A[2][2]*x[2]*y[2]
sage: b
1/2*x1*y2 + 3/2*x2*y1 + 3/2*x2*y3 + 1/2*x3*y2
sage: Qq(x1,x2,x3)=x1*x2+x2*x3
sage: Qa(y1,y2,y3)=expand(Qq(x1=y1+y2,x2=y1-y2,x3=y3));Qa
(y1, y2, y3) |--> y1^2 + y1*y3 - y2^2 - y2*y3
sage: Q(y1,y2,y3)=(y1+y3/2)^2+expand(Qa-(y1+y3/2)^2);Q
(y1, y2, y3) |--> 1/4*(2*y1 + y3)^2 - y2^2 - y2*y3 - 1/4*y3^2
sage: Qq1(z1,z2,z3)=Q(z1-1/2*z3,z2,z3);expand(Qq1)
(z1, z2, z3) |--> z1^2 - z2^2 - z2*z3 - 1/4*z3^2
sage: Qq2(z1,z2,z3)=(z2+z3/2)^2+expand(Qq1-(z2+z3/2)^2);Qq2
(z1, z2, z3) |--> 1/4*(2*z2 + z3)^2 + z1^2 - 2*z2^2 - 2*z2*z3 - 1/2*z3^2
sage: Qq3(t1,t2,t3)=Qq2(t1,t2-t3/2,t3);expand(Qq3)
(t1, t2, t3) |--> t1^2 - t2^2
sage: n=A.rank();n
2
sage: p=1;q=1;d=n-(p+q);s=(p,q,d);s
(1, 1, 0)
```

2. Determine an orthonormal basis of vector space $\mathbb{R}^3$ relative to which the quadratic form

$$f(\overline{x}) = -x_1^2 + x_2^2 - 5x_3^2 + 6x_1x_3 + 4x_2x_3$$

has a canonical expression.

3. Let $f : \mathbb{R}^4 \to \mathbb{R}$ be a quadratic form whose analytical expression form relative to the canonical basis of $\mathbb{R}^4$ is

$$f(\overline{x}) = x_1^2 + 5x_2^2 + 4x_3^2 - x_4^2 + 6x_1x_2 - 4x_1x_3 - 12x_2x_3 - 4x_2x_4 - 8x_3x_4.$$

Use the Gauss method to determine a canonical expression for $f$ and a basis of $\mathbb{R}^4$, relative to which $f$ has this canonical expression.

4. Let be the bilinear functional

$$b(\overline{x},\overline{y}) = x_1y_2 - x_2y_1 + x_1y_3 - x_3y_1 + x_1y_4 - x_4y_1 + x_2y_3 - x_3y_2 + x_2y_4 - x_4y_2 + x_3y_4 - x_4y_3, \ (\forall)\ \overline{x},\overline{y} \in \mathbb{R}^4 \tag{5.22}$$

(a) Prove that $b$ is a antisymmetic bilinear functional.
(b) Find the matrix corresponding to the bilinear functional $b : \mathbb{R}^4 \times \mathbb{R}^4 \to \mathbb{R}$ relative to the basis

$$B' = \left\{\overline{f}_1 = (1,1,1,0)\,,\ \overline{f}_2 = (0,1,1,1)\,,\ \overline{f}_3 = (1,1,0,1)\,,\ \overline{f}_4 = (1,0,1,1)\right\}.$$

**Solution**
Using Sage, we shall have:

```
sage: var("x1 x2 x3 x4 y1 y2 y3 y4")
(x1, x2, x3, x4, y1, y2, y3, y4)
sage: b(x1,x2,x3,x4,y1,y2,y3,y4)=x1*y2-x2*y1+x1*y3-x3*y1+x1*y4-x4*y1+x2*y3-x3*y2+x2*y4-
x4*y2+x3*y4-x4*y3
sage: expand(b(x1,x2,x3,x4,y1,y2,y3,y4)+b(y1,y2,y3,y4,x1,x2,x3,x4))
0
sage: c00=b(1,1,1,0,1,1,1,0);c01=b(1,1,1,0,0,1,1,1);c02=b(1,1,1,0,1,1,0,1);c03=b(1,1,1,0,1,0,1,1)
sage: c10=b(0,1,1,1,1,1,1,0);c11=b(0,1,1,1,0,1,1,1);c12=b(0,1,1,1,1,1,0,1);c13=b(0,1,1,1,1,0,1,1)
sage: c20=b(1,1,0,1,1,1,1,0);c21=b(1,1,0,1,0,1,1,1);c22=b(1,1,0,1,1,1,0,1);c23=b(1,1,0,1,1,0,1,1)
sage: c30=b(1,0,1,1,1,1,1,0);c31=b(1,0,1,1,0,1,1,1);c32=b(1,0,1,1,1,1,0,1);c33=b(1,0,1,1,1,0,1,1)
sage: C=matrix([[c00,c01,c02,c03],[c10,c11,c12,c13],[c20,c21,c22,c23],[c30,c31,c32,c33]])
sage: C
[ 0  5  1  3]
[-5  0 -3 -1]
[-1  3  0  1]
[-3  1 -1  0]
```

5. Let $B = \{\overline{e}_1, \overline{e}_2, \overline{e}_3\}$ be the canonical basis of the arithmetic vector space $\mathbb{R}^3$ and let $b : \mathbb{R}^3 \times \mathbb{R}^3 \to \mathbb{R}$ be the bilinear form for which:

$$\begin{cases} b(\overline{e}_1,\overline{e}_1) = -1,\ b(\overline{e}_2,\overline{e}_2) = 3,\ b(\overline{e}_3,\overline{e}_3) = -6 \\ b(\overline{e}_1 - \overline{e}_2,\overline{e}_2) = 2,\ b(\overline{e}_2,\overline{e}_1 + 2\overline{e}_2) = 5,\ b(\overline{e}_3 - \overline{e}_1,\overline{e}_1) = 4 \\ b(2\overline{e}_1 + \overline{e}_2,\overline{e}_3) = -7,\ b(\overline{e}_1 + \overline{e}_3,\overline{e}_2) = 4,\ b(\overline{e}_1 - 2\overline{e}_2,\overline{e}_3) = -1. \end{cases}$$

(a) Write the matrix corresponding to the bilinear functional $b$, relative to the basis $B$.
(b) Is $b$ a symmetic bilinear functional?

**Solution**
With Sage, we achieve:

```
sage: b00=-1;b11=3;b22=-6;var("x y x1 x2 x3 y1 y2 y3")
(x, y, x1, x2, x3, y1, y2, y3)
sage: u=solve([x-b11==2],x);b01=u[0].right()
sage: u1=solve([x+2*b11==5],x);b10=u1[0].right()
sage: u2=solve([x-b00==4],x);b20=u2[0].right()
sage: u3=solve([x+b01==4],x);b21=u3[0].right()
sage: u4=solve([2*x+y==-7,x-2*y==-1],x,y);b02=u4[0][0].right();b12=u4[0][1].right()
sage: B=matrix([[b00,b01,b02],[b10,b11,b12],[b20,b21,b22]]);B
[-1 5 -3]
[-1 3 -1]
[ 3 -1 -6]
sage: x=vector([x1,x2,x3]);y=vector([y1,y2,y3])
sage: b=b00*x[0]*y[0]+b01*x[0]*y[1]+b02*x[0]*y[2]
sage: b+=b10*x[1]*y[0]+b11*x[1]*y[1]+b12*x[1]*y[2]
sage: b+=b20*x[2]*y[0]+b21*x[2]*y[1]+b22*x[2]*y[2]
sage: b
-x1*y1 + 5*x1*y2 - 3*x1*y3 - x2*y1 + 3*x2*y2 - x2*y3 + 3*x3*y1 - x3*y2 - 6*x3*y3
sage: bb=b.subs(x1=y1,x2=y2,x3=y3,y1=x1,y2=x2,y3=x3);bb
-x1*y1 + 3*x1*y2 + 3*x1*y3 + 3*x2*y1 + x2*y2 + x2*y3 - 13/5*x3*y1 - 9/5*x3*y2 + 2*x3*y3
sage: expand(b-bb)
0
```

6. (a) Write the analytical expression of the quadratic form $f : \mathbb{R}^3 \to \mathbb{R}$, defined by $f(\overline{x}) = b(\overline{x}, \overline{x})$, $(\forall)\ \overline{x} \in \mathbb{R}^3$, $b$ being the bilinear functional from the previous problem.
   (b) Use the Jacobi method to determine a canonical expression for the quadratic form $f$ from the previous problem and that basis of $\mathbb{R}^3$, relative towhich $f$ has this canonical expression.
   (c) Prove that $f$ is negative definite.

**Solution**

We shall solve in Sage this problem:

```
sage: f=b.subs(y1=x1,y2=x2,y3=x3);f
-x1^2 + 4*x1*x2 + 3*x2^2 - 2*x2*x3 - 6*x3^2
sage: B1=B[0:2,0:2]
sage: d1=1;d2=B[0][0];d3=B1.determinant();d4=B.determinant()
sage: ff=d1/d2*y1^2+d2/d3*y2^2+d3/d4*y3^2;ff
-y1^2 - 1/2*y2^2 - y3^2
```

```
sage: I=identity_matrix(RR,3)
sage: s=solve([x1*B[0][0]==1],x1);c11=s[0].right()
sage: v=vector(SR,[x1,x2]);
sage: v1=B1*v.transpose()
sage: ss=solve([v1[0][0],v1[1][0]==1],x1,x2)
sage: c21=ss[0][0].right();c22=ss[0][1].right();
sage: vv1=B*x.transpose()
sage: ss1=solve([vv1[0][0],vv1[1][0],vv1[2][0]==1],x1,x2,x3)
sage: c31=ss1[0][0].right();c32=ss1[0][1].right();c33=ss1[0][2].right()
sage: ep1=c11*I[0];ep1.n(digits=2)
(-1.0, 0.00, 0.00)
sage: ep2=c21*I[0]+c22*I[1];ep2.n(digits=2)
(-2.5, -0.50, 0.00)
sage: ep3=c31*I[0]+c32*I[1]+c33*I[2];ep3.n(digits=2)
(-2.0, -1.0, -1.0)
```

7. Let $b : M_2(\mathbb{R}) \times M_2(\mathbb{R}) \to \mathbb{R}$ be a bilinear functional, defined by

$$b(A, B) = 2 \cdot \text{trace}(AB) - \text{trace}(A)\,\text{trace}(B), \quad (\forall)\ A, B \in M_2(\mathbb{R}).$$

(a) Prove that $b$ is symmetrically.
(b) Write the analytical expression of $b$, relative to the canonical basis of $M_2(\mathbb{R})$.
(c) Build the matrix associated to $b$, relative to the canonical basis of $M_2(\mathbb{R})$.
(d) Indicate the signature of the quadratic form $f : M_2(\mathbb{R}) \to \mathbb{R}$, defined by $f(A) = b(A, B)$, $(\forall)\ A \in M_2(\mathbb{R})$.

**Solution**

The solution in Sage of this problem is:

```
sage: var("a1 a2 a3 a4 b1 b2 b3 b4 c1 c2 c3 c4 la")
(a1, a2, a3, a4, b1, b2, b3, b4, c1, c2, c3, c4, la)
sage: A=matrix([[a1,a2],[a3,a4]]);B=matrix([[b1,b2],[b3,b4]])
sage: b(a1,a2,a3,a4,b1,b2,b3,b4)=2*(A*B).trace()-A.trace()*B.trace()
sage: expand(b(a1,a2,a3,a4,b1,b2,b3,b4)-b(b1,b2,b3,b4,a1,a2,a3,a4))
0
sage: expand(b(a1,a2,a3,a4,b1,b2,b3,b4))
a1*b1 - a1*b4 + 2*a2*b3 + 2*a3*b2 - a4*b1 + a4*b4
sage: u00=b(a1,a2,a3,a4,b1,b2,b3,b4).coeff(a1).coeff(b1)
sage: u01=b(a1,a2,a3,a4,b1,b2,b3,b4).coeff(a1).coeff(b2)
sage: u02=b(a1,a2,a3,a4,b1,b2,b3,b4).coeff(a1).coeff(b3)
sage: u03=b(a1,a2,a3,a4,b1,b2,b3,b4).coeff(a1).coeff(b4)
sage: u10=b(a1,a2,a3,a4,b1,b2,b3,b4).coeff(a2).coeff(b1)
sage: u11=b(a1,a2,a3,a4,b1,b2,b3,b4).coeff(a2).coeff(b2)
sage: u12=b(a1,a2,a3,a4,b1,b2,b3,b4).coeff(a2).coeff(b3)
sage: u13=b(a1,a2,a3,a4,b1,b2,b3,b4).coeff(a2).coeff(b4)
sage: u20=b(a1,a2,a3,a4,b1,b2,b3,b4).coeff(a3).coeff(b1)
sage: u21=b(a1,a2,a3,a4,b1,b2,b3,b4).coeff(a3).coeff(b2)
```

```
sage: u22=b(a1,a2,a3,a4,b1,b2,b3,b4).coeff(a3).coeff(b3)
sage: u23=b(a1,a2,a3,a4,b1,b2,b3,b4).coeff(a3).coeff(b4)
sage: u30=b(a1,a2,a3,a4,b1,b2,b3,b4).coeff(a4).coeff(b1)
sage: u31=b(a1,a2,a3,a4,b1,b2,b3,b4).coeff(a4).coeff(b2)
sage: u32=b(a1,a2,a3,a4,b1,b2,b3,b4).coeff(a4).coeff(b3)
sage: u33=b(a1,a2,a3,a4,b1,b2,b3,b4).coeff(a4).coeff(b4)
sage: U=matrix([[u00,u01,u02,u03],[u10,u11,u12,u13],[u20,u21,u22,u23],[u30,u31,u32,u33]]);U
[1 0 0-1]
[0 0 2 0]
[0 2 0 0]
[-1 0 0 1]
sage: f=b(a1,a2,a3,a4,a1,a2,a3,a4);expand(f)
a1^2 - 2*a1*a4 + 4*a2*a3 + a4^2
sage: A=matrix([[u00,u01/2,u02/2,u03],[u10,u11/2,u12/2,u13],[u20,u21/2,u22/2,u23],[u30,u31/2,u32/2,u33]]);A
[1 0 0-1]
[0 0 1 0]
[0 1 0 0]
[-1 0 0 1]


sage: I=identity matrix(RR,4);x=vector(SR,[c1^2,c2^2,c3^2,c4^2])
sage: s=solve([(A-la*I).determinant()],la,multiplicities=True)
sage: s1=s[0][0].right();s2=s[0][1].right();s3=s[0][2].right();s4=s[0][3].right()
sage: K=vector([s1,s2,s3,s4])
sage: ff=x*K.transpose();ff[0]
2*c1^2 - c2^2 + c3^2
sage: n=4;p=2;q=1;d=n-(p+q);sig=(p,q,d);sig
(2, 1, 1)
```

8. Use the eigenvalue method to determine the canonical expression and the basis in which makes this for the quadratic form:

$$f:\mathbb{R}^3\to\mathbb{R},\ f(\overline{x})=5x_1^2+6x_2^2+4x_3^2-4x_1x_2-4x_1x_3,\ (\forall)\ \overline{x}\in\mathbb{R}^3.$$

   Indicate the signature of $f$.

9. Use the Jacobi method to find the canonical expression and the basis in which makes this for the quadratic form:

$$f:\mathbb{R}^3\to\mathbb{R},\ f(\overline{x})=4x_1^2+2x_2^2-2x_1x_2+2x_1x_3+2x_2x_3,\ (\forall)\ \overline{x}\in\mathbb{R}^3.$$

10. Reduce to the canonical expression through the three method, the following quadratic form:

$$f:\mathbb{R}^4\to\mathbb{R},\ f(\overline{x})=\sum_{\substack{i<j\\ i=1}}^{4}x_ix_j.$$

# References

1. I. Vladimirescu, M. Popescu (eds.), *Algebră liniară şi geometrie analitică* (Universitaria, Craiova, 1993)
2. I. Vladimirescu, M. Popescu (eds.), *Algebră liniară şi geometrie n- dimensională* (Radical, Craiova, 1996)
3. V. Balan (ed.), *Algebră liniară, geometrie analitică* (Fair Partners, Bucureşti, 1999)
4. I. Vladimirescu, M. Popescu, M. Sterpu, *Algebră liniară şi geometrie analitică*. Note de curs şi aplicaţii (Universitatea din Craiova, 1993)
5. V. Postelnicu, S. Coatu (ed.), *Mică enciclopedie matematică* (Tehnică, Bucureşti, 1980)

# Chapter 6
# Differential Geometry of Curves and Surfaces

## 6.1 Analytical Definition of Curves in the Space

**Definition 6.1** (see [5], p. 544). The **regular arc of a curve** is defined as the set $\Gamma$ of the points $M(x, y, z)$ from the real three-dimensional Euclidean space $\mathbb{R}^3$, whose coordinates $x, y, z$ check one of the following systems of equations:

$$\begin{cases} F(x, y, z) = 0 \\ G(x, y, z) = 0 \end{cases}, \quad (\forall)\ (x, y, z) \in D \subseteq \mathbb{R}^3 \ (\textit{the implicit representation}) \tag{6.1}$$

$$\begin{cases} z = f(x, y) \\ z = g(x, y) \end{cases}, \quad (\forall)\ (x, y) \in D \subseteq \mathbb{R}^2 \ (\textit{the explicit representation}) \tag{6.2}$$

$$\begin{cases} x = f_1(t) \\ y = f_2(t), \\ z = f_3(t) \end{cases} \quad (\forall)\ t \in (a, b) \ (\textit{the parametric representation}), \tag{6.3}$$

where the functions $F, G, f, g, f_1, f_2, f_3$ satisfy the following conditions of regularity:

(a) are some real and continuous functions,
(b) the functions $f_1, f_2, f_3$ establish a biunivocal correspondence between the values of the parameter $t \in (a, b)$ and the points $M \in \Gamma$,
(c) allow first order derivatives, not all null,
(d) at least one of the functional determinants

$$\frac{\mathrm{D}(F, G)}{\mathrm{D}(y, z)}, \frac{\mathrm{D}(F, G)}{\mathrm{D}(z, x)}, \frac{\mathrm{D}(F, G)}{\mathrm{D}(x, y)}$$

is not equal to 0.

G. A. Anastassiou and I. F. Iatan, *Intelligent Routines II*,
Intelligent Systems Reference Library 58, DOI: 10.1007/978-3-319-01967-3_6,

**Definition 6.2** (see [5], p. 545). The **regular curve** is the reunion of the regular curve arcs.

**Definition 6.3** (see [5], p. 545). Let $\Gamma$ be a curve given by its parametric equations

$$\begin{cases} x = x(t) \\ y = y(t), (\forall)\, t \in (a, b) \\ z = z(t) \end{cases} \tag{6.4}$$

and let $M(x, y, z) \in \Gamma$ be a current point.

If we consider a system of rectangular axes of versors $\overline{i}, \overline{j}, \overline{k}$ and if $\overline{r}$ is the position vector of the point $M$, then the relation

$$\overline{r} = x(t)\,\overline{i} + y(t)\,\overline{j} + z(t)\,\overline{k}, (\forall)\, t \in (a, b) \tag{6.5}$$

is called the **vector equation** of the curve $\Gamma$.

## 6.2 Tangent and Normal Plane to a Curve in the Space

**Definition 6.4** (see [5], p. 552). One considers the regular curve $\Gamma$ and let be the points $M, M_1 \in \Gamma$. The limit position of the chord $M, M_1$ when $M_1$ tends to $M$ is called the **tangent to the curve** at the point $M$ (see Fig. 6.1).

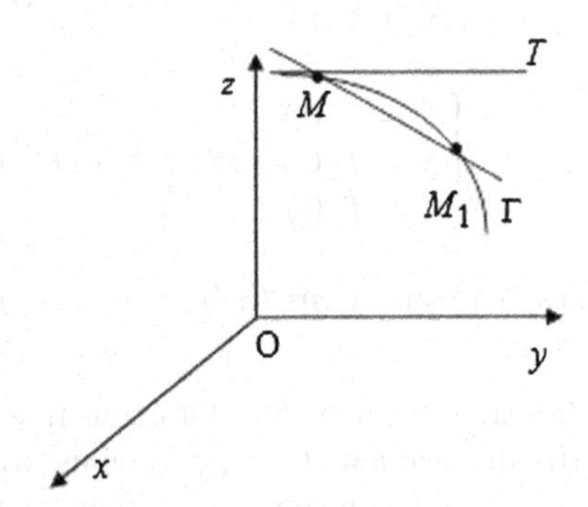


**Fig. 6.1** Tangent to a curve

The equations of the tangent to the curve at a point $M_0(x_0, y_0, z_0) \in \Gamma$ in the case when:

(a) the curve is given parametrically (as in the relation (6.4)) are :

$$T : \frac{x - x_0}{x'(t_0)} = \frac{y - y_0}{y'(t_0)} = \frac{z - z_0}{z'(t_0)}, \tag{6.6}$$

where

$$\begin{cases} x = x(t_0) \\ y = y(t_0), \quad (\forall)\ t_0 \in (a, b); \\ z = z(t_0) \end{cases}$$

(b) the curve is given implicitly (see the relation (6.1)) are

$$T : \frac{x - x_0}{\frac{D(F,G)}{D(y_0,z_0)}} = \frac{y - y_0}{\frac{D(F,G)}{D(z_0,x_0)}} = \frac{z - x_0}{\frac{D(F,G)}{D(x_0,y_0)}}; \tag{6.7}$$

(c) the curve is given explicitly (see the relation (6.2)) are

$$T : \frac{x - x_0}{g'_{y_0} - f'_{y_0}} = \frac{y - y_0}{f'_{x_0} - g'_{x_0}} = \frac{z - x_0}{f'_{x_0} g'_{y_0} - f'_{y_0} g'_{x_0}}. \tag{6.8}$$

**Definition 6.5** (see [5], p. 553). One considers the regular curve $\Gamma$ and let be $M \in \Gamma$. The **normal plane** to the curve $\Gamma$ in the point $M$ is the plane $\pi_N$ perpendicular to the tangent $T$ to the curve $\Gamma$ in the point $M$.

The equation of the normal plane $\pi_N$ to the curve $\Gamma$ in the point $M$, in the case when:

(a) the curve is given parametrically (as in the relation (6.4)) is:

$$\pi_N : x'(t_0)(x - x(t_0)) + y'(t_0)(y - y(t_0)) + z'(t_0)(z - z(t_0)) = 0; \tag{6.9}$$

(b) the curve is given implicitly (see the relation (6.1)) is

$$\pi_N : \frac{D(F, G)}{D(y_0, z_0)}(x - x(t_0)) + \frac{D(F, G)}{D(z_0, x_0)}(y - y(t_0)) + \frac{D(F, G)}{D(x_0, y_0)}(z - z(t_0)) = 0; \tag{6.10}$$

(c) the curve is given explicitly (see the relation (6.2)) is

$$\begin{aligned} \pi_N : \left(g'_{y_0} - f'_{y_0}\right)(x - x(t_0)) + \left(f'_{x_0} - g'_{x_0}\right)(y - y(t_0)) \\ + \left(f'_{x_0} g'_{y_0} - f'_{y_0} g'_{x_0}\right)(z - z(t_0)) = 0. \end{aligned} \tag{6.11}$$

## 6.3 Frenet Trihedron. Frenet Formulas

**Definition 6.6** (see [5], p. 576). Let $\Gamma$ be a regular curve and the point $M_0 \in \Gamma$. The **Frenet Trihedron** attached to the curve $\Gamma$ in the point $M_0$ is a right trihedron determined by the versors $\overline{\tau}, \overline{\beta}, \overline{\nu}$ (Fig. 6.2).

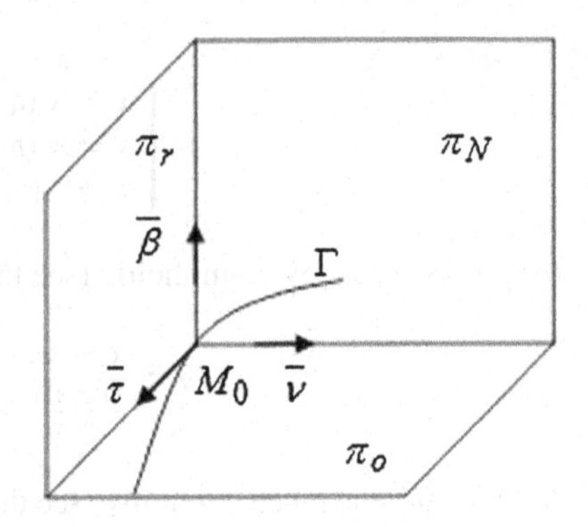


**Fig. 6.2** Frenet trihedral

The pair of versors $\overline{\tau}, \overline{\beta}, \overline{\nu}$ determine the following planes:

- *the normal plane* $\pi_N$, determined by $\overline{\nu}$ and $\overline{\beta}$;
- *the osculator plane* $\pi_0$, determined by $\overline{\tau}$ and $\overline{\nu}$;
- *the rectified plane* $\pi_r$, determined by $\overline{\tau}$ and $\overline{\beta}$.

The equation of the normal plane to the curve $\Gamma$ in a point $M_0(x_0, y_0, z_0) \in \Gamma$ can be expressed with the relations (6.9), (6.10) or (6.11).

The equation of the osculator plane to the curve $\Gamma$ in a point $M_0(x_0, y_0, z_0) \in \Gamma$ in the case when:

(a) the curve is given parametrically (as in the relation (6.4)) is

$$\pi_0 : \begin{vmatrix} x - x(t_0) & y - y(t_0) & z - z(t_0) \\ x'(t_0) & y'(t_0) & z'(t_0) \\ x''(t_0) & y''(t_0) & z''(t_0) \end{vmatrix} = 0; \tag{6.12}$$

(b) the curve is given implicitly (see the relation (6.1)) is

$$\pi_0 : \begin{vmatrix} x - x_0 & y - y_0 & z - z_0 \\ \frac{D(F,G)}{D(y_0,z_0)} & \frac{D(F,G)}{D(z_0,x_0)} & \frac{D(F,G)}{D(x_0,y_0)} \\ \frac{D^2(F,G)}{D^2(y_0,z_0)} & \frac{D^2(F,G)}{D^2(z_0,x_0)} & \frac{D^2(F,G)}{D^2(x_0,y_0)} \end{vmatrix} = 0; \tag{6.13}$$

(c) the curve is given explicitly (see the relation (6.2)) is

$$\pi_0 : \begin{vmatrix} x - x_0 & y - y_0 & z - z_0 \\ g'_{y_0} - f'_{y_0} & f'_{x_0} - g'_{x_0} & f'_{x_0} g'_{y_0} - f'_{y_0} g'_{x_0} \\ g''_{y_0} - f''_{y_0} & f''_{x_0} - g''_{x_0} & f''_{x_0} g''_{y_0} - f''_{y_0} g''_{x_0} \end{vmatrix} = 0. \tag{6.14}$$

If

$$\overline{v} = a_1\overline{i} + a_2\overline{j} + a_3\overline{k}, \tag{6.15}$$

then the equation of the rectified plane to the curve $\Gamma$ in the point $M_0(x_0, y_0, z_0) \in \Gamma$ is:

$$\pi_r : a_1 (x - x(t_0)) + a_2 (y - y(t_0)) + a_3 (z - z(t_0)) = 0. \tag{6.16}$$

The Frenet's trihedron axes in the point $M_0(x_0, y_0, z_0) \in \Gamma$ are:

1. *the tangent* in $M_0$ to the curve (the tangent versor is denoted by $\overline{\tau}$); the equations of the tangent to the curve $\Gamma$ in a point $M_0(x_0, y_0, z_0) \in \Gamma$ can be expressed with the relations (6.6), (6.7) or (6.8);
2. *the binormal* of a curve in $M_0$ (the normal which is perpendicular to the osculator plane, that passes through the point $M_0$; the binormal versor is denoted by $\overline{\beta}$) has the equations:

$$\left(\overline{\beta}\right) : \frac{x - x_0}{\begin{vmatrix} y'(t_0) & z'(t_0) \\ y''(t_0) & z''(t_0) \end{vmatrix}} = \frac{y - y_0}{\begin{vmatrix} z'(t_0) & x'(t_0) \\ z''(t_0) & x''(t_0) \end{vmatrix}} = \frac{z - z_0}{\begin{vmatrix} x'(t_0) & y'(t_0) \\ x''(t_0) & y''(t_0) \end{vmatrix}}.$$

3. *the principal normal* of a curve in $M_0$ (the straight line contained in the normal plane and in the osculator plane passing through $M_0$; the principal normal versor is denoted by $\overline{\nu}$) has the equations:

$$(\overline{\nu}) : \frac{x - x(t_0)}{a_1} = \frac{y - y(t_0)}{a_2} = \frac{z - z(t_0)}{a_3}, \tag{6.17}$$

$\overline{\nu}$ being expressed in (6.15).

Since the rectified plane is that plane which passes through $M_0(x_0, y_0, z_0)$ and contains the directions of the tangent and of the binormal it results that the equation of the rectified plane is:

$$\pi_r : \begin{vmatrix} x - x_0 & y - y_0 & z - z_0 \\ t_1 & t_2 & t_3 \\ b_1 & b_2 & b_3 \end{vmatrix} = 0, \tag{6.18}$$

- $\overline{\tau} = t_1\overline{i} + t_2\overline{j} + t_3\overline{k}$ is the director vector of the tangent in the point $M_0(x_0, y_0, z_0)$;
- $\overline{\beta} = b_1\overline{i} + b_2\overline{j} + b_3\overline{k}$ is the director vector of the binormale in the point $M_0(x_0, y_0, z_0)$.

**Definition 6.7** (see [5], p. 550). **The arc element of a curve** $\Gamma$ is the differential $ds$ of the function $s = s(t)$, which signifies the length of the respective arc, from the curve $\Gamma$.

If the curve $\Gamma$ is given by the parametric equations (6.4), then

$$ds = \sqrt{dx^2 + dy^2 + dz^2}. \tag{6.19}$$

If we dispose by the vector equation from the relation (6.5) of the curve $\Gamma$, then

$$ds = \|d\overline{r}\|. \tag{6.20}$$

The versors of the Frenet trihedron are:

1. *the tangent versor*:

$$\overline{\tau} = \frac{d\overline{r}}{ds} = \frac{\overline{r}'}{\|\overline{r}'\|} = \frac{x'(t)\,\overline{i} + y'(t)\,\overline{j} + z'(t)\,\overline{k}}{\sqrt{x'(t)^2 + y'(t)^2 + z'(t)^2}} \tag{6.21}$$

2. *the binormal versor*:

$$\overline{\beta} = \frac{\overline{r}' \times \overline{r}''}{\|\overline{r}' \times \overline{r}''\|} \tag{6.22}$$

3. *the principal normal versor*:

$$\overline{\nu} = \overline{\beta} \times \overline{\tau}, \tag{6.23}$$

where $\overline{r}$ is expressed in the relation (6.5).

**Theorem 6.8** (see [5], p. 576). Let $\Gamma$ be a regular curve and $\left(\overline{\tau}, \overline{\beta}, \overline{\nu}\right)$ be the Frenet trihedron attached to the curve $\Gamma$ in the point $M \in \Gamma$. There are the following relations among the versors $\overline{\tau}, \overline{\beta}, \overline{\nu}$:

$$\begin{cases} \overline{\tau} = \overline{\nu} \times \overline{\beta} \\ \overline{\nu} = \overline{\beta} \times \overline{\tau} \\ \overline{\beta} = \overline{\tau} \times \overline{\nu} \\ \|\overline{\tau}\| = \left\|\overline{\beta}\right\| = \|\overline{\nu}\| = 1 \\ \overline{\tau} \cdot \overline{\nu} = \overline{\tau} \cdot \overline{\beta} = \overline{\nu} \cdot \overline{\beta} = 0. \end{cases} \tag{6.24}$$

**Example 6.9** (see [1], p. 137). Let be the curve:

$$\overline{r}(t) = 2t\overline{i} + t^2\overline{j} + \ln t\overline{k}, t > 0.$$

Find the equations of the edges (the tangent, the principal normal and the binormal) and of the planes (the normal plane, the rectified plane and the osculator plane) corresponding to the Frenet trihedron in the point $t = 1$.

**Solution**

The tangent equations are:

$$(\mathcal{T}) : \frac{x - 2t}{2} = \frac{y - t^2}{2t} = \frac{z - \ln t}{\frac{1}{t}}.$$

The tangent equations in the point $t = 1$:

$$(\mathcal{T}) : \frac{x - 2}{2} = \frac{y - 1}{2} = \frac{z}{1}.$$

The binormale equations in the point $t = 1$:

$$\left(\overline{\beta}\right) : \frac{x - x(1)}{\begin{vmatrix} y'(1) & z'(1) \\ y''(1) & z''(1) \end{vmatrix}} = \frac{y - y(1)}{\begin{vmatrix} z'(1) & x'(1) \\ z''(1) & x''(1) \end{vmatrix}} = \frac{z - z(1)}{\begin{vmatrix} x'(1) & y'(1) \\ x''(1) & y''(1) \end{vmatrix}},$$

i.e.

$$\left(\overline{\beta}\right) : \frac{x - 2}{\begin{vmatrix} 2 & 1 \\ 2 & -1 \end{vmatrix}} = \frac{y - 1}{\begin{vmatrix} 1 & 2 \\ -1 & 0 \end{vmatrix}} = \frac{z}{\begin{vmatrix} 2 & 2 \\ 0 & 2 \end{vmatrix}}$$

or

$$\left(\overline{\beta}\right) : \frac{x - 2}{-4} = \frac{y - 1}{2} = \frac{z}{4}.$$

The equation of the normal plane is:

$$\pi_N : 2(x - 2t) + 2t\left(y - t^2\right) + \frac{1}{t}(z - \ln t) = 0;$$

in the point $t = 1$, the equation of the normal plane becomes:

$$\pi_N : 2(x - 2) + 2(y - 1) + z = 0$$

or

$$\pi_N : 2x + 2y + z - 6 = 0.$$

Using (6.12), the equation of the normal plane in the point $t = 1$:

$$\pi_0 : \begin{vmatrix} x - x(1) & y - y(1) & z - z(1) \\ x'(1) & y'(1) & z'(1) \\ x''(1) & y''(1) & z''(1) \end{vmatrix} = 0;$$

i.e

$$\pi_0 : \begin{vmatrix} x-2 & y-1 & z-1 \\ 2 & 2 & 1 \\ 0 & 2 & -1 \end{vmatrix} = 0$$

or

$$\pi_o : 2x - y - 2z = 3.$$

The principal normal is the intersection between the osculator plane and the normal plane:

$$(\overline{\nu}) : \begin{cases} 2x - y - 2z = 3 \\ 2x + 2y + z = 6. \end{cases}$$

As the rectified plane is the plane that passes through the point $t = 1$ and contains the director vectors of the tangent and of the binormal in this point, based on (6.18), the equation of the rectified plane will be:

$$\pi_r : \begin{vmatrix} x - x(1) & y - y(1) & z - z(1) \\ a_1 & a_2 & a_3 \\ b_1 & b_2 & b_3 \end{vmatrix} = 0,$$

where:

- $\overline{a} = (a_1, a_2, a_3)$ is the director vector of the tangent in the point $t = 1$, i.e. $\overline{a} = (2, 2, 1)$,
- $\overline{b} = (b_1, b_2, b_3)$ is the director vector of the binormal in the point $t = 1$, i.e. $\overline{b} = (-4, 2, 4)$;

it will result:

$$\pi_r : \begin{vmatrix} x-2 & y-1 & z \\ 2 & 2 & 1 \\ -4 & 2 & 4 \end{vmatrix} = 0.$$

A solution in Sage will be given, too:

```
sage: x,y,z,t,a,b,c,d=var('x,y,z,t,a,b,c,d');(t>0).assume()
sage: r=vector([2*t,t^2,ln(t)]);r0=r.subs(t=1)
sage: rp=diff(r,t);rp0=rp.subs(t=1)
sage: (x-r0[0])/rp0[0]==(y-r0[1])/rp0[1];(y-r0[1])/rp0[1]==(z-r0[2])/rp0[2]
1/2*x - 1 == 1/2*y - 1/2
1/2*y - 1/2 == z
sage: A=matrix([[a,b],[c,d]])
sage: rs=diff(r,t,2);rs0=rs.subs(t=1)
sage: A1=A.subs(a=rp0[1],b=rp0[2],c=rs0[1],d=rs0[2])
sage: d1=A1.determinant()
```

```
sage: A2=A.subs(a=rp0[2],b=rp0[0],c=rs0[2],d=rs0[0])
sage: d2=A2.determinant()
sage: A3=A.subs(a=rp0[0],b=rp0[1],c=rs0[0],d=rs0[1])
sage: d3=A3.determinant()
sage: (x-r0[0])/d1==(y-r0[1])/d2;(y-r0[1])/d2==(z-r0[2])/d3
-1/4*x + 1/2 == 1/2*y - 1/2
1/2*y - 1/2 == 1/4*z
sage: u=rp0[0]*(x-r0[0])+rp0[1]*(y-r0[1])+rp0[2]*(z-r0[2]);u==0
2*x + 2*y + z - 6 == 0
sage: B=matrix([[x-r0[0],y-r0[1],z-r0[2]],[rp0[0],rp0[1],rp0[2]],[rs0[0],rs0[1],rs0[2]]])
sage: v=B.determinant();v==0
-4*x + 2*y + 4*z + 6 == 0
sage: n1=vector([u.coeff(x),u.coeff(y),u.coeff(z)])
sage: n2=vector([v.coeff(x),v.coeff(y),v.coeff(z)])
sage: n=n1.cross_product(n2)
sage: (x-r0[0])/n[0]==(y-r0[1])/n[1];(y-r0[1])/n[1]==(z-r0[2])/n[2]
1/6*x - 1/3 == -1/12*y + 1/12
-1/12*y + 1/12 == 1/12*z
sage: C=matrix([[x-r0[0],y-r0[1],z-r0[2]],[rp0[0],rp0[1],rp0[2]],[d1,d2,d3]])
sage: C.determinant()==0
6*x - 12*y + 12*z == 0
```

The Frenet's formulas establish some relations between the edge versors that bears its name and their derivatives.

**Theorem 6.10** (see [5], p. 582). Let $\Gamma$ be a regular curve and the point $M \in \Gamma$ be a current point, having the position vector $\overline{r}$. Let $\overline{\tau}, \overline{\beta}, \overline{\nu}$ be the versors of the tangent, binormal and the principale normal in $M$. If $\mathrm{d}s$ is the arc element on the curve $\Gamma$, then the following relations are satisfied, called *Frenet's formulas*:

1. *the first Frenet formula*:

$$\frac{\mathrm{d}\overline{\tau}}{\mathrm{d}s} = \frac{\overline{\nu}}{R} \tag{6.25}$$

2. *the second Frenet formula*:

$$\frac{\mathrm{d}\overline{\beta}}{\mathrm{d}s} = -\frac{\overline{\nu}}{T} \tag{6.26}$$

3. *the third Frenet formula*:

$$\frac{\mathrm{d}\overline{\nu}}{\mathrm{d}s} = -\left(\frac{\overline{\tau}}{R} - \frac{\overline{\beta}}{T}\right). \tag{6.27}$$

## 6.4 Curvature and Torsion of the Space Curves

The scalars $1/R$ and $1/T$, that are introduced through the Frenet formulas are called *the curvature* and *the torsion* of the given curve in the point $M$.

We shall examine the geometric interpretation of these scalars.

**Definition 6.11** (see [5], p. 579). The **mean curvature** is the variation of the tangent direction, per arc unit; the **curvature** at a point of a curve is the limit of the mean curvature, when the considered arc element tends to 0 (see Fig. 6.3), i.e.

$$K = \lim_{\Delta s \to 0} \left| \frac{\Delta \overline{\tau}}{\Delta s} \right| . \tag{6.28}$$

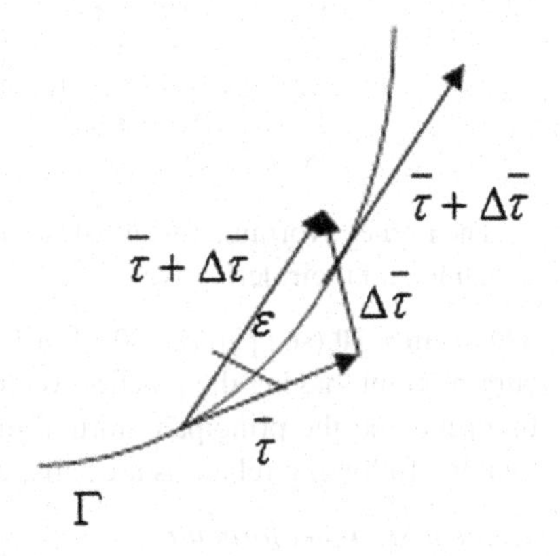


**Fig. 6.3** Mean curvature

**Definition 6.12** (see [5], p. 579). The **curvature radius** in a point from a curve is equal to the inverse of curvature at that point and is denoted with $R$.

**Remark 6.13** (see [2], p. 30). The curvature of the curve indicates the speed with which the curve moves away from the tangent.

**Theorem 6.14** (see [5], p. 586). Let $\Gamma$ be a regular curve. The necessary and sufficient condition that this curve to be a straight line is that $K = 0$.

**Definition 6.15** (see [5], p. 581). The **torsion** in a point of a curve is the variation of the binormal direction, per unit of arc, when the considered arc element tends to 0.

**Definition 6.16** (see [5], p. 581). The **torsion radius** in a point of a curve is equal to the inverse of the torsion at that point and is denoted by $T$.

**Remark 6.17** (see [2], p. 30). The torsion of the curve indicates the speed with which the curve moves away from the osculator plane.

**Theorem 6.18** (see [5], p. 587). Let $\Gamma$ be a regular curve. The necessary and sufficient condition that this curve to be a plane curve is that $\frac{1}{T} = 0$.

**Theorem 6.19 (computing the curvature**, see [5], p. 589). Let $\Gamma$ be a regular curve and $M \in \Gamma$ be the current point, with position vector $\overline{r}$, d$s$ be the arc element on the curve $\Gamma$ and $R$ be the radius of curvature of the the curve $\Gamma$ in the point $M$. Then

$$K = \frac{1}{R} = \frac{\left\| \overline{r'} \times \overline{r''} \right\|}{\left\| \overline{r'} \right\|^3}. \tag{6.29}$$

**Proof**
Using the first Frenet formula we have:

$$\frac{\mathrm{d}\overline{\tau}}{\mathrm{d}s} = \frac{\overline{\nu}}{R} \Leftrightarrow \frac{\mathrm{d}^2\overline{r}}{\mathrm{d}s^2} = \frac{\overline{\nu}}{R} \Rightarrow$$

$$\overline{\nu} = R\frac{\mathrm{d}^2\overline{r}}{\mathrm{d}s^2}. \tag{6.30}$$

We shall achieve:

$$\left\| \frac{\mathrm{d}^2\overline{r}}{\mathrm{d}s^2} \right\| = \frac{1}{R} \cdot \underbrace{\|\overline{\nu}\|}_{=1} \Leftrightarrow \left\| \frac{\mathrm{d}^2\overline{r}}{\mathrm{d}s^2} \right\| = \frac{1}{R};$$

therefore

$$\frac{1}{R} = \sqrt{\left(\frac{\mathrm{d}^2x}{\mathrm{d}s^2}\right)^2 + \left(\frac{\mathrm{d}^2y}{\mathrm{d}s^2}\right)^2 + \left(\frac{\mathrm{d}^2z}{\mathrm{d}s^2}\right)^2}.$$

Using the third formula from (6.24) and the relation (6.30) we deduce:

$$\left\| \overline{\beta} \right\| = \left\| \frac{\mathrm{d}\overline{r}}{\mathrm{d}s} \times R\frac{\mathrm{d}^2\overline{r}}{\mathrm{d}s^2} \right\| = 1 \Rightarrow$$

$$\frac{1}{R} = \left\| \frac{\mathrm{d}\overline{r}}{\mathrm{d}s} \times \frac{\mathrm{d}^2\overline{r}}{\mathrm{d}s^2} \right\|. \tag{6.31}$$

As

$$\frac{\mathrm{d}\overline{r}}{\mathrm{d}s} = \frac{\mathrm{d}\overline{r}}{\mathrm{d}t} \cdot \frac{\mathrm{d}t}{\mathrm{d}s} \tag{6.32}$$

we have:

$$\frac{d^2\overline{r}}{ds^2}=\frac{d}{ds}\left(\frac{d\overline{r}}{ds}\right)=\frac{d}{ds}\left(\frac{d\overline{r}}{dt}\cdot\frac{dt}{ds}\right)=\frac{d}{ds}\left(\frac{d\overline{r}}{dt}\right)\cdot\frac{dt}{ds}+\frac{d\overline{r}}{dt}\cdot\frac{d^2t}{ds^2}$$
$$=\left[\frac{d}{dt}\left(\frac{d\overline{r}}{dt}\right)\cdot\frac{dt}{ds}\right]\cdot\frac{dt}{ds}+\frac{d\overline{r}}{dt}\cdot\frac{d^2t}{ds^2},$$

i.e.

$$\frac{d^2\overline{r}}{ds^2}=\frac{d^2\overline{r}}{dt^2}\cdot\left(\frac{dt}{ds}\right)^2+\frac{d\overline{r}}{dt}\cdot\frac{d^2t}{ds^2}. \tag{6.33}$$

We deduce

$$\frac{d\overline{r}}{ds}\times\frac{d^2\overline{r}}{ds^2}=\left(\frac{d\overline{r}}{dt}\cdot\frac{dt}{ds}\right)\times\left[\frac{d^2\overline{r}}{dt^2}\cdot\left(\frac{dt}{ds}\right)^2+\frac{d\overline{r}}{dt}\cdot\frac{d^2t}{ds^2}\right]$$
$$=\left(\frac{dt}{ds}\right)^3\cdot\left(\frac{d\overline{r}}{dt}\times\frac{d^2\overline{r}}{dt^2}\right)+\frac{dt}{ds}\cdot\frac{d^2t}{ds^2}\cdot\underbrace{\left(\frac{d\overline{r}}{dt}\times\frac{d\overline{r}}{dt}\right)}_{=0}.$$

Therefore
$$\frac{d\overline{r}}{ds}\times\frac{d^2\overline{r}}{ds^2}=\left(\frac{dt}{ds}\right)^3\cdot\left(\frac{d\overline{r}}{dt}\times\frac{d^2\overline{r}}{dt^2}\right)=\frac{\frac{d\overline{r}}{dt}\times\frac{d^2\overline{r}}{dt^2}}{\left(\frac{ds}{dt}\right)^3},$$

i.e.

$$\frac{d\overline{r}}{ds}\times\frac{d^2\overline{r}}{ds^2}\overset{(6.20)}{=}\frac{\frac{d\overline{r}}{dt}\times\frac{d^2\overline{r}}{dt^2}}{\left\|\frac{d\overline{r}}{dt}\right\|^3}. \tag{6.34}$$

From (6.31) and (6.34) we obtain

$$\frac{1}{R}=\frac{\left\|\frac{d\overline{r}}{dt}\times\frac{d^2\overline{r}}{dt^2}\right\|}{\left\|\frac{d\overline{r}}{dt}\right\|^3}=\frac{\left\|\overline{r}'\times\overline{r}''\right\|}{\left\|\overline{r}'\right\|^3}.$$

**Theorem 6.20 (computing the torsion**, see [5], p. 593). Let $\Gamma$ be a regular curve and $M\in\Gamma$ be the current point, with position vector $\overline{r}$, $ds$ be the arc element on the curve $\Gamma$ and $R$ be the radius of curvature of the the curve $\Gamma$ in the point $M$. Then

$$\frac{1}{T}=\frac{\overline{r}'\cdot\left(\overline{r}''\times\overline{r}'''\right)}{\left\|\overline{r}'\times\overline{r}''\right\|^2}. \tag{6.35}$$

**Proof**

Taking into account the relations (6.32) and (6.33), we have

$$\begin{aligned}
\frac{\mathrm{d}^3\overline{r}}{\mathrm{d}s^3} &= \frac{\mathrm{d}}{\mathrm{d}s}\left(\frac{\mathrm{d}^2\overline{r}}{\mathrm{d}t^2}\cdot\left(\frac{\mathrm{d}t}{\mathrm{d}s}\right)^2+\frac{\mathrm{d}\overline{r}}{\mathrm{d}t}\cdot\frac{\mathrm{d}^2t}{\mathrm{d}s^2}\right)=\frac{\mathrm{d}}{\mathrm{d}s}\left(\frac{\mathrm{d}^2\overline{r}}{\mathrm{d}t^2}\cdot\left(\frac{\mathrm{d}t}{\mathrm{d}s}\right)^2\right)+\frac{\mathrm{d}}{\mathrm{d}s}\left(\frac{\mathrm{d}\overline{r}}{\mathrm{d}t}\cdot\frac{\mathrm{d}^2t}{\mathrm{d}s^2}\right)\\
&=\frac{\mathrm{d}}{\mathrm{d}s}\left(\frac{\mathrm{d}^2\overline{r}}{\mathrm{d}t^2}\right)\cdot\left(\frac{\mathrm{d}t}{\mathrm{d}s}\right)^2+\frac{\mathrm{d}^2\overline{r}}{\mathrm{d}t^2}\cdot\frac{\mathrm{d}}{\mathrm{d}s}\left(\left(\frac{\mathrm{d}t}{\mathrm{d}s}\right)^2\right)+\frac{\mathrm{d}}{\mathrm{d}s}\left(\frac{\mathrm{d}\overline{r}}{\mathrm{d}t}\right)\cdot\frac{\mathrm{d}^2t}{\mathrm{d}s^2}+\frac{\mathrm{d}\overline{r}}{\mathrm{d}t}\cdot\frac{\mathrm{d}^3t}{\mathrm{d}s^3}\\
&=\left[\frac{\mathrm{d}}{\mathrm{d}t}\left(\frac{\mathrm{d}^2\overline{r}}{\mathrm{d}t^2}\right)\cdot\frac{\mathrm{d}t}{\mathrm{d}s}\right]\cdot\left(\frac{\mathrm{d}t}{\mathrm{d}s}\right)^2+\frac{\mathrm{d}^2\overline{r}}{\mathrm{d}t^2}\cdot\frac{\mathrm{d}}{\mathrm{d}s}\left(\frac{\mathrm{d}t}{\mathrm{d}s}\cdot\frac{\mathrm{d}t}{\mathrm{d}s}\right)\\
&\quad+\left[\frac{\mathrm{d}}{\mathrm{d}t}\left(\frac{\mathrm{d}\overline{r}}{\mathrm{d}t}\right)\cdot\frac{\mathrm{d}t}{\mathrm{d}s}\right]\cdot\frac{\mathrm{d}^2t}{\mathrm{d}s^2}+\frac{\mathrm{d}\overline{r}}{\mathrm{d}t}\cdot\frac{\mathrm{d}^3t}{\mathrm{d}s^3}\\
&=\frac{\mathrm{d}^3\overline{r}}{\mathrm{d}t^3}\cdot\left(\frac{\mathrm{d}t}{\mathrm{d}s}\right)^3+\frac{\mathrm{d}^2\overline{r}}{\mathrm{d}t^2}\cdot\frac{\mathrm{d}^2t}{\mathrm{d}s^2}\cdot\frac{\mathrm{d}t}{\mathrm{d}s}+\frac{\mathrm{d}^2\overline{r}}{\mathrm{d}t^2}\cdot\frac{\mathrm{d}t}{\mathrm{d}s}\cdot\frac{\mathrm{d}^2t}{\mathrm{d}s^2}+\frac{\mathrm{d}^2\overline{r}}{\mathrm{d}t^2}\cdot\frac{\mathrm{d}t}{\mathrm{d}s}\cdot\frac{\mathrm{d}^2t}{\mathrm{d}s^2}\\
&\quad+\frac{\mathrm{d}\overline{r}}{\mathrm{d}t}\cdot\frac{\mathrm{d}^3t}{\mathrm{d}s^3},
\end{aligned}$$

i.e.

$$\frac{\mathrm{d}^3\overline{r}}{\mathrm{d}s^3}=\frac{\mathrm{d}^3\overline{r}}{\mathrm{d}t^3}\cdot\left(\frac{\mathrm{d}t}{\mathrm{d}s}\right)^3+3\frac{\mathrm{d}^2\overline{r}}{\mathrm{d}t^2}\cdot\frac{\mathrm{d}t}{\mathrm{d}s}\cdot\frac{\mathrm{d}^2t}{\mathrm{d}s^2}+\frac{\mathrm{d}\overline{r}}{\mathrm{d}t}\cdot\frac{\mathrm{d}^3t}{\mathrm{d}s^3}. \tag{6.36}$$

Using the first Frenet formula we achieve:

$$\frac{\mathrm{d}^3\overline{r}}{\mathrm{d}s^3}=\frac{\mathrm{d}}{\mathrm{d}s}\left(\frac{1}{R}\right)\overline{\nu}+\frac{1}{R}\cdot\frac{\mathrm{d}\overline{\nu}}{\mathrm{d}s}. \tag{6.37}$$

Using the third Frenet formula, from (6.37) we shall have:

$$\frac{\mathrm{d}^3\overline{r}}{\mathrm{d}s^3}=\frac{\mathrm{d}}{\mathrm{d}s}\left(\frac{1}{R}\right)\overline{\nu}-\frac{1}{R}\left(\frac{\overline{\tau}}{R}-\frac{\overline{\beta}}{T}\right)=\frac{\mathrm{d}}{\mathrm{d}s}\left(\frac{1}{R}\right)\overline{\nu}-\frac{1}{R^2}\overline{\tau}+\frac{1}{R\cdot T}\overline{\beta}. \tag{6.38}$$

By multiplying with $\overline{\beta}$ in (6.38), it will result:

$$\frac{\mathrm{d}^3\overline{r}}{\mathrm{d}s^3}\cdot\overline{\beta}=\frac{\mathrm{d}}{\mathrm{d}s}\left(\frac{1}{R}\right)\underbrace{\left(\overline{\nu}\cdot\overline{\beta}\right)}_{=0}-\frac{1}{R^2}\underbrace{\left(\overline{\tau}\cdot\overline{\beta}\right)}_{=0}+\frac{1}{R\cdot T}\underbrace{\left(\overline{\beta}\cdot\overline{\beta}\right)}_{=1}\overset{\text{Theorem 6.8}}{=}\frac{1}{R\cdot T}$$

$$\Rightarrow\frac{1}{T}=R\frac{\mathrm{d}^3\overline{r}}{\mathrm{d}s^3}\cdot\left(\overline{\tau}\times\overline{\nu}\right);$$

i.e.

$$\frac{1}{T}=R\frac{\mathrm{d}^3\overline{r}}{\mathrm{d}s^3}\cdot\left(\frac{\mathrm{d}\overline{r}}{\mathrm{d}s}\times R\frac{\mathrm{d}^2\overline{r}}{\mathrm{d}s^2}\right)=R^2\frac{\mathrm{d}^3\overline{r}}{\mathrm{d}s^3}\cdot\left(\frac{\mathrm{d}\overline{r}}{\mathrm{d}s}\times\frac{\mathrm{d}^2\overline{r}}{\mathrm{d}s^2}\right)=R^2\frac{\mathrm{d}\overline{r}}{\mathrm{d}s}\cdot\left(\frac{\mathrm{d}^2\overline{r}}{\mathrm{d}s^2}\times\frac{\mathrm{d}^3\overline{r}}{\mathrm{d}s^3}\right).$$

We obtain:

$$\frac{1}{T}=\frac{\frac{d\overline{r}}{ds}\cdot\left(\frac{d^2\overline{r}}{ds^2}\times\frac{d^3\overline{r}}{ds^3}\right)}{\frac{1}{R^2}}\overset{(6.31)}{=}\frac{\frac{d\overline{r}}{ds}\cdot\left(\frac{d^2\overline{r}}{ds^2}\times\frac{d^3\overline{r}}{ds^3}\right)}{\left\|\frac{d\overline{r}}{ds}\times\frac{d^2\overline{r}}{ds^2}\right\|^2}=\frac{\frac{d^3\overline{r}}{ds^3}\cdot\left(\frac{d\overline{r}}{ds}\times\frac{d^2\overline{r}}{ds^2}\right)}{\left\|\frac{d\overline{r}}{ds}\times\frac{d^2\overline{r}}{ds^2}\right\|^2}. \tag{6.39}$$

We have:

$$\frac{d\overline{r}}{ds}\times\frac{d^2\overline{r}}{ds^2}\overset{(6.34)}{=}\frac{\frac{d\overline{r}}{ds}\times\frac{d^2\overline{r}}{ds^2}}{\left\|\frac{d\overline{r}}{dt}\right\|^3};$$

i.e

$$\frac{d^3\overline{r}}{ds^3}\cdot\left(\frac{d\overline{r}}{ds}\times\frac{d^2\overline{r}}{ds^2}\right)\overset{(6.34)}{=}\frac{1}{\left\|\frac{d\overline{r}}{dt}\right\|^3}\left[\frac{d^3\overline{r}}{ds^3}\cdot\left(\frac{d\overline{r}}{ds}\times\frac{d^2\overline{r}}{ds^2}\right)\right]$$

$$=\frac{1}{\left\|\frac{d\overline{r}}{dt}\right\|^3}\left[\left(\frac{d^3\overline{r}}{dt^3}\cdot\left(\frac{dt}{ds}\right)^3+3\frac{d^2\overline{r}}{dt^2}\cdot\frac{d^2t}{ds^2}\cdot\frac{dt}{ds}+\frac{d\overline{r}}{dt}\cdot\frac{d^3t}{ds^3}\right)\cdot\left(\frac{d\overline{r}}{dt}\times\frac{d^2\overline{r}}{dt^2}\right)\right]$$

$$=\frac{1}{\left\|\frac{d\overline{r}}{dt}\right\|^3}\left[\left(\frac{dt}{ds}\right)^3\cdot\frac{d^3\overline{r}}{dt^3}\cdot\left(\frac{d\overline{r}}{dt}\times\frac{d^2\overline{r}}{dt^2}\right)+3\frac{d^2t}{ds^2}\cdot\frac{dt}{ds}\cdot\underbrace{\frac{d^2\overline{r}}{dt^2}\cdot\left(\frac{d\overline{r}}{dt}\times\frac{d^2\overline{r}}{dt^2}\right)}_{=0}\right.$$

$$\left.+\frac{d^3t}{ds^3}\cdot\underbrace{\frac{d\overline{r}}{dt}\cdot\left(\frac{d\overline{r}}{dt}\times\frac{d^2\overline{r}}{dt^2}\right)}_{=0}\right].$$

Finally, we get

$$\frac{d^3\overline{r}}{ds^3}\cdot\left(\frac{d\overline{r}}{ds}\times\frac{d^2\overline{r}}{ds^2}\right)=\frac{\left(\frac{dt}{ds}\right)^3\cdot\frac{d^3\overline{r}}{dt^3}\cdot\left(\frac{d\overline{r}}{dt}\times\frac{d^2\overline{r}}{dt^2}\right)}{\left\|\frac{d\overline{r}}{dt}\right\|^3}=\frac{\frac{d^3\overline{r}}{dt^3}\cdot\left(\frac{d\overline{r}}{dt}\times\frac{d^2\overline{r}}{dt^2}\right)}{\left\|\frac{d\overline{r}}{dt}\right\|^3\cdot\left(\frac{ds}{dt}\right)^3},$$

i.e.

$$\frac{d^3\overline{r}}{ds^3}\cdot\left(\frac{d\overline{r}}{ds}\times\frac{d^2\overline{r}}{ds^2}\right)=\frac{\frac{d^3\overline{r}}{dt^3}\cdot\left(\frac{d\overline{r}}{dt}\times\frac{d^2\overline{r}}{dt^2}\right)}{\left\|\frac{d\overline{r}}{dt}\right\|^3\cdot\left\|\frac{d\overline{r}}{dt}\right\|^3}=\frac{\frac{d^3\overline{r}}{dt^3}\cdot\left(\frac{d\overline{r}}{dt}\times\frac{d^2\overline{r}}{dt^2}\right)}{\left\|\frac{d\overline{r}}{dt}\right\|^6}. \tag{6.40}$$

Substituting (6.40) into (6.39) we achieve:

$$\frac{1}{T}=\frac{\dfrac{\frac{d\overline{r}}{dt}\cdot\left(\frac{d^2\overline{r}}{dt^2}\times\frac{d^3\overline{r}}{dt^3}\right)}{\left\|\frac{d\overline{r}}{dt}\right\|^6}}{\dfrac{\left\|\frac{d\overline{r}}{ds}\times\frac{d^2\overline{r}}{ds^2}\right\|^2}{\left\|\frac{d\overline{r}}{dt}\right\|^6}}=\frac{\frac{d\overline{r}}{dt}\cdot\left(\frac{d^2\overline{r}}{dt^2}\times\frac{d^3\overline{r}}{dt^3}\right)}{\left\|\frac{d\overline{r}}{ds}\times\frac{d^2\overline{r}}{ds^2}\right\|^2}=\frac{\overline{r}'\cdot\left(\overline{r}''\times\overline{r}'''\right)}{\left\|\overline{r}'\times\overline{r}''\right\|^2}.$$

**Example 6.21** (see [4], p. 37). Let be the curve:

$$\begin{cases} x=3\cos t\\ y=3\sin t\\ z=4t.\end{cases}$$

Determine:

(a) the tangent versor in any point $M$;
(b) the curvature of the curve in the point $M$;
(c) the principal normal versor in $M$;
(d) the equations of the principal normal;
(e) the binormal versor in $M$;
(f) the equations of the binormal;
(g) the torsion of the curve in $M$;
(h) the equation of the osculator plane in $M$.

**Solution**

(a) Using (6.21), the tangent versor is:

$$\overline{\tau}=\frac{-3\sin t\,\overline{i}+3\cos t\,\overline{j}+4\overline{k}}{\sqrt{9+16}}=\frac{1}{5}\left(-3\sin t\,\overline{i}+3\cos t\,\overline{j}+4\overline{k}\right).$$

We can get the tangent in Sage, too:

```
sage: t=var('t');x=3 * cos(t);y=3 * sin(t);z=4*t
sage: ta=vector([diff(x),diff(y),diff(z)])
sage: tn=ta[0]^2+ta[1]^2+ta[2]^2
sage: tau=ta/sqrt(tn);tau.simplify_trig()
(-3/5*sin(t), 3/5*cos(t), 4/5)
```

We need the following Sage code to plot the curve:

```
sage: show(parametric_plot3d( (x, y, z), (t, 0, 2*pi)))
```

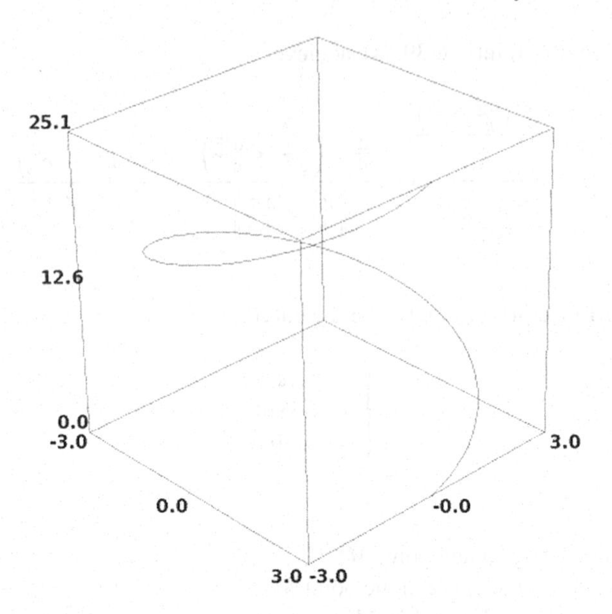


(b) + (c) The given curve has the vector equation

$$\overline{r} = 3\cos t\,\overline{i} + 3\sin t\,\overline{j} + 4t\,\overline{k}.$$

We compute:

$$\frac{d\overline{\tau}}{ds} = \frac{d\overline{\tau}}{dt}\cdot\frac{dt}{ds} = \frac{1}{\frac{dt}{ds}}\cdot\frac{d\overline{\tau}}{dt} = \frac{1}{\left\|\frac{d\overline{r}}{dt}\right\|}\cdot\frac{d\overline{\tau}}{dt} = \frac{1}{\left\|\overline{r}'\right\|}\cdot\frac{d\overline{\tau}}{dt} = \frac{1}{25}\left(-3\cos t\,\overline{i} - 3\sin t\,\overline{j}\right).$$

Using the first Frenet formula, it results:

$$\frac{1}{R}\overline{\nu} = \frac{3}{25}\left(-\cos t\,\overline{i} - \sin t\,\overline{j}\right);$$

therefore:

- the curvature of the curve is
$$\frac{1}{R} = \frac{3}{25};$$
- the principal normal versor is:
$$\overline{\nu} = -\cos t\,\overline{i} - \sin t\,\overline{j}.$$

The curvature of the curve can be also determined using the formula (6.29). Solving (b) + (c) + (e) with Sage, we have:

```
sage: tap=vector([diff(x,2),diff(y,2),diff(z,2)])
sage: tt=ta.cross_product(tap).simplify_trig()
sage: n1=sqrt(tt[0]^2+tt[1]^2+tt[2]^2).simplify_trig()
sage: K=n1/(tn^(3/2)).simplify_trig();K
3/25
sage: K=n1/(tn^(3/2)).simplify_trig()
sage: be=tt/n1;be
(4/5*sin(t), -4/5*cos(t), 3/5)
sage: nu=be.cross_product(tau);nu.simplify_trig()
(-cos(t), -sin(t), 0)
```

(d) The equations of the principal normal are:

$$(\overline{\nu}):\begin{cases}\frac{x-3\cos t}{-\cos t}=\frac{y-3\sin t}{-\sin t}\\ z=4t\end{cases}.$$

(e) the binormal versor can be determined using one of the two formulas: (6.22) or the third relation from (6.24). We achieve:

$$\overline{\beta}=\frac{1}{5}\left(4\sin t\,\overline{i}-4\cos t\,\overline{j}+3\overline{k}\right).$$

(f) The binormal equations are determined using the formulas (6.16); it results:

$$\left(\overline{\beta}\right):\frac{x-3\cos t}{4\sin t}=\frac{y-3\sin t}{-4\cos t}=\frac{z-4t}{3}.$$

(g) As

$$\frac{\mathrm{d}\overline{\beta}}{\mathrm{d}s}=\frac{\mathrm{d}\overline{\beta}}{\mathrm{d}t}\cdot\frac{\mathrm{d}t}{\mathrm{d}s}=\frac{1}{\frac{\mathrm{d}s}{\mathrm{d}t}}\cdot\frac{\mathrm{d}\overline{\beta}}{\mathrm{d}t}=\frac{1}{\left\|\frac{\mathrm{d}\overline{r}}{\mathrm{d}t}\right\|}\cdot\frac{\mathrm{d}\overline{\beta}}{\mathrm{d}t}$$
$$=\frac{1}{\left\|\overline{r}'\right\|}\cdot\frac{\mathrm{d}\overline{\beta}}{\mathrm{d}t}=\frac{1}{25}\left(4\cos t\,\overline{i}+4\sin t\,\overline{j}\right),$$

using the second Frenet formula we shall have:

$$-\frac{1}{T}\left(-\cos t\,\overline{i}-\sin t\,\overline{j}\right)=\frac{4}{25}\left(\cos t\,\overline{i}+\sin t\,\overline{j}\right);$$

hence, the torsion of the curve is:

$$\frac{1}{T} = \frac{4}{25}.$$

The same value can be found using Sage:

```
sage: tac=vector([diff(x,3),diff(y,3),diff(z,3)])
sage: g=tap.cross_product(tac).simplify_trig(); g
(0, 0, 9)
sage: iT=ta.dot_product(g)/n1^2;iT
4/25
```

(h) The equation of the osculator plane in $M$ can be determined with (6.12):

$$\pi_0 : \begin{vmatrix} x - 3\cos t & y - 3\sin t & z - 4t \\ -3\sin t & 3\cos t & 4 \\ -3\cos t & -3\sin t & 0 \end{vmatrix} = 0.$$

Using Sage, we achieve:

```
sage: X,Y,Z=var('X,Y,Z')
sage: A=matrix([[X-x,Y-y,Z-z],[ta[0],ta[1],ta[2]],[tap[0],tap[1],tap[2]]])
sage: expand(A.determinant())
9*Z*sin(t)^2 + 9*Z*cos(t)^2 - 36*t*sin(t)^2 - 36*t*cos(t)^2 + 12*X*sin(t) - 12*Y*cos(t)
```

## 6.5 Envelope of a Family of Curves in Plane

**Definition 6.22** (see [3], p. 121). A relation of the form

$$F(x, y, \lambda) = 0, \tag{6.41}$$

where $\lambda$ is a real parameter represents a **family of curves** in the plane $xOy$, each curve from the family being determined by the value of the respective parameter $\lambda$.

**Definition 6.23** (see [3], p. 121). The **envelope of a family of curves** is the tangent curve in every of its points, to a curve from that family (see Fig. 6.4).

As the considered family of curves depends on the parameter $\lambda$ it results that as well as the envelope points will depend on the values of $\lambda$. Therefore, the envelope of a family of curves has the parametric representation:

$$I : \begin{cases} x = x(\lambda) \\ y = y(\lambda). \end{cases} \tag{6.42}$$

The common points of the envelope and of the curves that belong to the respective family check the equation:

$$F\left(x\left(\lambda\right), y\left(\lambda\right), \lambda\right) = 0. \tag{6.43}$$

By differentiating with respect to $\lambda$ the relation (6.43) we achieve:

$$\frac{\partial F}{\partial x} \cdot \frac{\partial x}{\partial \lambda} + \frac{\partial F}{\partial y} \cdot \frac{\partial y}{\partial \lambda} + \frac{\partial F}{\partial \lambda} = 0, \tag{6.44}$$

i.e.

$$F_x' x_\lambda' + F_y' y_\lambda' + F_\lambda' = 0. \tag{6.45}$$

The equation of the tangent in a point $M\left(x, y\right)$ to a curve from the family of curves that has the Eq. (6.41) is

$$Y - y = -\frac{F_x'}{F_y'}\left(X - x\right), \tag{6.46}$$

where $X, Y$ are some current coordinates on the straight line.

The equation of the tangent in a point $M\left(x, y\right)$ to the envelope of a family of curves, characterized by the Eq. (6.42) is

$$Y - y = \frac{y_\lambda'}{x_\lambda'}\left(X - x\right). \tag{6.47}$$

As the envelope and the family of curves are tangent in the common points it results that the slopes of tangents coincide, i.e. from the relations (6.46) and (6.47) we deduce

$$\frac{y_\lambda'}{x_\lambda'} = -\frac{F_x'}{F_y'} \Leftrightarrow F_x' x_\lambda' + F_y' y_\lambda' = 0. \tag{6.48}$$

Substituting (6.48) into (6.45) we achieve

$$F_\lambda'\left(x, y, \lambda\right) = 0. \tag{6.49}$$

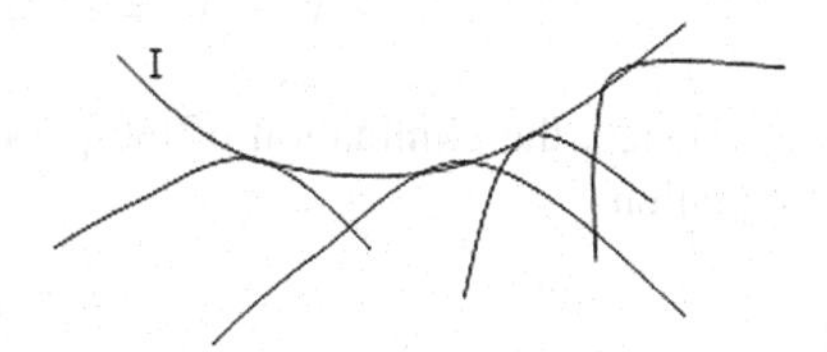


**Fig. 6.4** The envelope of a family of curves

Therefore, in conclusion, to get the equation of the considered family envelope it must that the parameter $\lambda$ to be eliminated between the Eqs. (6.41) and (6.49).

If the parameter $\lambda$ can't be eliminated, then from the Eqs. (6.41) and (6.49) one determines the coordinates of a current point of the envelope, depending on $\lambda$ and these will constitute the parametric equations of the envelope.

**Example 6.24** (see [4], p. 331). Determine the envelope of the family of straight lines :

$$F_{\lambda}(x, y, \lambda) = x \cos \lambda + y \sin \lambda - a = 0,$$

$\lambda$ being a real parameter.

**Solution**

We have

$$F'_{\lambda}(x, y, \lambda) = -x \sin \lambda + y \cos \lambda.$$

Solving the system, which contains the Eqs. (6.41) and (6.49), i.e.:

$$\begin{cases} F_{\lambda}(x, y, \lambda) = 0 \\ F'_{\lambda}(x, y, \lambda) = 0 \end{cases}$$

it will result:

$$\begin{cases} x \cos \lambda + y \sin \lambda - a = 0 \\ -x \sin \lambda + y \cos \lambda = 0. \end{cases} \tag{6.50}$$

By multiplying the first equation of the system (6.50) with $\sin \lambda$ and the second equation with $\cos \lambda$ we achieve

$$y = a \sin \lambda.$$

By multiplying the first equation of the system (6.50) with $\cos \lambda$ and the second equation with $-\sin \lambda$ we achieve

$$x = a \cos \lambda.$$

We shall deduce

$$x^2 + y^2 = a^2 \sin^2 \lambda + a^2 \cos^2 \lambda = a^2.$$

Hence, the elimination of the parameter $\lambda$ from the system (6.50) gives us the equation

$$x^2 + y^2 = a^2,$$

i.e. the envelope is a circle centered in the origin and having the radius $a$.

Solving this problem in Sage, we obtain:

```
sage: var("x,y,la,a")
(x, y, la, a)
sage: F=x*cos(la)+y*sin(la)-a
sage: Fp=diff(F,la)
sage: s=solve([F,Fp],x,y)
sage: x=s[0][0].right().simplify_trig();y=s[0][1].right().simplify_trig()
sage: x;y
a*cos(la)
a*sin(la)
sage: (x^2+y^2).simplify_trig()
a^2
```

## 6.6 Analytic Definition of Surfaces

**Definition 6.25** (see [5], p. 602). A **regular portion of a surface** is the set $\Sigma$ of the points $M(x, y, z)$ from the three-dimensional Euclidean real space $\mathbb{R}^3$, whose coordinates $x, y, z$ check one of the following systems of equations:

$$F(x, y, z) = 0, \quad (x, y, z) \in D \subseteq \mathbb{R}^3 \text{ (the implicit representation)} \tag{6.51}$$

$$z = f(x, y), \quad (x, y) \in D \subseteq \mathbb{R}^2 \text{ (the explicit representation)} \tag{6.52}$$

$$\begin{cases} x = f_1(u, v) \\ y = f_2(u, v) \\ z = f_3(u, v) \end{cases} = 0, (u, v) \in D \subseteq \mathbb{R}^2 \text{ (the parametric representation)}, \tag{6.53}$$

where the functions $F$, $f_1$, $f_2$, $f_3$ satisfy the following regularity conditions:

(a) they are real and continuous functions;
(b) the functions $f_1$, $f_2$, $f_3$ establish a biunivocal correspondence between the points $M \in \Sigma$ and the ordered pairs $(u, v)$, where $u$ and $v$ are some real parameters;
(c) they admit first order derivatives, continuous, that aren't all null;
(d) at least one of the functional determinants $\frac{D(f_1, f_2)}{D(u,v)}$, $\frac{D(f_2, f_3)}{D(u,v)}$ and $\frac{D(f_3, f_1)}{D(u,v)}$ isn't equal to 0.

**Definition 6.26** (see [5], p. 604). Let $\Sigma$ be a regular portion of surface, given by its parametric equations

$$\begin{cases} x = x(u, v) \\ y = y(u, v), \quad (u, v) \in D \subseteq \mathbb{R}^2 \\ z = z(u, v) \end{cases} \tag{6.54}$$

and $M(x, y, z) \in \Sigma$ be a current point.

If we consider a system of rectangular axes of versors $\bar{i}$, $\bar{j}$, $\bar{k}$ and if $\bar{r}$ is the position vector of the point $M$, then the relation

$$\bar{r} = x(u, v)\,\bar{i} + y(u, v)\,\bar{j} + z(u, v)\,\bar{k} \tag{6.55}$$

constitutes the **vector representation** of $\Sigma$.

**Definition 6.27** (see [5], p. 603). A **regular surface** is the union of the regular surface portions.

**Definition 6.28** (see [6], p. 708). Let $\Sigma$ be a regular surface, given by its parametric equations (6.54). The ordered pairs $(u, v)$, that determines the position of a point from the surface, are called the **curvilinear coordinates** on the surface $\Sigma$.

**Definition 6.29** (see [5], p. 647). Let $\Sigma$ be a regular surface, given by its parametric equations (6.54). The set of the poins $M(x, y, z) \in \Sigma$, whose coordinates verify the equations

$$\begin{cases} x = x(u(t), v(t)) \\ y = y(u(t), v(t)), \quad t \in (a, b) \\ z = z(u(t), v(t)) \end{cases} \tag{6.56}$$

forms a curve $\Gamma$, called a **curve traced on the surface** $\Sigma$, $\Gamma \subset \Sigma$.

The Eq. (6.56) are called the **parametric equations** of the curve $\Gamma$, traced on the surface $\Sigma$.

**Theorem 6.30** (see [5], p. 609). Let $\Sigma$ be a regular surface. If $(u, v)$ is a curvilinear coordinate system on the surface $\Sigma$, then any curve $\Gamma \subset \Sigma$ can be analytically represented by one of the following equation:

$$\begin{cases} u = u(t) \\ v = v(t) \end{cases} \tag{6.57}$$

$$f(u, v) = 0 \tag{6.58}$$

$$u = g(v). \tag{6.59}$$

## 6.7 Tangent Plane and Normal to a Surface

**Definition 6.31** (see [5], p. 615). The **tangent plane** in a point of the surface $\Sigma$ is set of the tangents pursued to all of the curves from the surface, passing through that point (see Fig. 6.5).

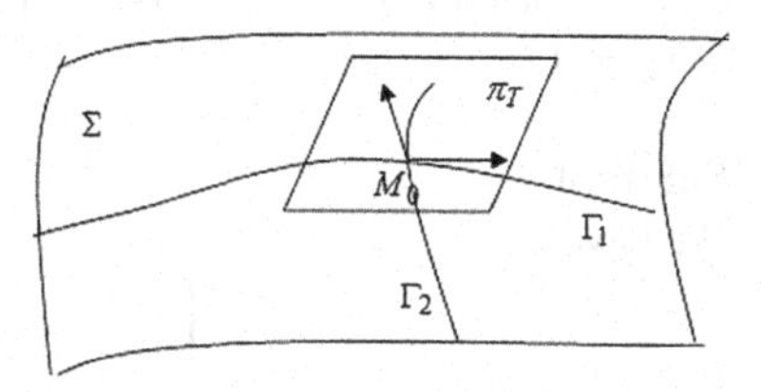


**Fig. 6.5** Tangent plane

The equation of the tangent plane to the surface $\Sigma$ in a point $M_0(x_0, y_0, z_0) \in \Sigma$, in the case when:

(a) the surface is given parametrically (as in the relation (6.54)) is

$$\pi_T : \begin{vmatrix} x - x_0 & y - y_0 & z - z_0 \\ x_{u_0} & y_{u_0} & z_{u_0} \\ x_{v_0} & y_{v_0} & z_{v_0} \end{vmatrix} = 0, \tag{6.60}$$

where

$$\begin{cases} x_0 = x(u_0, v_0), \ y_0 = y(u_0, v_0), \ z_0 = z(u_0, v_0) \\ x_{u_0} = \frac{\partial x}{\partial u}|_{u=u_0, v=v_0}, \ y_{u_0} = \frac{\partial y}{\partial u}|_{u=u_0, v=v_0}, \ z_{u_0} = \frac{\partial z}{\partial u}|_{u=u_0, v=v_0} \\ x_{v_0} = \frac{\partial x}{\partial v}|_{u=u_0, v=v_0}, \ y_{u_0} = \frac{\partial y}{\partial v}|_{u=u_0, v=v_0}, \ z_{v_0} = \frac{\partial z}{\partial v}|_{u=u_0, v=v_0}. \end{cases}$$

(b) the surface is given implicitly (see the relation (6.51)) is

$$\pi_T : (x - x_0) \frac{\partial F}{\partial x}|_{x=x_0, y=y_0, z=z_0} + (y - y_0) \frac{\partial F}{\partial y}|_{x=x_0, y=y_0, z=z_0} + (z - z_0) \frac{\partial F}{\partial z}|_{x=x_0, y=y_0, z=z_0} = 0 \tag{6.61}$$

(c) the surface is given explicitly (see the relation (6.52)) is

$$\pi_T : (X - x_0) \frac{\partial z}{\partial x}|_{x=x_0, y=y_0} + (Y - y_0) \frac{\partial z}{\partial y}|_{x=x_0, y=y_0} - (Z - z_0) = 0. \tag{6.62}$$

**Example 6.32** (see [5], p. 618). Let be the surface

$$\Sigma : \begin{cases} x = u + v \\ y = u - v \\ z = uv. \end{cases}$$

Find the equation of the tangent plane to the surface $\Sigma$ in the point $M$, for $u_0 = 2, v_0 = 1$.

**Solution**
We have

$$\begin{cases} x_u = 1, \ y_u = 1, \ z_u = v \\ x_v = 1, \ y_v = -1, \ z_v = u. \end{cases}$$

Considering $u_0 = 2, v_0 = 1$ it results $x_0 = 2, y_0 = 1, z_0 = 2$; by replacing them in the equation of the tangent plane (6.60) we have

$$\pi_T : \begin{vmatrix} x-3 & y-1 & z-2 \\ 1 & 1 & 1 \\ 1 & -1 & 2 \end{vmatrix} = 0,$$

i.e

$$\pi_T : 3(x-3) - (y-1) - 2(z-2) = 0 \Leftrightarrow$$
$$\pi_T : 3x - y - 2z - 4 = 0.$$

A solution in Sage will be given, too:

```
sage: x,y,z=var('x,y,z')
sage: u=2;v=1;xu=1;yu=1;zu=v;xv=1;yv=-1;zv=u
sage: x0=u+v;y0=u-v;z0=u*v
sage: A=matrix([[x-x0,y-y0,z-z0],[xu,yu,zu],[xv,yv,zv]])
sage: A.determinant()
3*x - y - 2*z - 4
```

**Example 6.33** (see [5], p. 619). Let be the surface

$$\Sigma : x^2 + 2xy + y^2 + 4xz + z^2 + 2x + 4y - 6z + 8 = 0.$$

Find the equation of the tangent plane to the surface in the point $M(0, 0, 2)$.

**Solution.**
We have

$$F(x, y, z) = x^2 + 2xy + y^2 + 4xz + z^2 + 2x + 4y - 6z + 8.$$

Whereas

$$\begin{cases} \frac{\partial F}{\partial x} = 2x + 2y + 4z + 2 \\ \frac{\partial F}{\partial y} = 2x + 2y + 4 \\ \frac{\partial F}{\partial z} = 4x + 2z - 6, \end{cases}$$

with (6.61), the equation of the tangent plane to the surface in the point $M\,(0, 0, 2)$ will be

$$\pi_T : 10x + 4y - 2\,(z - 2) = 0,$$

i.e

$$\pi_T : 5x + 2y - z + 2 = 0.$$

The problem can be also solved in Sage:

```
sage: x,y,z=var('x,y,z');x0=0;y0=0;z0=2
sage: F=x^2+2*x*y+y^2+4*x*z+z^2+2*x+4*y-6*z+8
sage: f1=diff(F,x);f2=diff(F,y);f3=diff(F,z)
sage: ff1=f1.subs(x=x0,y=y0,z=z0)
sage: ff2=f2.subs(x=x0,y=y0,z=z0)
sage: ff3=f3.subs(x=x0,y=y0,z=z0)
sage: (x-x0)*ff1+(y-y0)*ff2+(z-z0)*ff3
10*x + 4*y - 2*z + 4
```

**Definition 6.34** (see [5], p. 621). A **normal** in a point of a surface is the straight line perpendicular to the tangent plane to surface in that point (Fig. 6.6).

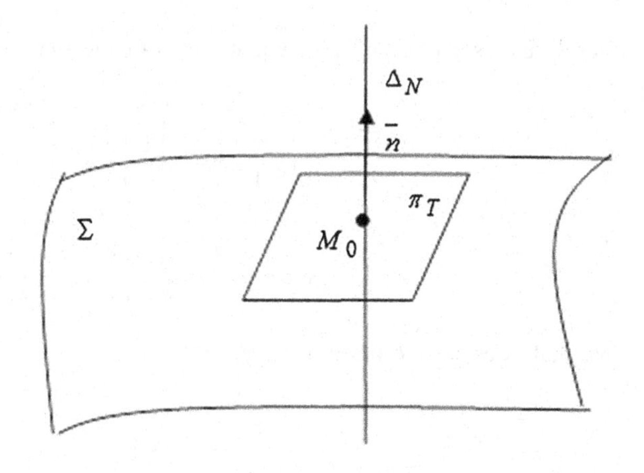


**Fig. 6.6** Normal in a point of surface

The equations of the normal to a surface $\Sigma$ in a point $M_0\,(x_0, y_0, z_0) \in \Sigma$, in the case when:

(a) the surface is given parametrically (as in the relation (6.54)) is

$$\Delta_N : \frac{x - x_0}{\begin{vmatrix} y_{u_0} & z_{u_0} \\ y_{v_0} & z_{v_0} \end{vmatrix}} = \frac{y - y_0}{\begin{vmatrix} z_{u_0} & x_{u_0} \\ z_{v_0} & x_{v_0} \end{vmatrix}} = \frac{z - z_0}{\begin{vmatrix} x_{u_0} & y_{u_0} \\ x_{v_0} & y_{v_0} \end{vmatrix}}; \tag{6.63}$$

(b) the surface is given implicitly (see the relation (6.51)) is

$$\Delta_N : \frac{x - x_0}{\frac{\partial F}{\partial x}|_{x=x_0,y=y_0,z=z_0}} = \frac{y - y_0}{\frac{\partial F}{\partial y}|_{x=x_0,y=y_0,z=z_0}} = \frac{z - z_0}{\frac{\partial F}{\partial z}|_{x=x_0,y=y_0,z=z_0}}; \tag{6.64}$$

(c) the surface is given explicitly (see the relation (6.52)) is

$$\Delta_N : \frac{X - x_0}{\frac{\partial z}{\partial x}|_{x=x_0,y=y_0}} = \frac{Y - y_0}{\frac{\partial z}{\partial y}|_{x=x_0,y=y_0}} = \frac{Z - z_0}{-1}. \tag{6.65}$$

**Example 6.35** (see [5], p. 623). Find the equation of the normal to the surface from the Example 6.32, in the point $u_0 = 2$, $v_0 = 1$.

**Solution**

For $u_0 = 2$, $v_0 = 1$ it results that $x_0 = 2$, $y_0 = 1$, $z_0 = 2$.

We shall have:

$$\begin{cases} \frac{\partial x}{\partial u} = 1, \frac{\partial y}{\partial u} = 1, \frac{\partial z}{\partial u} = v \\ \frac{\partial x}{\partial v} = 1, \frac{\partial y}{\partial v} = -1, \frac{\partial z}{\partial v} = u; \end{cases}$$

therefore, based on (6.63), the equations of the normal to the surface will be:

$$\Delta_N : \frac{x - 3}{\begin{vmatrix} 1 & 1 \\ -1 & 2 \end{vmatrix}} = \frac{y - y_0}{\begin{vmatrix} 1 & 1 \\ 2 & 1 \end{vmatrix}} = \frac{z - z_0}{\begin{vmatrix} 1 & 1 \\ 1 & -1 \end{vmatrix}}$$

or

$$\Delta_N : \frac{x - 3}{3} = \frac{y - 1}{-1} = \frac{z - 2}{-2}.$$

We shall present a solution in Sage, too:

```
sage: x,y,z,a,b,c,d=var('x,y,z,a,b,c,d')
sage: u=2;v=1;xu=1;yu=1;zu=v;xv=1;yv=-1;zv=u
sage: x0=u+v;y0=u-v;z0=u*v
sage: A=matrix([[a,b],[c,d]])
sage: A1=A.subs(a=yu,b=zu,c=yv,d=zv);a1=A1.determinant()
sage: A2=A.subs(a=zu,b=xu,c=zv,d=xv);a2=A2.determinant()
sage: A3=A.subs(a=xu,b=yu,c=xv,d=yv);a3=A3.determinant()
sage: (x-x0)/a1==(y-y0)/a2;(y-y0)/a2==(z-z0)/a3
1/3*x - 1 == -y + 1
-y + 1 == -1/2*z + 1
```

**Example 6.36** (see [5], p. 623). Find the equation of the normal $\Delta_N$ to the surface from the **Example 6.33**, in the point $M_0\,(0, 0, 2) \in \Sigma$.

**Solution**
As in the Example 6.33 we computed: $\frac{\partial F}{\partial x}, \frac{\partial F}{\partial y}, \frac{\partial F}{\partial z}$, using the relation (6.64) we achieve

$$\Delta_N : \frac{x-3}{10} = \frac{y}{4} = \frac{z-2}{-2} \Leftrightarrow \Delta_N : \frac{x}{5} = \frac{y}{2} = \frac{z-2}{-1}.$$

The solution with Sage is:

```
sage: x,y,z=var('x,y,z');x0=0;y0=0;z0=2
sage: F=x^2+2*x*y+y^2+4*x*z+z^2+2*x+4*y-6*z+8
sage: f1=diff(F,x);f2=diff(F,y);f3=diff(F,z)
sage: ff1=f1.subs(x=x0,y=y0,z=z0)
sage: ff2=f2.subs(x=x0,y=y0,z=z0)
sage: ff3=f3.subs(x=x0,y=y0,z=z0)
sage: (x-x0)/ff1==(y-y0)/ff2;(y-y0)/ff2==(z-z0)/ff3
1/10*x == 1/4*y
1/4*y == -1/2*z + 1
```

## 6.8 First Fundamental Form of a Surface. Curves on a Surface

**Definition 6.37**(see [5], p. 637). The **first fundamental form of a surface** $\Sigma$, denoted by $\Phi_1$ is the square of the arc element corresponding to an arbitrary arc from the surface:

$$\Phi_1 = \mathrm{d}s^2. \tag{6.66}$$

**Remark 6.38** (see [5], p. 637). The first fundamental form $\Phi_1$ is also called the **metric** of the surface $\Sigma$.
**Theorem 6.39 (expression of the first fundamental form**, see [5], p. 637). Let $\Sigma$ be a regular surface, given through its parametric equations (6.54).

Then

$$\Phi_1 = E\mathrm{d}u^2 + 2F\mathrm{d}u\mathrm{d}v + G\mathrm{d}v^2, \tag{6.67}$$

where

$$\begin{cases} E = x_u^2 + y_u^2 + z_u^2 \\ F = x_u x_v + y_u y_v + z_u z_v \\ G = x_v^2 + y_v^2 + z_v^2 \end{cases} \tag{6.68}$$

and

$$\begin{cases} x_u = \frac{\partial x}{\partial u},\ y_u = \frac{\partial y}{\partial u},\ z_u = \frac{\partial z}{\partial u} \\ x_v = \frac{\partial x}{\partial u},\ y_u = \frac{\partial y}{\partial u},\ z_u = \frac{\partial z}{\partial u}. \end{cases} \tag{6.69}$$

**Definition 6.40** (see [5], p. 643). Let $\Gamma \subset \Sigma$ be the curve given by (6.54). If $M_1$, $M_2$ are two points corresponding to the values of $t = t_1$ and $t = t_2$ then the **length of the arc** $M_1M_2$ is

$$s = \left| \int_{t_1}^{t_2} \mathrm{d}s \right| = \left| \int_{t_1}^{t_2} \sqrt{E\left(\frac{\mathrm{d}u}{\mathrm{d}t}\right)^2 + 2F\frac{\mathrm{d}u}{\mathrm{d}t} \cdot \frac{\mathrm{d}v}{\mathrm{d}t} + G\left(\frac{\mathrm{d}v}{\mathrm{d}t}\right)^2}\,\mathrm{d}t \right|, \tag{6.70}$$

$\mathrm{d}s$ being the arc element on the curve $\Gamma$.

**Example 6.41** (see [5], p. 643). One considers the surface

$$\begin{cases} x = u\cos v \\ y = u\sin v \\ z = av,\ a > 0. \end{cases}$$

Let $\Gamma_1, \Gamma_2, \Gamma_3$ be three curves traced on the surface $\Sigma$, as shown below

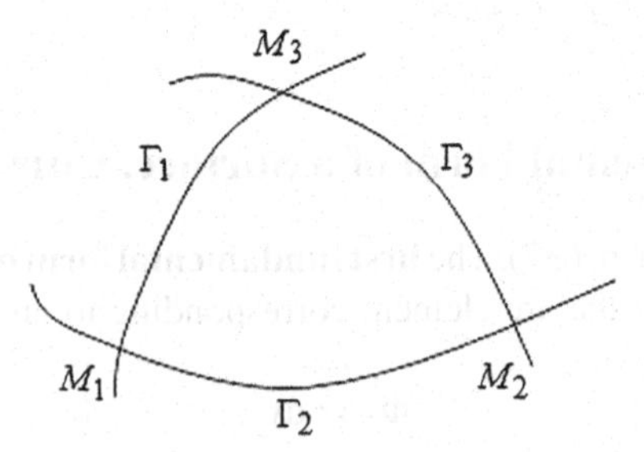


given by the parametric equations in the curvilinear coordinates

$$\begin{cases} \Gamma_1 : v = 1 \\ \Gamma_2 : u = \frac{1}{2}av^2 \\ \Gamma_3 : u = -\frac{1}{2}av^2. \end{cases}$$

If $M_1 \in \Gamma_1 \cap \Gamma_2$, $M_2 \in \Gamma_2 \cap \Gamma_3$, $M_3 \in \Gamma_1 \cap \Gamma_3$ and $s_1, s_2, s_3$ are the lengths of the arcs $M_1M_2$, $M_2M_3$, $M_3M_1$ find the perimeter of the curvilinear triangle $M_1M_2M_3$.

**Solution**

The curvilinear coordinates of the points are $M_1$, $M_2$, $M_3$ are

$$M_1\left(u = \frac{1}{2}a, v = 1\right), M_2\left(u = 0, v = 0\right), M_3\left(u = -\frac{1}{2}a, v = 1\right).$$

The parametric equations of the curves $\Gamma_1, \Gamma_2, \Gamma_3$ traced on the surface $\Sigma$ are:

$$\Gamma_1 : \begin{cases} x = u\cos 1 \\ y = u\sin 1 \\ z = a \end{cases} \Rightarrow$$

$$\begin{cases} \mathrm{d}x = \mathrm{d}u\cos 1 \\ \mathrm{d}y = \mathrm{d}u\sin 1 \\ \mathrm{d}z = \mathrm{d}a \end{cases} \Rightarrow \mathrm{d}s \overset{(6.19)}{=} \mathrm{d}u;$$

$$\Gamma_2 : \begin{cases} x = \frac{1}{2}av^2\cos v \\ y = \frac{1}{2}av^2\sin v \\ z = av \end{cases} \Rightarrow$$

$$\begin{cases} \mathrm{d}x = \left(\frac{1}{2}a\cdot 2v\cos v - \frac{1}{2}av^2\sin v\right)\mathrm{d}v \\ \mathrm{d}y = \left(\frac{1}{2}a\cdot 2v\sin v + \frac{1}{2}av^2\cos v\right)\mathrm{d}v \\ \mathrm{d}z = a\mathrm{d}v \end{cases} \Rightarrow$$

$$\mathrm{d}s \overset{(6.19)}{=} a\left(\frac{1}{2}v^2 + 1\right)\mathrm{d}v;$$

$$\Gamma_3 : \begin{cases} x = -\frac{1}{2}av^2\cos v \\ y = -\frac{1}{2}av^2\sin v \\ z = av \end{cases} \Rightarrow$$

$$\begin{cases} \mathrm{d}x = \left(-\frac{1}{2}a\cdot 2v\cos v + \frac{1}{2}av^2\sin v\right)\mathrm{d}v \\ \mathrm{d}y = \left(-\frac{1}{2}a\cdot 2v\sin v - \frac{1}{2}av^2\cos v\right)\mathrm{d}v \\ \mathrm{d}z = a\mathrm{d}v \end{cases} \Rightarrow$$

$$\mathrm{d}s \overset{(6.19)}{=} a\left(\frac{1}{2}v^2 + 1\right)\mathrm{d}v.$$

We shall have

$$\begin{cases} s_1 = \left| \int_1^0 a\left(\frac{1}{2}v^2 + 1\right) \mathrm{d}v \right| = \frac{7}{6}a, \\ s_2 = \left| \int_0^1 a\left(\frac{1}{2}v^2 + 1\right) \mathrm{d}v \right| = \frac{7}{6}a \\ s_3 = \left| \int_{-\frac{1}{2}a}^{\frac{1}{2}a} \mathrm{d}u \right| = a. \end{cases}$$

The perimeter of the curvilinear triangle $M_1M_2M_3$ will be

$$P = s_1 + s_2 + s_3 = \frac{7}{6}a + \frac{7}{6}a + a = \frac{20}{6}a = \frac{10}{3}a.$$

This problem will be also solve in Sage:

```
sage: a,u,v=var('a u v');(a > 0).assume()
sage: aa=1;x=u*cos(v);y=u*sin(v);z=aa*v;
sage: S=parametric_plot3d( (x, y, z), (u,-0.6,0.6),(v, 0, 1.2),color='yellow')
sage: x1=u*cos(1);y1=u*sin(1);z1=aa
sage: x2=1/2*aa*v^2*cos(v);y2=1/2*aa*v^2*sin(v);z2=aa*v
sage: x3=-1/2*aa*v^2*cos(v);y3=-1/2*aa*v^2*sin(v);z3=aa*v
sage: S1=parametric_plot3d( (x1, y1, z1), (u, -0.5, 0.5),color='black')
sage: S2=parametric_plot3d((x2, y2, z2),(v, 0, 1),color='black')
sage: S3=parametric_plot3d((x3, y3, z3),(v, 0, 1),color='black')
sage: S3+S2+S1+S
```

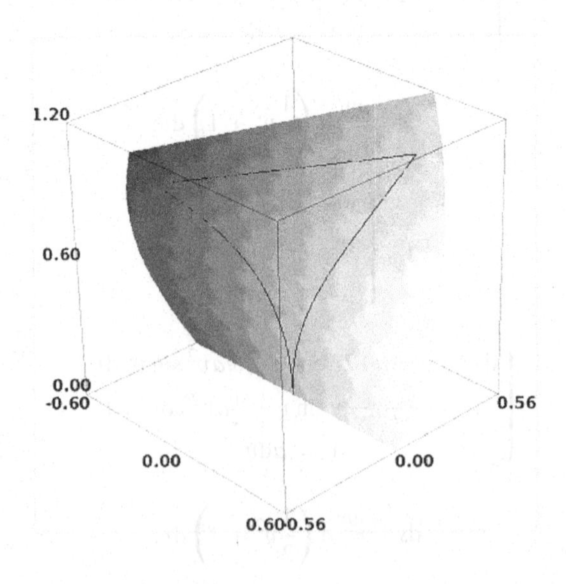

```
sage: I1=integral(a*(1/2*v^2+1), v, 1, 0)
sage: I2=integral(a*(1/2*v^2+1), v, 0, 1)
sage: I3=integral(1, u, -1/2*a, 1/2*a)
sage: I1.abs().simplify_exp()+I2.abs().simplify_exp()+I3.abs().simplify_exp()
10/3*a
```

**Definition 6.42** (see [5], p. 644). Let $\Sigma$ be a regular surface and $\Gamma_1 \subset \Sigma, \Gamma_2 \subset \Sigma$ be two curves traced on the surface $\Sigma$. The **angle** between the curves $\Gamma_1, \Gamma_2$ is the angle of the tangents to the two curves in their point of intersection.

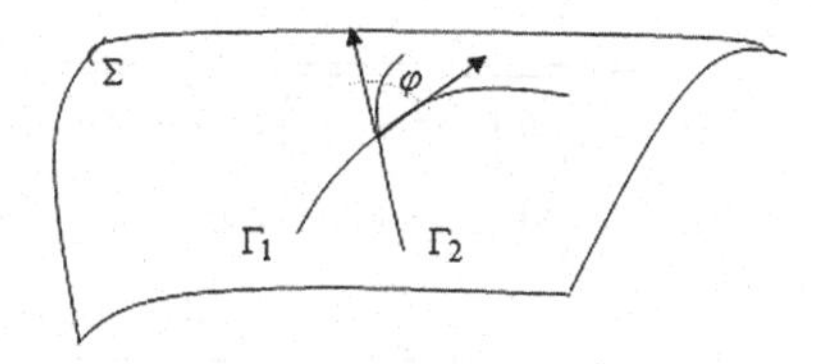


**Fig. 6.7** The angle between the two curves

**Theorem 6.43** (see [5], p. 645). Let the surface $\Sigma$ given by (6.54) and $\Gamma_1 \subset \Sigma, \Gamma_2 \subset \Sigma$ be two curves traced on the surface $\Sigma$. The **angle** between the curves $\Gamma_1, \Gamma_2$ in the point $M \in \Gamma_1 \cap \Gamma_2$ is given by the relation

$$\cos\varphi = \frac{E du \partial u + F\,(du\partial u + dv\partial v) + G dv\partial v}{\sqrt{Edu^2 + 2F du dv + G dv^2} \cdot \sqrt{E\partial u^2 + 2F\partial u\partial v + G\partial v^2}}, \tag{6.71}$$

where

- $E, F, G$ are the coefficients from the metric of the surface $\Sigma$ computed in $M$,
- $du, dv$ are two differentials along the curve $\Gamma_1$, and $\partial u, \partial v$ are two differentials along the curve $\Gamma_2$.

**Remark 6.44** (see [3], p. 146). If $\Gamma_1 : u = C_1, \Gamma_2 : v = C_2, C_1, C_2 \in \mathbb{R}$ (the curves have constant coordinates); as $du = 0, \partial v = 0$ it results that $\cos\varphi = \frac{F}{\sqrt{EG}}$.

**Example 6.45** (see [3], p. 147). Compute the angle between the curves $\Gamma_1 : u = v$ and $\Gamma_2 : u + v = 2$, traced on the surface

$$\Sigma : \overline{r} = \left(u^2 + v^2\right)\overline{i} + \left(u^2 - v^2\right)\overline{j} + 2uv\overline{k}.$$

**Solution**

The parametric equations of the surface $\Sigma$ will be

$$\begin{cases} x = u^2 + v^2 \\ y = u^2 - v^2 \\ z = 2uv. \end{cases}$$

Using (6.68), we shall compute the coefficients of the first fundamental form:

$$\begin{cases} E = 4u^2 + 4u^2 + 4v^2 = 8u^2 + 4v^2 \\ F = 2u \cdot 2v - 2u \cdot 2v + 2v \cdot 2u = 4uv \\ G = 4v^2 + 4v^2 + 4u^2 = 4u^2 + 8v^2. \end{cases}$$

As $\Gamma_1 : u = v$ and $\Gamma_2 : u + v = 2$ it results that d$u$ =d$v$ and $\partial u = -\partial v$.
We shall achieve:

$$\cos\varphi \overset{(6.71)}{=} \left[-4\left(2u^2+v^2\right)\mathrm{d}v\partial v + 4uv\left(\mathrm{d}v\partial v - \mathrm{d}v\partial v\right) + 4\left(2v^2+u^2\right)\mathrm{d}v\partial v\right]\cdot$$

$$= \frac{1}{\sqrt{4\left(2u^2+v^2\right)\mathrm{d}v^2 + 8uv\mathrm{d}v^2 + 4\left(2v^2+u^2\right)\mathrm{d}v^2}\cdot\sqrt{4\left(2u^2+v^2\right)\partial v^2 - 8uv\partial v^2 + 4\left(2v^2+u^2\right)\partial v^2}}$$

$$= \frac{4\left(-u^2+v^2\right)\mathrm{d}v\partial v}{\sqrt{4\left(3u^2+2uv+3v^2\right)\mathrm{d}v^2}\cdot\sqrt{4\left(3u^2-2uv+3v^2\right)\partial v^2}}$$

$$= \frac{-u^2+v^2}{\sqrt{\left(3u^2+2uv+3v^2\right)\left(3u^2-2uv+3v^2\right)}} = \frac{-u^2+v^2}{\sqrt{9\left(u^2+v^2\right)^2 - 4u^2v^2}}.$$

Taking into account that the point of intersection of two curves is $u = 1, v = 1$; therefore

$$\cos\varphi = \frac{0}{\sqrt{36-4}},$$

i.e. the two curves are orthogonal.

The same result can be achieved if we shall solve the problem in Sage:

```
sage: u,v,du,dv,ddu,ddv=var('u,v,du,dv,ddu,ddv')
sage: (dv>0).assume();(ddv>0).assume()
sage: x=u^2+v^2;y=u^2-v^2;z=2*u*v
sage: E=diff(x,u)^2+diff(y,u)^2+diff(z,u)^2
sage: F=diff(x,u)*diff(x,v)+diff(y,u)*diff(y,v)+diff(z,u)*diff(z,v)
sage: G=diff(x,v)^2+diff(y,v)^2+diff(z,v)^2
sage: du=dv;ddu=-ddv
sage: f=E*du^2+2*F*du*dv+G*dv^2;g=E*ddu^2+2*F*ddu*ddv+G*ddv^2
sage: h=sqrt(factor(f*g)).simplify_exp()
sage: t=(E*du*ddu+F*(du*ddv+dv*ddu)+G*dv*ddv).simplify_exp()
sage: hh=t.subs(du=dv,ddu=-ddv).factor()
sage: c=hh/h
sage: c.subs(u=1,v=1)
0
```

**Theorem 6.46** (see [5], p. 655). Let $\Sigma$ be a regular surface. The *area of the surface* $\Sigma$ is calculated using the following surface integral

$$\sigma = \int\int_{\Sigma} \sqrt{EG - F^2} \mathrm{d}u \mathrm{d}v, \tag{6.72}$$

where $E, F, G$ are the coefficients of the first fundamental form.

**Definition 6.47** (see [5], p. 655). Let $\Sigma$ be a regular surface and $\sigma$ from (6.72) be its area. The **surface area element** corresponding to $\Sigma$, denoted by $\mathrm{d}\sigma$ is the expression

$$\mathrm{d}\sigma = \sqrt{EG - F^2} \mathrm{d}u \mathrm{d}v. \tag{6.73}$$

**Theorem 6.48** (see [5], p. 655). Let $\Sigma$ be a regular surface. The *area* and the *surface area element* have respectively the expressions from bellow; if

(a) the surface is given parametrically (as in the relation (6.54)), then

$$\begin{cases} \sigma = \int\int_{\Sigma} \sqrt{A^{*2} + B^{*2} + C^{*2}} \mathrm{d}u \mathrm{d}v \\ \mathrm{d}\sigma = \sqrt{A^{*2} + B^{*2} + C^{*2}} \mathrm{d}u \mathrm{d}v, \end{cases} \tag{6.74}$$

where

$$\begin{cases} A^* = \begin{vmatrix} y_u & z_u \\ y_v & z_v \end{vmatrix} \\ B^* = \begin{vmatrix} z_u & x_u \\ z_v & x_v \end{vmatrix} \\ C^* = \begin{vmatrix} x_u & y_u \\ x_v & y_v \end{vmatrix}; \end{cases} \tag{6.75}$$

(b) the surface is given implicitly (see the relation (6.54)), then

$$\begin{cases} \sigma = \int\int_{\Sigma} \frac{\sqrt{\left(\frac{\partial F}{\partial x}\right)^2 + \left(\frac{\partial F}{\partial y}\right)^2 + \left(\frac{\partial F}{\partial z}\right)^2}}{\frac{\partial F}{\partial z}} \mathrm{d}x \mathrm{d}y \\ \mathrm{d}\sigma = \sqrt{1 + \left(\frac{\partial z}{\partial x}\right)^2 + \left(\frac{\partial z}{\partial y}\right)^2} \mathrm{d}x \mathrm{d}y. \end{cases} \tag{6.76}$$

## 6.9 Problems

1. Let be the curve

$$\Gamma : \overline{r} = e^{-t}\overline{i} + e^{t}\overline{j} + \sqrt{2}t\overline{k}.$$

   (a) Determine the analytical expressions of the versors of the Frenet trihedron in an arbitrary point of the curve.
   (b) Write the equations of the edges and of the planes of the Frenet trihedron in an arbitrary point of the curve.
   (c) Find the curve and the torsion of the curve in an arbitrary point.

2. Find the versors of the Frenet trihedron in the origin for the curve:

$$\Gamma : \overline{r} = t\,\overline{i} + t^2\,\overline{j} + t^3\overline{k}.$$

**Solution**
Solving in Sage, we get:

```
sage: t=var('t')
sage: r=vector([t,t^2,t^3])
sage: rp=diff(r,t);rs=diff(r,t,2)
sage: n=rp.norm().simplify_exp()
sage: tau=(rp/n).subs(t=0);tau
(1, 0, 0)
sage: v=rp.cross_product(rs)
sage: vn=v.norm().simplify_exp()
sage: be=(v/vn).subs(t=0);be
(0, 0, 1)
sage: nu=be.cross_product(tau).subs(t=0);nu
(0, 1, 0)
sage: c=parametric_plot3d((r[0],r[1],r[2]),(t,-1.3,1.3),thickness=5)
sage: a1=arrow3d((0,0,0),(tau[0],tau[1],tau[2]))
sage: a2=arrow3d((0,0,0),(be[0],be[1],be[2]))
sage: a3=arrow3d((0,0,0),(nu[0],nu[1],nu[2]))
sage: c+a1+a2+a3
```

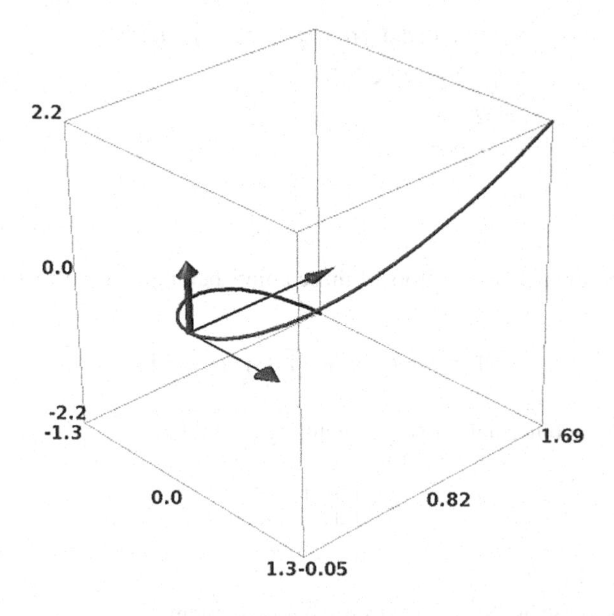


3. Write the tangent equations and the equation of the normal plane to the curve:

$$\Gamma : \begin{cases} x(t) = \mathrm{e}^t \cos 3t \\ y(t) = \mathrm{e}^t \sin 3t \\ z(t) = \mathrm{e}^{-2t} \end{cases}$$

   in the point $t = 0$.
4. Find the relation between the curvature and the torsion of the curve:

$$\Gamma : \overline{r} = t\,\overline{i} + \frac{t^2}{2}\,\overline{j} + \frac{t^3}{6}\overline{k}.$$

**Solution**
We shall use Sage to determine the asked relation:

```
sage: t=var('t')
sage: r=vector([t,t^2/2,t^3/6])
sage: rp=diff(r,t);rs=diff(r,t,2);rt=diff(r,t,3)
sage: rpn=sqrt(rp[0]^2+rp[1]^2+rp[2]^2)
sage: v=rp.cross_product(rs)
sage: vn=sqrt(v[0]^2+v[1]^2+v[2]^2)
sage: C=vn/rpn^3;C.factor()
4/(t^2 + 2)^2
```

```
sage: w=rp.dot_product(rs.cross_product(rt))
sage: iT=w/vn^2;iT.factor()
4/(t^2 + 2)^2
sage: C/iT
1
```

5. Determine the position vectors of those points belonging to the curve

$$\Gamma : \overline{r} = t^{-1}\,\overline{i} + t\,\overline{j} + \left(2t^2 - 1\right)\overline{k}$$

for which the binormale is perpendicularly on the line

$$d : \begin{cases} x + y = 0 \\ 4x - z = 0. \end{cases}$$

**Solution**
We need the following Sage code to solve this problem:

```
sage: n1=vector([1,1,0]);n2=vector([4,0,-1])
sage: n=n1.cross_product(n2)
sage: t=var('t');(t!=0).assume()
sage: r=vector([1/t,t,2*t^2-1])
sage: r1=diff(r,t);r2=diff(r,t,2)
sage: rr=r1.cross_product(r2)
sage: rn=sqrt(rr[0]^2+rr[1]^2+rr[2]^2)
sage: w=rr/rn.simplify_exp()
sage: s=solve([n.dot_product(w).simplify_exp()],t);s
[t == 2, t == -1]
sage: t1=s[0].right();t2=s[1].right()
sage: v1=r.subs(t=t1);v2=r.subs(t=t2);v1;v2
(1/2, 2, 7)
(-1, -1, 1)
```

6. Let be the surface

$$\Sigma : \begin{cases} x\,(u, v) = u \cos v \\ y\,(u, v) = u \sin v \\ z\,(u, v) = av. \end{cases}$$

Write the first fundamental form of the surface $\Sigma$.

**Solution**
The solution in sage of this problem is:

```
sage: a,u,v,du,dv=var('a,u,v,du,dv')
sage: x=u*cos(v);y=u*sin(v);z=a*v
sage: E=diff(x,u)^2+diff(y,u)^2+diff(z,u)^2
sage: F=diff(x,u)*diff(x,v)+diff(y,u)*diff(y,v)+diff(z,u)*diff(z,v)
sage: G=diff(x,v)^2+diff(y,v)^2+diff(z,v)^2
sage: Phi=E*du^2+2*F*du*dv+G*dv^2;Phi.simplify_trig()
a^2*dv^2 + dv^2*u^2 + du^2
```

7. Let be the surface

$$\Sigma : \begin{cases} x(u,v) = u\mathrm{e}^{v} \\ y(u,v) = u\mathrm{e}^{-v} \\ z(u,v) = 4uv. \end{cases}$$

(a) Find the equation of the tangent plane to the surface in the point $u = 0$, $v = 2$;
(b) Find the equation of the normal to the surface in that point.

8. Let be the surface

$$\Sigma : z = 5x^2 + 4y - 3.$$

(a) Find the equation of the tangent plane in the point $M(1, 0, 2)$.
(b)Write the equation of the of normal to the surface $\Sigma$ in the point $M(1, 0, 2)$.

**Solution**
The solution in Sage of this problem is:

```
sage: x,y,z=var('x,y,z');x0=1;y0=0;z0=2
sage: zz=5*x^2+4*y-3
sage: p=diff(zz,x).subs(x=x0,y=y0);q=diff(zz,y).subs(x=x0,y=y0)
sage: expand((x-x0)*p+(y-y0)*q-(z-z0))==0
10*x + 4*y - z - 8 == 0
sage: (x-x0)/p==(y-y0)/q;(y-y0)/q==(z-z0)/(-1)
1/10*x - 1/10 == 1/4*y
1/4*y == -z + 2
```

9. Calculate the angle between the curves $\Gamma_1 : u - \mathrm{e}^{v} = 0$ and $\Gamma_2 : u^2 + u + 1 - \mathrm{e}^{-v} = 0$, traced on the surface

$$\Sigma : \overline{r} = u \cos v\, \overline{i} + u \sin v\, \overline{j} + (u + v)\, \overline{k}.$$

10. Find the area of the quadrilateral, bounded by the curves: $u = 0$, $u = a$, $v = 0$, $v = \pi$, traced on the surface

$$\Sigma : \overline{r} = u \cos v\, \overline{i} + u \sin v\, \overline{j} + av\, \overline{k}.$$

## References


1. G.h. Atanasiu, G.h. Munteanu, M. Postolache, *Algebră liniară, geometrie analitică şi diferenţială, ecuaţii diferenţiale, ed* (ALL, Bucureşti) (1998)
2. I. Bârză, *Elemente de geometrie diferenţială, ed* (MatrixRom, Bucureşti) (2007)
3. T. Didenco, *Geometrie analitică şi diferenţială* (Academia Militară, Bucureşti) (1977)
4. C. Ionescu- Bujor, O. Sacter, *Exerciţii şi probleme de geometrie analitică şi diferenţială, vol II, ed* (Didactică şi pedagogică, Bucureşti) (1963)
5. E. Murgulescu, S. Flexi, O. Kreindler, O. Sacter, M. Tîrnoveanu, *Geometrie analitică şi diferenţială, ed* (Didactică şi pedagogică, Bucureşti) (1965)
6. V. Postelnicu, S. Coatu, *Mică enciclopedie matematică, ed* (Tehnică, Bucureşti) (1980)

# Chapter 7
# Conics and Quadrics

## 7.1 General Equation of a Conic

The analytical geometry replaces the definition and the geometrical study of the curves and the surfaces with that algebraic: a curve respectively a surface is defined by an algebraic equation and the study of the *curves* and of the *surfaces* is reduced to the study of the equation corresponding to each of them.

**Definition 7.1** (see [1], p. 158). We consider the function

$$f : \mathbb{R}^2 \to \mathbb{R},\ f(x, y) = a_{11}x^2 + 2a_{12}xy + a_{22}y^2 + 2b_1x + 2b_2y + c. \tag{7.1}$$

A **second order algebraic curve** or a **conic** is the set $\Gamma$ of the points $M(x, y)$ from the plane, whose coordinates relative to an orthonormal Cartesian reference check thegeneral equation

$$f(x, y) = 0 \tag{7.2}$$

where the coefficients $a_{11}, a_{12}, a_{22}, b_1, b_2, c$ are some real constants, with $a_{11}^2 + a_{12}^2 + a_{22}^2 \neq 0$; therefore

$$\Gamma = \left\{ M(x, y) \mid (x, y) \in \mathbb{R}^2,\ f(x, y) = 0 \right\}. \tag{7.3}$$

**Definition 7.2** (see [2]). The **invariants of a conic** are those expressions made with the coefficients of the conic equation, that keep the same value to the changes of an orthonormal reference.

**Proposition 7.3** (see [2] and [3], p. 56). We can assign three invariants of the conics from (7.2):

$$1)\ I = a_{11} + a_{22}, \tag{7.4}$$

$$2)\ \delta = \begin{vmatrix} a_{11} & a_{12} \\ a_{12} & a_{22} \end{vmatrix},$$

G. A. Anastassiou and I. F. Iatan, *Intelligent Routines II*,
Intelligent Systems Reference Library 58, DOI: 10.1007/978-3-319-01967-3_7,

$$3)\ \Delta = \begin{vmatrix} a_{11} & a_{12} & b_1 \\ a_{12} & a_{22} & b_2 \\ b_1 & b_2 & c \end{vmatrix};$$

the first invariant is linear, the second is quadratic, while the third is a cubical one.

The invariant $\Delta$ determines the nature of a conic. Thus, if:

- $\Delta \neq 0$, we say that $\Gamma$ is a *non-degenerate conic* (the circle, the ellipse, the hyperbola and the parabola)
- $\Delta = 0$, we say that $\Gamma$ is a *degenerate* conic.

With the help of $\delta$ one establishes the type of a conic. Thus, if:

- $\delta > 0$, we say that $\Gamma$ has an *elliptic type*
- $\delta < 0$, we say that $\Gamma$ is of a *hyperbolic type*
- $\delta = 0$, we say that $\Gamma$ has a *parabolic type.*

**Definition 7.4** (see [1], p. 168). The **center of symmetry of a conic** $\Gamma$ (in the case when this exists, the conic is called a *conic with center*) is a point $C$ from the plane, which has the property that for any point $M \in \Gamma$, the reflection of $M$ with respect to $C$ satisfies the equation of the conic $\Gamma$, too.

**Theorem 7.5** (see [1], p. 168). The conic $\Gamma$ from (7.2) admits a unique center of symmetry $C(x_0, y_0)$ if and only if its invariant $\delta$ is non-null; in this case, its coordinates are the solutions of the linear system

$$\begin{cases} \frac{\partial f}{\partial x} = 0 \\ \frac{\partial f}{\partial y} = 0 \end{cases} \Leftrightarrow \begin{cases} a_{11}x + a_{12}y + b_1 = 0 \\ a_{12}x + a_{22}y + b_2 = 0. \end{cases} \tag{7.5}$$

The non-degenerate conics

- with center are: the circle, the ellipse, the hyperbola
- without center is: the parabola.

**The General Table of the Conic Discussion**

(I) If $\Delta \neq 0$ then for:

(1) $\delta > 0$, the conic is
(a) a real ellipse when $I \cdot \Delta < 0$
(b) an imaginary ellipse when $I \cdot \Delta > 0$
(2) $\delta = 0$, the conic is a parabola
(3) $\delta < 0$, the conic is a hyperbola

(II) If $\Delta = 0$ then for:

(1) $\delta > 0$, we achieve two imaginary concurrent lines with a real intersection
(2) $\delta = 0$, we can obtain:
(a) two parallel lines if $\delta_1 < 0$
(b) two confounded lines if $\delta_1 = 0$

(c) two imaginary parallel lines if $\delta_1 > 0$

(3) $\delta < 0$, we achieve two real concurrent lines,

where

$$\delta_1 = a_{11}c - b_1^2. \tag{7.6}$$

**Theorem 7.6** (see [4], p. 174). Each conic has one of the following canonical forms:

(1) imaginary ellipse:

$$\frac{x^2}{a^2} + \frac{y^2}{b^2} + 1 = 0; \tag{7.7}$$

(2) real ellipse:

$$\frac{x^2}{a^2} + \frac{y^2}{b^2} - 1 = 0; \tag{7.8}$$

(3) hyperbola:

$$\frac{x^2}{a^2} - \frac{y^2}{b^2} - 1 = 0; \tag{7.9}$$

(4) two imaginary concurrent lines having a real intersection

$$\frac{x^2}{a^2} + \frac{y^2}{b^2} = 0; \tag{7.10}$$

(5) two real concurrent lines

$$\frac{x^2}{a^2} - \frac{y^2}{b^2} = 0; \tag{7.11}$$

(6) parabola

$$y^2 = 2px; \tag{7.12}$$

(7) two imaginary parallel lines

$$\frac{x^2}{a^2} + 1 = 0; \tag{7.13}$$

(8) two real parallel lines

$$\frac{x^2}{a^2} - 1 = 0; \tag{7.14}$$

(9) a pair of confounded lines

$$x^2 = 0. \tag{7.15}$$

## 7.2 Conics on the Canonical Equations

### 7.2.1 Circle

**Definition 7.7** (see [5], p. 117). The **circle** is the set of the equally spaced points from the plane by a fixed point called the *center*, the distance from the center to the points of the circle, being the *radius*.

We report the circle plane to an orthogonal Cartesian reference $\{O, \bar{i}, \bar{j}\}$. Let $C(a, b)$ be the center of the circle and $M(x, y)$ be an arbitrary point of it (see Fig. 7.1).

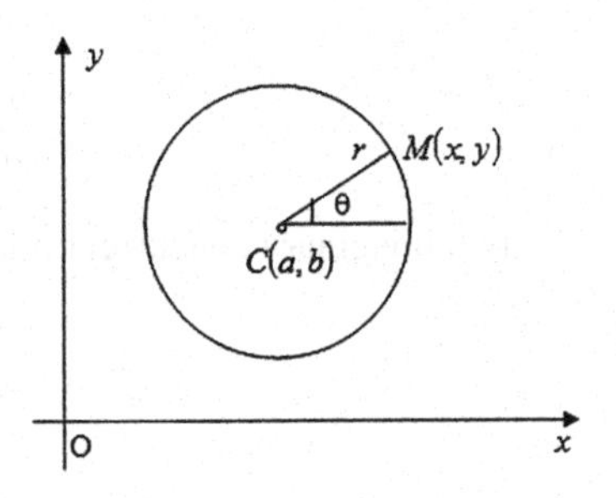


**Fig. 7.1** Circle

From the Definiton 7.7, it results that the distance from $C$ and $M$ is constant and equal to the radius $r$ of the circle, i.e.:

$$\left\| \overline{CM} \right\| = r;$$

hence

$$\sqrt{(x-a)^2 + (y-b)^2} = r.$$

By raising to the power of 2, from the previous relation we get:

$$(x-a)^2 + (y-b)^2 = r^2. \tag{7.16}$$

If we open the squares from (7.16) we achieve the circle equation in the form:

$$x^2 + y^2 - 2ax - 2by + a^2 + b^2 - r^2 = 0. \tag{7.17}$$

Denoting by

$$\begin{cases} m = -a \\ n = -b \\ p = a^2 + b^2 - r^2, \end{cases} \tag{7.18}$$

the Eq. (7.17) becomes

$$x^2 + y^2 + 2mx + 2ny + p = 0. \tag{7.19}$$

In the case of the circle, the set $\Gamma$ will be:

$$\Gamma = \left\{ M(x, y) \mid (x, y) \in \mathbb{R}^2, x^2 + y^2 + 2mx + 2ny + p = 0 \right\}. \tag{7.20}$$

As the Eq. (7.20) can be written in the form

$$(x + m)^2 + (y + n)^2 = m^2 + n^2 - p$$

it results that:

1. if $m^2 + n^2 - p > 0$ then the circle $\Gamma$ will have the center $C(-m, -n)$ and the radius $r = \sqrt{m^2 + n^2 - p}$;
2. if $m^2 + n^2 - p = 0$ then the circle $\Gamma$ one reduces to the point $C(-m, -n)$;
3. if $m^2 + n^2 - p < 0$ then $\Gamma = \Phi$ (empty set).

For $m^2 + n^2 - p > 0$, the Eq. (7.19) is called the *general Cartesian equation* of the circle $\Gamma$.

If $\theta$ is that angle made by the radius with the positive direction of the axis $Ox$, then the parametric equations of the circle will be:

$$\begin{cases} x = a + r\cos\theta \\ y = b + r\sin\theta \end{cases}, \quad \theta \in [0, 2\pi].$$

**Example 7.8** (see [5], p. 121). Find the equation of the circle determined by the points: $M(-1, 1)$, $N(2, -1)$, $P(1, 3)$.

**Solution**

Using the Eq. (7.19) we deduce

$$\begin{cases} 2m - 2n - p = 2 \\ 4m - 2n + p = -5 \\ 2m + 6n + p = -10; \end{cases} \Rightarrow \begin{cases} m = -\frac{11}{10} \\ n = -\frac{9}{10} \\ p = -\frac{12}{5}. \end{cases}$$

The equation of the circle will be:

$$x^2 + y^2 - \frac{22}{10}x - \frac{18}{10}y - \frac{12}{5} = 0$$

or

$$5\left(x^2+y^2\right)-11x-9y-12=0.$$

The solution in Sage is:

```
sage: x,y,m,n,p=var('x,y,m,n,p')
sage: f(x,y)=x^2+y^2+2*m*x+2*n*y+p
sage: s=solve([f(-1,1),f(2,-1),f(1,3)],m,n,p)
sage: s1=s[0][0].right();s2=s[0][1].right();s3=s[0][2].right()
sage: f.subs(m=s1,n=s2,p=s3)==0
(x, y) |--> x^2 + y^2 - 11/5*x - 9/5*y - 12/5 == 0
sage: a=-s1;b=-s2;r=sqrt(a^2+b^2-s3);r.n(digits=3)
2.10
sage: M=point2d((-1,1),size=22);t1=text("M",(-1.1,1.1))
sage: N=point2d((2,-1),size=22);t2=text("N",(2,-1.1))
sage: P=point2d((1,3),size=22);t3=text("P",(1,3.1));C=circle((a,b), r)
sage: (M+N+P+C+t1+t2+t3).show(aspect_ratio=1)
```

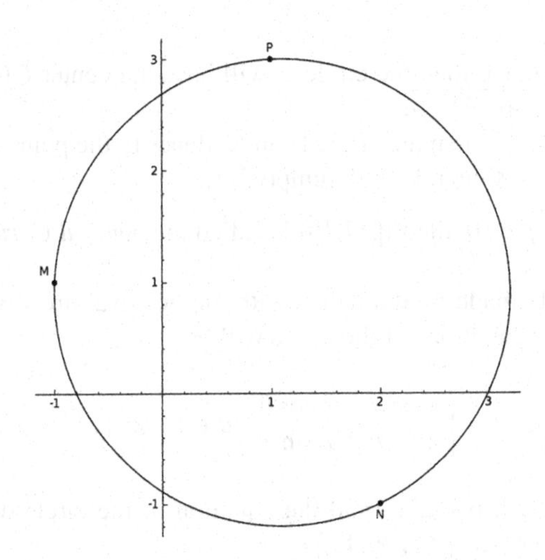


## 7.2.2 Ellipse

Let $c>0$ be a positive real number and $F$, $F'$ be two fixed points of the plane such that $FF'=2c$.

**Definition 7.9** (see [5], p. 150). The **ellipse** is the set of the points $M$ of the plane that satisfy the relation

$$MF+MF'=2a=ct, \tag{7.21}$$

which has a constant sum of the distances to two fixed points.

We choose $FF'$ as O$x$ axis and the mediator of the segment $FF'$ as O$y$ axis (see Fig. 7.2). We denote $FF' \cap BB' = \{O\}$.

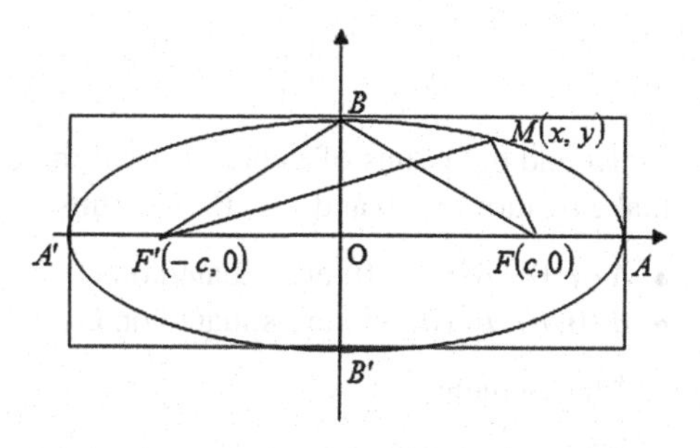


**Fig. 7.2** Ellipse

Therefore $F(c, 0)$ and $F'(-c, 0)$; the points $F$ and $F'$ are called the *foci* of the ellipse and the distance $FF'$ constitutes the *focus distance* of the ellipse. $MF$, $MF'$ are the *focus radiuses (focus radii)* of the point $M$. The ellipse admits a unique center of symmetry O and two axes of symmetry O$x$,O$y$. The ellipse is a bounded curve (there is a rectangle that contains all its points).

To find the ellipse equation we shall transform analytically the Eq. (7.21). If $M(x, y)$, then the relation (7.21) becomes

$$\sqrt{(x-c)^2+y^2}+\sqrt{(x+c)^2+y^2}=2a. \tag{7.22}$$

As we want to simplify the relation (7.22), we write:

$$(x-c)^2+y^2=4a^2-4a\sqrt{(x+c)^2+y^2}+(x+c)^2+y^2$$

or

$$a\sqrt{(x+c)^2+y^2}=a^2+cx;$$

therefore

$$\left(a^2-c^2\right)x^2+a^2y^2-a^2\left(a^2-c^2\right)=0. \tag{7.23}$$

In the triangle $MFF'$ it is known that $MF+MF' > FF'$ or $2a > 2c$, hence a > c; so that $a^2-c^2>0$.

Therefore, we can denote

$$a^2-c^2=b^2. \tag{7.24}$$

Dividing by $a^2b^2$ in (7.23) it results the *implicit Cartesian equation* (7.8) of the ellipse. If $a=b=r$, the Eq. (7.8) becomes

$$x^2+y^2=r^2 \tag{7.25}$$

and it represents a circle centered in origin and having the radius $r$. Therefore, the circle is a special case of an ellipse.

Thereby:

$$\Gamma = \left\{ M\,(x, y) \mid (x, y) \in \mathbb{R}^2, \frac{x^2}{a^2} + \frac{y^2}{b^2} = 1 \right\}. \tag{7.26}$$

To find the points of intersection of the curve with the coordinate axes we shall make by turn $y = 0$ and $x = 0$. It results:

- $A\,(a, 0)$, $A'\,(-a, 0)$ are situated on $Ox$,
- $B\,(0, b)$, $B'\,(0, -b)$ are situated on $Oy$.

The segment

- $AA' = 2a$ is called the *major axis* of the ellipse;
- $BB' = 2b$ is called the *minor axis* of the ellipse.

Their halves, i.e. $OA = a$ and $OB = b$ are the *semi-axes* of the ellipse. The points $A, A', B, B'$ are called the *vertices of the ellipse.*

From the Eq. (7.8) one deduces the *explicit Cartesian equations of the ellipse*:

$$y = \pm \frac{b}{a} \sqrt{a^2 - x^2}, x \in [-a, a]. \tag{7.27}$$

To obtain the parametric equations of the ellipse, one proceeds (see [6], p. 378) as follows (see Fig. 7.3):

(1) build two concentric circles with the radii $a$ and respectively $b$, $a > b$;
(2) trace a semi- line through origin, which intersects the two circles in the points $A$ and respectively $B$;
(3) bulid through the points $A$ and $B$ some lines, that are parallel with the axes; the intersection of these points will be a point $M$ of the ellipse;
(4) denote by $\theta$ the angle formed by the radius $OA$ with the axis $Ox$.

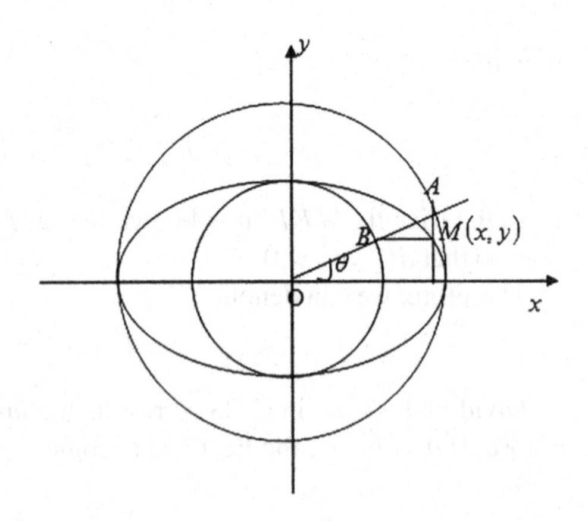


**Fig. 7.3** Concentric circles

We shall deduce that the parametric equations of the ellipse are:

$$\begin{cases} x = a\cos\theta \\ y = b\sin\theta \end{cases}, \ \theta \in [0, 2\pi]. \tag{7.28}$$

## 7.2.3 Hyperbola

As in the case of the ellipse, we shall also consider $c > 0$ a real positive number and $F$, $F'$ two fixed points from the plane such that $FF' = 2c$.

**Definition 7.10** (see [5], p. 156). The **hyperbola** is the set of the points $M$ of the plane that satisfy the relation

$$\left|MF - MF'\right| = 2a = ct, \tag{7.29}$$

i.e. that have constantly the difference of the distances to two fixed points.

We choose $FF'$ as O$x$ axis and the mediator of the segment $\overrightarrow{FF'}$ as O$y$ axis (see Fig. 7.4).

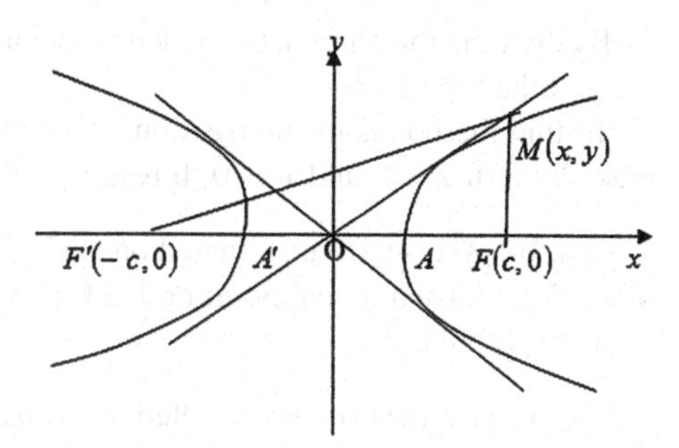


**Fig. 7.4** Hyperbola

$MF$, $MF'$ are the *focus radii* of the point $M$. The hyperbola admits a unique center of symmetry O and two axes of symmetry O$x$, O$y$.

From Fig. 7.4 we can note that the hyperbola is an unbounded curve.

The points $F(c, 0)$ and $F'(c, 0)$ are called the *foci* of the hyperbola and the distance $FF'$ constitutes the *focus distance* of the hyperbola.

To find the hyperbola equation we shall transform analytically the Eq. (7.29). If $M(x, y)$, then the relation (7.29) becomes:

$$\sqrt{(x-c)^2 + y^2} - \sqrt{(x+c)^2 + y^2} = \pm 2a. \tag{7.30}$$

We shall obtain:

$$(x+c)^2+y^2=(x-c)^2+y^2\pm 4a\sqrt{(x-c)^2+y^2}+4a^2.$$

By reducing the similar terms and passing in the first part of all those terms that don't contain radical signs, we have:

$$cx-a^2=\pm a\sqrt{(x-c)^2+y^2};$$

hence

$$\left(c^2-a^2\right)x^2-a^2y^2-a^2\left(c^2-a^2\right)=0. \tag{7.31}$$

By multiplying (7.30) with $-1$ and by changing the signs in the parentheses, we note that we have obtained the same Eq. (7.23) from the case of ellipse. The difference of the hyperbola towards the ellipse (where we have $a>c$) comes from the fact that in the triangle $MFF'$: from $FF'>\left|MF-MF'\right|=2a$ we have $c>a$; so that $c^2-a^2=b^2$. Therefore, we can denote

$$c^2-a^2=b^2. \tag{7.32}$$

By dividing the relation (7.31) with $a^2b^2$ it results the *implicit Cartesian equation* (7.9) of the hiperbola.

To find the points of intersection of the curve with the coordinate axes we shall make by turn $y=0$ and $x=0$. It results that

- $A(a,0)$, $A'(-a,0)$ are situated on $Ox$,
- we don't have any real point on the $Oy$ axis; therefore the $Oy$ axis doesn't cross the hyperbola.

That is why, the $Ox$ axis is called the *transverse axis* and the $Oy$ axis is called the *untransverse axis*.

The points $A$, $A'$ represent the *vertices of the hyperbola*.

From the Eq. (7.9) one deduces the *explicit Cartesian equations of the hyperbola*:

$$y=\pm\frac{b}{a}\sqrt{x^2-a^2},\ x\in(-\infty,-a]\cup[a,\infty). \tag{7.33}$$

The hyperbola admits two oblique asymptotes:

$$y=\pm\frac{b}{a}x. \tag{7.34}$$

From (7.9) we shall have:

$$\left(\frac{x}{a}+\frac{y}{b}\right)\left(\frac{x}{a}-\frac{y}{b}\right)=1.$$

If $x \in [a, \infty)$, by denoting

$$\left.\begin{array}{l}\frac{x}{a}+\frac{y}{b}=\mathrm{e}^{t} \\ \frac{x}{a}-\frac{y}{b}=\mathrm{e}^{-t}\end{array}\right\} \Rightarrow x=a \frac{\mathrm{e}^{t}+\mathrm{e}^{-t}}{2};$$

it results that the *parametric equations of the hyperbola* are:

$$\begin{cases} x = a \cdot \cosh t \\ y = b \cdot \sinh t \end{cases}, t \in \mathbb{R}.$$

If $x \in (-\infty, -a]$, by denoting

$$\left.\begin{array}{l}\frac{x}{a}+\frac{y}{b}=-\mathrm{e}^{t} \\ \frac{x}{a}-\frac{y}{b}=-\mathrm{e}^{-t}\end{array}\right\} \Rightarrow x=-a \frac{\mathrm{e}^{t}+\mathrm{e}^{-t}}{2};$$

it results that the *parametric equations of the hyperbola* are:

$$\begin{cases} x = -a \cdot \cosh t \\ y = -b \cdot \sinh t \end{cases}, t \in \mathbb{R}.$$

**Example 7.11** (see [5], p. 158). Determine the vertices, the foci and the asymptotes of the hyperbola

$$2x^2 - 5y^2 - 8 = 0.$$

**Solution**

Writing the equation of the hyperbola in the form

$$\frac{x^2}{4} - \frac{y^2}{\frac{8}{5}} - 1 = 0,$$

we deduce:

$$\begin{cases} a^2 = 4 \\ b^2 = \frac{8}{5} \\ c^2 = a^2 + b^2 = \frac{28}{5}. \end{cases}$$

The hyperbola vertices are: $A(2, 0)$, $A'(-2, 0)$ and its foci: $F\left(2\sqrt{\frac{7}{5}}, 0\right)$, $F'\left(-2\sqrt{\frac{7}{5}}, 0\right)$. The equations of the hyperbola asymptotes are:

$$y = \pm\sqrt{\frac{2}{5}}x.$$

Using Sage, we achieve:

```
sage: R.<x,y> = PolynomialRing(QQ,2)
sage: I = R.ideal([x^2/4-y^2/(8/5)- 1])
sage: F=point2d((2*sqrt(7/5),0),size=22);t1=text("F",(2*sqrt(7/5)+0.3,0.2))
sage: Fp=point2d((-2*sqrt(7/5),0),size=22);t2=text("Fp",(-2*sqrt(7/5)-0.3,0.2))
sage: A=point2d((2,0),size=22);t3=text("A",(2-0.3,0.2))
sage: Ap=point2d((-2,0),size=22);t4=text("Ap",(-2+0.5,0.1))
sage: a1=plot(sqrt(2/5)*x, (-10, 10))
sage: a2=plot(-sqrt(2/5)*x, (-10, 10))
sage: I.plot()+F+Fp+A+Ap+t1+t2+t3+t4+a1+a2
```

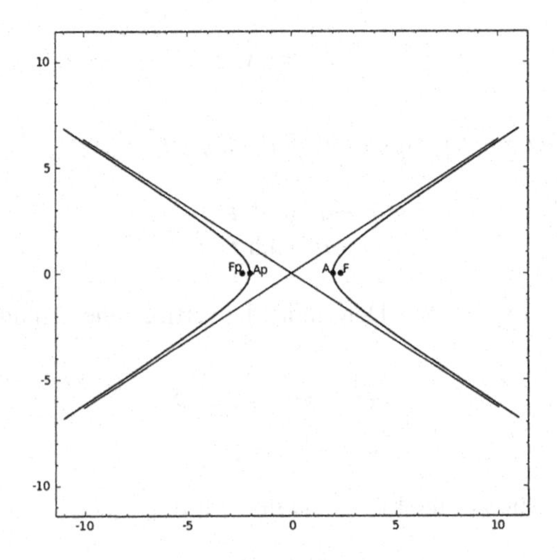


## 7.2.4 Parabola

**Definition 7.12** (see [5], p. 182). The **parabola** is the set of the points from plane, that are equidistant from a fixed line and a fixed point.

The fixed line is called the *directrix of the parabola* and the fixed point is called the *focus of the parabola*.

To find the equation of the parabola we choose a Cartesian reference whose axes are:

- the perpendicular from the focus $F$ on the directrix $d$ as $Ox$ axis,
- the parallel to $d$ going to the midway between the focus and the directrix $d$ as $Oy$ axis.

We denote $A = d \cap Ox$. Let $M$ be a point of the parabola and $N$ be its projection on the directrix (see Fig. 7.5).

**Fig. 7.5** Parabola

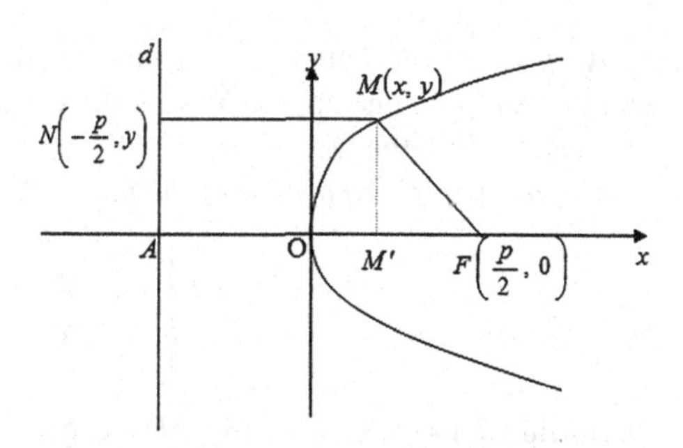


The parabola has no center of symmetry and it has a single axis of symmetry, O$x$. It is an unbounded curve.

We denote $AF = p$; it results $F\left(\frac{p}{2},0\right)$ and $A\left(-\frac{p}{2},0\right)$. If $M$ is an arbitrary point of the parabola, then within the **Definition 7.12**, the point $M$ has to satisfy the relation:

$$MF = MN. \tag{7.35}$$

As

$$MN = x - \left(-\frac{p}{2}\right) = x + \frac{p}{2},$$

the relation (7.35) becomes:

$$\sqrt{\left(x - \frac{p}{2}\right)^2 + y^2} = x + \frac{p}{2};$$

hence one achieves *the implicit Cartesian equation of the parabola*:

$$y^2 = 2px. \tag{7.36}$$

**Remark 7.13** (see [1], p. 165). In the case when $x \le 0$, the *implicit Cartesian equation of the parabola* will become:

$$y^2 = -2px.$$

The O$x$ axis cuts the parabola in the point O(0, 0) called the *parabola vertex*. From the Eq. (7.36) one deduces the *explicit Cartesian equations of the parabola*:

$$y = \pm\sqrt{2px}, x \ge 0, \tag{7.37}$$

$p$ being a positive number called the *parabola parameter*, which indicates its form.

As $p$ is smaller, both the focus and the directrix one approach [6] by the O$y$ axis, and the parabola one approaches by the O$x$ axis (when $p \to 0$ then the parabola degenerates into the O$x$ axis).

As $p$ is greater, both the focus and the directrix one depart [6] by the $Oy$ axis, and the parabola one approaches by the $Oy$ axis (when $p \to \infty$ then the parabola degenerates into the $Oy$ axis).

The *parametric equations of the parabola* are:

$$\begin{cases} x = \frac{t^2}{2p} \\ y = t \end{cases}, t \in \mathbb{R}. \tag{7.38}$$

**Definition 7.14** (see [5], p. 187). The eccentricity denoted by $e$ is an element which characterizes the circle, the ellipse, the hiperbola and the parabola, representing the ratio of the distances from an arbitrary point of the respective conic to the focus and the directrix, namely:

- $e = 0$ characterizes the circle $\left(e = \frac{c}{a} = \frac{0}{a}\right)$,
- $0 < e < 1$ characterizes the ellipse $\left(e = \frac{c}{a}\right)$,
- $e > 1$ characterizes the hyperbola $\left(e = \frac{c}{a}\right)$,
- $e = 1$ characterizes the parabola $\left(e = \frac{MF}{MN}\right)$.

## 7.3 Reducing to the Canonical Form of a Conic Equation

We propose to determine an orthonormal Cartesian reference, relative to which the general equation from (7.2) of has one of the canonical forms from (7.8), (7.9) or (7.12).

We can distinguish [1], [2] the following situations:

*Case 1.* $\delta \neq 0$, i.e. the conic admits an unique center of symmetry $C(x_0, y_0)$.

The stages that are necessary in this case to obtain a canonical equation of a conic are:

(1) attach the square form

$$Q(\overline{v}) = a_{11}x^2 + 2a_{12}xy + a_{22}y^2, (\forall)\ \overline{v} = (x, y) \in \mathbb{R}^2 \tag{7.39}$$

to the Eq. (7.2)

(2) write the matrix

$$A = \begin{pmatrix} a_{11} & a_{12} \\ a_{12} & a_{22} \end{pmatrix}$$

associated to the form $Q$ relative to the canonical basis $B$ of $\mathbb{R}^2$ and then build the characteristic polynomial:

$$P(\lambda) = \begin{vmatrix} a_{11} - \lambda & a_{12} \\ a_{12} & a_{22} - \lambda \end{vmatrix} = \lambda^2 - I\lambda + \delta. \tag{7.40}$$

(3) make a change of an orthonormal reference such that the center of symmetry to be the origin of the new reference.

The transition from the coordinates $(x, y)$ to the coordinates $\left(x', y'\right)$ in the new reference is achieved by a translation of the vector $\overline{OC}$, characterized by the equations:

$$\begin{cases} x = x_0 + x' \\ y = y_0 + y'. \end{cases} \tag{7.41}$$

By this transformation, the general Eq. (7.2) of the conic becomes:

$$a_{11}\left(x_0 + x'\right)^2 + 2a_{12}\left(x_0 + x'\right)\left(y_0 + y'\right) + a_{22}^2\left(y_0 + y'\right)^2 + 2b_1\left(x_0 + x'\right) + 2b_2\left(y_0 + y'\right) + c = 0,$$

i.e.

$$a_{11}x'^2 + 2a_{12}x'y' + a_{22}^2 y'^2 + c' = 0, c' = f\left(x_0, y_0\right). \tag{7.42}$$

(4) determine an orthonormal basis $B'$, composed of the eigenvectors corresponding to the eigenvalues $\lambda_1$ and $\lambda_2$ of the matrix $A$.
(5) rewrite the Eq. (7.42) relative to the basis $B'$; it will become

$$\lambda_1 x''^2 + \lambda_2 y''^2 + c' = 0. \tag{7.43}$$

The transition to the new coordinates $\left(x'', y''\right)$ one achieves through the relation:

$$\begin{pmatrix} x' \\ y' \end{pmatrix} = \mathrm{M}_{(B,B')} \cdot \begin{pmatrix} x'' \\ y'' \end{pmatrix}. \tag{7.44}$$

We can note that for the Eq. (7.43) we have

$$\begin{cases} \delta = \begin{vmatrix} \lambda_1 & 0 \\ 0 & \lambda_2 \end{vmatrix} = \lambda_1\lambda_2 \\ \Delta = \begin{vmatrix} \lambda_1 & 0 & 0 \\ 0 & \lambda_2 & 0 \\ 0 & 0 & c' \end{vmatrix} = \lambda_1\lambda_2 c'. \end{cases}$$

As $\delta \neq 0$ it results that

$$\frac{\Delta}{\delta} = c'.$$

(6) obtain the canonical equation of the conic:

$$\lambda_1 x''^2 + \lambda_2 y''^2 + \frac{\Delta}{\delta} = 0. \tag{7.45}$$

**Remark 7.15** (see [2]). If $\lambda_1$ and $\lambda_2$ have the same sign and $\frac{\Delta}{\delta}$ has an opposite one, then from (7.45) one obtains an ellipse. If $\lambda_1$ and $\lambda_2$ have different signs then from (7.45) one obtains an hyperbola.

**Example 7.16** (see [4], p. 209). Let be the conic

$$\Gamma : 9x^2 - 4xy + 6y^2 + 16x - 8y - 2 = 0.$$

Bring it to the canonical form, pointing the necessary reference changes, recognize the obtained conic and plot its graph.

**Solution**

Using (7.4), we have:

$$\begin{aligned} I &= 9 + 6 = 15, \\ \delta &= \begin{vmatrix} 9 & -2 \\ -2 & 6 \end{vmatrix} = 50 > 0 \\ \Delta &= \begin{vmatrix} 9 & -2 & 8 \\ -2 & 6 & -4 \\ 8 & -4 & -2 \end{vmatrix} = -500 \neq 0. \end{aligned}$$

As

- $\delta > 0$, the conic $\Gamma$ has an elliptic type
- $\Delta \neq 0$, the conic $\Gamma$ is non-degenerate
- $\delta \neq 0$, the conic $\Gamma$ admits an unique center of symmetry.

The center of the conic is given by the system:

$$\begin{cases} 18x - 4y + 16 = 0 \\ -4x + 12y - 8 = 0, \end{cases}$$

i.e. it is the point $C\left(-\frac{4}{5}, \frac{2}{5}\right)$.

We shall make a change of an orthonormal reference so that the center of symmetry to be the origin of the new reference:

$$\begin{cases} x' = x + \frac{4}{5} \\ y' = y - \frac{2}{5} \end{cases} \Leftrightarrow \begin{cases} x = -\frac{4}{5} + x' \\ y = \frac{2}{5} + y'. \end{cases}$$

By this transformation, the equation of the conic becomes:

$$9\left(-\frac{4}{5} + x'\right)^2 - 4\left(-\frac{4}{5} + x'\right)\left(\frac{2}{5} + y'\right) + 6\left(\frac{2}{5} + y'\right)^2 + 16\left(-\frac{4}{5} + x'\right) - 8\left(\frac{2}{5} + y'\right) - 2 = 0$$

or

$$\Gamma : 9x'^2 - 4x'y' + 6y'^2 - 10 = 0. \quad (7.46)$$

We note that

$$c' \overset{(7.43)}{=} f\left(-\frac{4}{5}, \frac{2}{5}\right),$$

where

$$f(x, y) = 9x^2 - 4xy + 6y^2 + 16x - 8y - 2.$$

The matrix associated to the quadratic form (7.46) is

$$A = \begin{pmatrix} 9 & -2 \\ -2 & 6 \end{pmatrix}.$$

As the characteristic polynomial associated to the matrix $A$ is

$$P(\lambda) \overset{(7.40)}{=} \lambda^2 - 15\lambda + 50 = (\lambda - 10)(\lambda - 5)$$

it results the eigenvalues: $\lambda_1 = 5, \lambda_2 = 10$.

The eigensubspace associated to the eigenvalue $\lambda_1$ will be

$$V_{\lambda_1} = \left\{\overline{v} \in \mathbb{R}^2 | A\overline{v} = \lambda_1 \overline{v}\right\}.$$

We achieve:

$$\begin{pmatrix} 9 & -2 \\ -2 & 6 \end{pmatrix} \begin{pmatrix} v_1 \\ v_2 \end{pmatrix} = 5 \begin{pmatrix} v_1 \\ v_2 \end{pmatrix};$$

therefore

$$\begin{cases} 9v_1 - 2v_2 = 5v_1 \\ -2v_1 + 6v_2 = 5v_2 \end{cases} \Rightarrow 2v_1 = v_2;$$

it results

$$V_{\lambda_1} = \left\{\overline{v} \in \mathbb{R}^2 | \, \overline{v} = v_1 \underbrace{(1, 2)}_{\overline{w}}, (\forall)\, v_1 \in \mathbb{R}\right\}.$$

We shall obtain the orthonormal basis $B_1 = \{\overline{f}_1\}$, where:

$$\overline{f}_1 = \frac{\overline{w}}{\|\overline{w}\|} = \left(\frac{1}{\sqrt{5}}, \frac{2}{\sqrt{5}}\right).$$

The eigensubspace associated to the eigenvalue $\lambda_2$ will be

$$V_{\lambda_2} = \left\{ \overline{u} \in \mathbb{R}^2 | A\overline{u} = \lambda_2 \overline{u} \right\}.$$

We achieve:

$$\begin{pmatrix} 9 & -2 \\ -2 & 6 \end{pmatrix} \begin{pmatrix} u_1 \\ u_2 \end{pmatrix} = 10 \begin{pmatrix} u_1 \\ u_2 \end{pmatrix};$$

therefore

$$\begin{cases} 9u_1 - 2u_2 = 10u_1 \\ -2u_1 + 6u_2 = 10u_2 \end{cases} \Rightarrow u_1 = -2u_2;$$

it results

$$V_{\lambda_2} = \left\{ \overline{u} \in \mathbb{R}^2 | \overline{u} = u_2 \underbrace{(-2, 1)}_{\overline{z}}, (\forall)\, u_2 \in \mathbb{R} \right\}.$$

We shall obtain the orthonormal basis $B_2 = \left\{ \overline{f}_2 \right\}$, where:

$$\overline{f}_2 = \frac{\overline{z}}{\|\overline{z}\|} = \left( -\frac{2}{\sqrt{5}}, \frac{1}{\sqrt{5}} \right).$$

It will result

$$B = B_1 \cup B_2 = \left\{ \overline{f}_1, \overline{f}_2 \right\}.$$

The equation of the conic becomes:

$$\Gamma : 5x''^2 + 10y''^2 - 10 = 0.$$

The transition to the new coordinates $\left(x'', y''\right)$ one achieves through the relation (7.44), where

$$\mathrm{M}_{(B,B')} = \begin{pmatrix} \frac{1}{\sqrt{5}} & \frac{2}{\sqrt{5}} \\ -\frac{2}{\sqrt{5}} & \frac{1}{\sqrt{5}} \end{pmatrix};$$

therefore

$$\begin{cases} x' = \frac{1}{\sqrt{5}} x'' + \frac{2}{\sqrt{5}} y'' \\ y' = -\frac{2}{\sqrt{5}} x'' + \frac{1}{\sqrt{5}} y''. \end{cases} \tag{7.47}$$

The conic equation will have the canonical form

$$\Gamma : \frac{x''^2}{2} + y''^2 - 1 = 0$$

and it represents an ellipse.

The ellipse axes have the equations $x'' = 0$ and respectively $y'' = 0$.

Solving the system from (7.47) we deduce

$$\begin{cases} x'' = \frac{x'-2y'}{\sqrt{5}} \\ y'' = \frac{2x'+y'}{\sqrt{5}}. \end{cases} \tag{7.48}$$

Therefore, the axes of the ellipse have the equations:

$$\begin{cases} x' - 2y' = 0 \\ 2x' + y' = 0. \end{cases} \tag{7.49}$$

Taking into account that

$$\begin{cases} x' = x + \frac{4}{5} \\ y' = y - \frac{2}{5} \end{cases}$$

we deduce that the ellipse axes will have the equations:

$$\begin{cases} x + \frac{4}{5} - 2\left(y - \frac{2}{5}\right) = 0 \\ 2\left(x + \frac{4}{5}\right) + y - \frac{2}{5} = 0, \end{cases}$$

i.e.

$$\begin{cases} x - 2y + \frac{8}{5} = 0 \\ 2x + y + \frac{6}{5} = 0. \end{cases} \tag{7.50}$$

The ellipse center will be $O''\left(x'' = 0, y'' = 0\right)$. We have:

$$\begin{cases} x'' = 0 \\ y'' = 0 \end{cases} \overset{(7.48)}{\Leftrightarrow} \begin{cases} \frac{x'-2y'}{\sqrt{5}} = 0 \\ \frac{2x'+y'}{\sqrt{5}} = 0 \end{cases} \overset{(7.49)}{\Leftrightarrow}$$

$$\begin{cases} x' - 2y' = 0 \\ 2x' + y' = 0 \end{cases} \overset{(7.50)}{\Leftrightarrow} \begin{cases} x - 2y + \frac{8}{5} = 0 \\ 2x + y + \frac{6}{5} = 0 \end{cases}$$

$$\Leftrightarrow \begin{cases} x = -\frac{4}{5} \\ y = \frac{2}{5}; \end{cases}$$

hence the ellipse center will be $O''\left(x = -\frac{4}{5}, y = \frac{2}{5}\right)$.

The vertices of the ellipse will be:

$$A\left(x''=\sqrt{2}, y''=0\right), A'\left(x''=-\sqrt{2}, y''=0\right),$$
$$B\left(x''=0, y''=1\right), B'\left(x''=0, y''=-1\right).$$

We deduce

$$\begin{cases} x''=\sqrt{2} \\ y''=0 \end{cases} \overset{(7.48)}{\Leftrightarrow} \begin{cases} \frac{x'-2y'}{\sqrt{5}}=\sqrt{2} \\ \frac{2x'+y'}{\sqrt{5}}=0 \end{cases} \overset{(7.49)}{\Leftrightarrow}$$

$$\begin{cases} x'-2y'=\sqrt{10} \\ 2x'+y'=0 \end{cases} \overset{(7.50)}{\Leftrightarrow} \begin{cases} x-2y+\frac{8}{5}=\sqrt{10} \\ 2x+y+\frac{6}{5}=0 \end{cases}$$

$$\Leftrightarrow \begin{cases} x=\frac{\sqrt{10}-4}{5} \\ y=\frac{2\left(1-\sqrt{10}\right)}{5}; \end{cases}$$

$$\begin{cases} x''=-\sqrt{2} \\ y''=0 \end{cases} \Leftrightarrow \begin{cases} x=\frac{-\sqrt{10}-4}{5} \\ y=\frac{2\left(1+\sqrt{10}\right)}{5}; \end{cases}$$

$$\begin{cases} x''=0 \\ y''=1 \end{cases} \overset{(7.48)}{\Leftrightarrow} \begin{cases} \frac{x'-2y'}{\sqrt{5}}=0 \\ \frac{2x'+y'}{\sqrt{5}}=1 \end{cases} \overset{(7.49)}{\Leftrightarrow}$$

$$\begin{cases} x'-2y'=0 \\ 2x'+y'=\sqrt{5} \end{cases} \overset{(7.50)}{\Leftrightarrow} \begin{cases} x-2y+\frac{8}{5}=0 \\ 2x+y+\frac{6}{5}=\sqrt{5} \end{cases}$$

$$\Leftrightarrow \begin{cases} x=\frac{2\sqrt{5}-4}{5} \\ y=\frac{2+\sqrt{5}}{5}; \end{cases}$$

$$\begin{cases} x''=0 \\ y''=-1 \end{cases} \Leftrightarrow \begin{cases} x=\frac{-2\sqrt{5}-4}{5} \\ y=\frac{2-\sqrt{5}}{5}. \end{cases}$$

We obtain the following vertices of the ellipse:

$$A\left(x=\frac{\sqrt{10}-4}{5}, y=\frac{2\left(1-\sqrt{10}\right)}{5}\right), A'\left(x=\frac{-\sqrt{10}-4}{5}, y=\frac{2\left(1+\sqrt{10}\right)}{5}\right)$$

$$B\left(x=\frac{2\sqrt{5}-4}{5}, y=\frac{2+\sqrt{5}}{5}\right), B'\left(x=\frac{-2\sqrt{5}-4}{5}, y=\frac{2-\sqrt{5}}{5}\right).$$

The obtained ellipse has the following graphic representation:

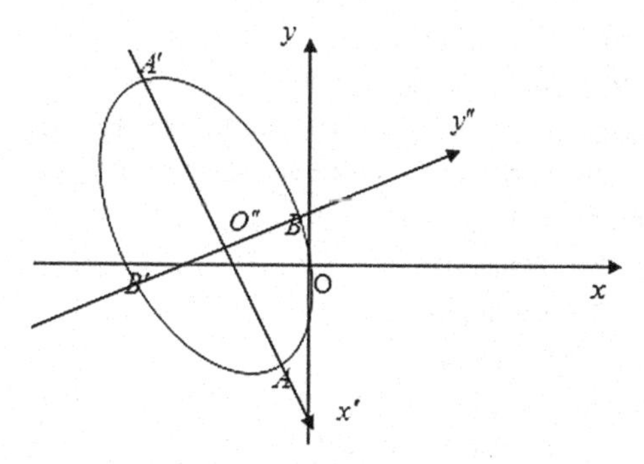


The solution in Sage is:

```
sage: a11=9;a12=-2;a22=6;b1=8;b2=-4;c=-2
sage: I=a11+a22;I
15
sage: de=matrix([[a11,a12],[a12,a22]]).det();de
50
sage: De=matrix([[a11,a12,b1],[a12,a22,b2],[b1,b2,c]]).det();De
-500
sage: x,y,xp,yp,xs,ys,la,v1,v2=var('x,y,xp,yp,xs,ys,la,v1,v2')
sage: f=a11*x^2+2*a12*x*y+a22*y^2+2*b1*x+2*b2*y+c
sage: s=solve([diff(f,x),diff(f,y)],x,y);s
[[x == (-4/5), y == (2/5)]]
sage: ff1=expand(f.subs(x=-4/5+xp,y=2/5+yp))
sage: aa11=ff1.coeff(xp^2);aa22=ff1.coeff(yp^2);aa12=ff1.coeff(xp).coeff(yp)
sage: Ap=matrix([[aa11,aa12/2],[aa12/2,aa22]])
sage: laa=solve([la^2-I*la+de],la)
sage: la1=laa[1].right();la2=laa[0].right()
sage: v=vector(SR,[v1,v2]);vv=v.transpose()
sage: m=Ap*vv-la1*vv;m1=Ap*vv-la2*vv
sage: ss=solve([m[0][0],m[1][0]],v1,v2);ss
[[v1 == 1/2*r1, v2 == r1]]
sage: f1=vector([ss[0][0].right().subs(r1=2),ss[0][1].right().subs(r1=2)])
sage: g1=f1/f1.norm()
sage: ss1=solve([m1[0][0],m1[1][0]],v1,v2);ss1
[[v1 == -2*r2, v2 == r2]]
sage: f2=vector([ss1[0][0].right().subs(r2=1),ss1[0][1].right().subs(r2=1)])
sage: g2=f2/f2.norm()
sage: M=matrix([g1,g2]);M
[ 1/5*sqrt(5) 2/5*sqrt(5)]
[-2/5*sqrt(5) 1/5*sqrt(5)]
sage: la1*xs^2+la2*ys^2+f.subs(x=s[0][0].right(),y=s[0][1].right())
5*xs^2 + 10*ys^2 - 10
sage: u0=M*vector(SR,[xs,ys])-vector(SR,[xp,yp])
```

```
sage: sol=solve([u0[0],u0[1]],xs,ys)
sage: eq=[sol[0][0].right(),sol[0][1].right()]
sage: eq1=expand(eq[0].subs(xp=x+4/5,yp=y-2/5)).simplify_exp()
sage: eq2=expand(eq[1].subs(xp=x+4/5,yp=y-2/5)).simplify_exp()
sage: eq1==0;eq2==0
1/25*(5*x - 10*y + 8)*sqrt(5) == 0
1/25*(10*x + 5*y + 6)*sqrt(5) == 0
sage: zz=solve([eq1,eq2],x,y)
sage: O=vector([zz[0][0].right().n(digits=3),zz[0][1].right().n(digits=3)])
sage: z=solve([eq1==sqrt(2),eq2],x,y)
sage: A=vector([z[0][0].right().n(digits=3),z[0][1].right().n(digits=3)])
sage: z1=solve([eq1==-sqrt(2),eq2],x,y)
sage: A1=vector([z1[0][0].right().n(digits=3),z1[0][1].right().n(digits=3)])
sage: z2=solve([eq1,eq2==1],x,y)
sage: B=vector([z2[0][0].right().n(digits=3),z2[0][1].right().n(digits=3)])
sage: z3=solve([eq1,eq2==-1],x,y)
sage: B1=vector([z3[0][0].right().n(digits=3),z3[0][1].right().n(digits=3)])
sage: OA=A-O;oa=OA.norm();OB=B-O;ob=OB.norm()
sage: e=ellipse((O[0], O[1]), oa, ob,asin(A[1]/ob))
sage: l1=implicit_plot(eq1, (x,-3,3), (y,-3,3))
sage: l2=implicit_plot(eq2, (x,-3,3), (y,-3,3))
sage: el=implicit_plot(f, (x,-2,2), (y,-2,2))
sage: Aa=point2d((A[0],A[1]),size=22);Aa1=point2d((A1[0],A1[1]),size=22)
sage: Bb=point2d((B[0],B[1]),size=22);Bb1=point2d((B1[0],B1[1]),size=22)
sage: t1=text("A",(A[0]+0.1,A[1]-0.1));t2=text("Ap",(A1[0]+0.1,A1[1]+0.1))
sage: t3=text("B",(B[0]+0.12,B[1]+0.12));t4=text("Bp",(B1[0]-0.12,B1[1]+0.12))
sage: Aa+l1+l2+Aa1+Bb+Bb1+t1+t2+t3+t4+e
```

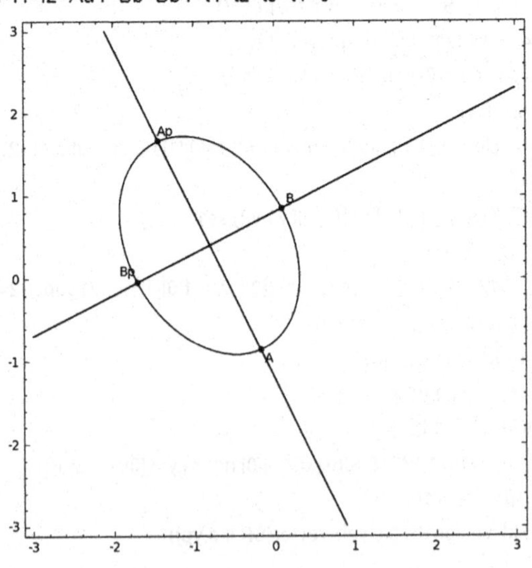

*Case 2*. $\delta = 0$, i.e. namely the conic hasn't an unique center of symmetry.
The stages that are necessary in this case to obtain a canonical equation of a conic are (1), (2), (4) from the *Case 1* followed by:

(1') relative to the basis $B'$, thc Eq. (7.1) will become

$$\lambda_1 x'^2 + \lambda_2 y'^2 + 2b'_1 x' + 2b'_2 y' + c' = 0. \tag{7.51}$$

The transition to the new coordinates $(x', y')$ one achieves through the relation:

$$\begin{pmatrix} x \\ y \end{pmatrix} = \mathrm{M}_{(B,B')} \cdot \begin{pmatrix} x' \\ y' \end{pmatrix}. \tag{7.52}$$

We can note that for the Eq. (7.51) we have

$$\delta = \begin{vmatrix} \lambda_1 & 0 \\ 0 & \lambda_2 \end{vmatrix} = \lambda_1 \lambda_2.$$

We suposse $\lambda_2 = 0$ for $\delta = 0$. We shall obtain the equation

$$\lambda_1 x'^2 + 2b'_1 x' + 2b'_2 y' + c' = 0. \tag{7.53}$$

(2') Form a perfect square

$$\lambda_1 \left( x' + \frac{b'_1}{\lambda_1} \right)^2 + 2b'_2 y' + c' - \frac{b'^2_1}{\lambda_1} = 0.$$

(3') Make the change of coordinates

$$\begin{cases} x'' = x' + \frac{b'_1}{\lambda_1} \\ y'' = y'; \end{cases}$$

it results the equation:

$$\lambda_1 x''^2 + 2b'_2 y'' + c'' = 0.$$

We note that:

$$\Delta = \begin{vmatrix} \lambda_1 & 0 & 0 \\ 0 & 0 & b'_2 \\ 0 & b'_2 & c'' \end{vmatrix} = -\lambda_1 b'^2_2.$$

(4') $\Delta \neq 0 \Leftrightarrow b'_2 \neq 0$; therefore, we can write

$$\lambda_1 x''^2 + 2b'_2 \left( y'' + \frac{c''}{2b'_2} \right) = 0.$$

(5') Making the change of coordinates

$$\begin{cases} X = x'' \\ Y = y'' + \frac{c''}{2b'_2} \end{cases}$$

it results the canonical equation

$$\lambda_1 X^2 + 2b'_2 Y = 0, \tag{7.54}$$

which corresponds to a parabola.

**Example 7.17** (see [2]). Let be the conic

$$\Gamma : 4x^2 - 4xy + y^2 - 3x + 4y - 7 = 0.$$

Bring it to the canonical form, pointing the necessary reference changes, recognize the obtained conic and plot its graph.

**Solution**

Using (7.4), we have:

$$I = 4 + 1 = 5,$$
$$\delta = \begin{vmatrix} 4 & -2 \\ -2 & 1 \end{vmatrix} = 0$$
$$\Delta = \begin{vmatrix} 4 & -2 & -3/2 \\ -2 & 1 & 2 \\ -3/2 & 2 & -7 \end{vmatrix} = -\frac{25}{4} \neq 0.$$

As

- $\Delta \neq 0$, the conic $\Gamma$ is non-degenerate
- $\delta = 0$, the conic $\Gamma$ doesn't admit an unique center of symmetry.

The matrix associated to the quadratic form is

$$A = \begin{pmatrix} 4 & -2 \\ -2 & 1 \end{pmatrix}.$$

As the characteristic polynomial associated to the matrix $A$ is

$$P(\lambda) \overset{(7.40)}{=} \lambda^2 - 5 = \lambda(\lambda - 5)$$

it results the eigenvalues: $\lambda_1 = 0$, $\lambda_2 = 5$.

The eigensubspace associated to the eigenvalue $\lambda_1$ will be

$$V_{\lambda_1} = \left\{ \overline{u} \in \mathbb{R}^2 | \, A\overline{u} = \lambda_1 \overline{u} \right\}.$$

We achieve:

$$\begin{pmatrix} 4 & -2 \\ -2 & 1 \end{pmatrix}\begin{pmatrix} u_1 \\ u_2 \end{pmatrix} = 0 \cdot \begin{pmatrix} u_1 \\ u_2 \end{pmatrix};$$

therefore

$$\begin{cases} 4u_1 - 2u_2 = 0 \\ -2u_1 + u_2 = 0 \end{cases} \Rightarrow u_2 = 2u_1;$$

it results

$$V_{\lambda_1} = \left\{ \overline{v} \in \mathbb{R}^2 | \overline{u} = u_1 \underbrace{(1, 2)}_{\overline{w}}, (\forall)\, u_1 \in \mathbb{R} \right\}.$$

We shall obtain the orthonormal basis $B_1 = \{\overline{f}_1\}$, where:

$$\overline{f}_1 = \frac{\overline{w}}{\|\overline{w}\|} = \left(\frac{1}{\sqrt{5}}, \frac{2}{\sqrt{5}}\right).$$

The eigensubspace associated to the eigenvalue $\lambda_2$ will be

$$V_{\lambda_2} = \left\{ \overline{v} \in \mathbb{R}^2 | A\overline{v} = \lambda_2 \overline{v} \right\}.$$

We achieve:

$$\begin{pmatrix} 4 & -2 \\ -2 & 1 \end{pmatrix}\begin{pmatrix} v_1 \\ v_2 \end{pmatrix} = 5 \begin{pmatrix} v_1 \\ v_2 \end{pmatrix};$$

therefore

$$\begin{cases} 4v_1 - 2v_2 = 5v_1 \\ -2v_1 + v_2 = 5v_2 \end{cases} \Rightarrow v_1 = -2v_2;$$

it results

$$V_{\lambda_2} = \left\{ \overline{v} \in \mathbb{R}^2 | \overline{v} = v_2 \underbrace{(-2, 1)}_{\overline{z}}, (\forall)\, u_2 \in \mathbb{R} \right\}.$$

We shall obtain the orthonormal basis $B_2 = \{\overline{f}_2\}$, where:

$$\overline{f}_2 = \frac{\overline{z}}{\|\overline{z}\|} = \left(-\frac{2}{\sqrt{5}}, \frac{1}{\sqrt{5}}\right).$$

It will result

$$B = B_1 \cup B_2 = \{\overline{f}_1, \overline{f}_2\}.$$

Relative to the basis $B'$, the conic equation will become:

$$\lambda_1 x'^2 + \lambda_2 y'^2 - 3\left(\frac{1}{\sqrt{5}}x' - \frac{2}{\sqrt{5}}y'\right) + 4\left(\frac{2}{\sqrt{5}}x' + \frac{1}{\sqrt{5}}y'\right) - 7 = 0$$

or

$$5y'^2 - \frac{11}{\sqrt{5}}x' - \frac{2}{\sqrt{5}}y' - 7 = 0. \tag{7.55}$$

The transition to the new coordinates $(x', y')$ one achieves through the relation (7.52), where

$$\mathrm{M}_{(B,B')} = \begin{pmatrix} \frac{1}{\sqrt{5}} & \frac{2}{\sqrt{5}} \\ -\frac{2}{\sqrt{5}} & \frac{1}{\sqrt{5}} \end{pmatrix};$$

therefore

$$\begin{cases} x = \frac{1}{\sqrt{5}}x' + \frac{2}{\sqrt{5}}y' \\ y = -\frac{2}{\sqrt{5}}x' + \frac{1}{\sqrt{5}}y'. \end{cases}$$

We will form a perfect square, writing the Eq. (7.55) in the form

$$5\left(y' - \frac{1}{5\sqrt{5}}\right)^2 - \frac{11}{\sqrt{5}}x' - \frac{176}{25} = 0.$$

Making the change of coordinates

$$\begin{cases} x'' = x' \\ y'' = y' - \frac{1}{5\sqrt{5}} \end{cases}$$

it results the equation

$$5y''^2 - \frac{11}{\sqrt{5}}x'' - \frac{176}{25} = 0.$$

We can write:

$$5y''^2 - \frac{11}{\sqrt{5}}\left(x'' - \frac{176}{25} \cdot \frac{\sqrt{5}}{11}\right) = 0.$$

Making the change of coordinates

$$\begin{cases} X = x'' - \frac{176}{25} \cdot \frac{\sqrt{5}}{11}, \\ Y = y'' \end{cases}$$

the conic equation will have the canonical form

$$\Gamma : Y^2 = \frac{11}{\sqrt{5}} X$$

and it represents a parabola.

Solving this problem in Sage we achieve:

```
sage: a11=4;a12=-2;a22=1;b1=-3/2;b2=2;c=-7
sage: I=a11+a22;I
5
sage: de=matrix([[a11,a12],[a12,a22]]).det();de
0
sage: De=matrix([[a11,a12,b1],[a12,a22,b2],[b1,b2,c]]).det();De
-25/4
sage: x,y,xp,yp,xs,ys,la,v1,v2,X,Y=var('x,y,xp,yp,xs,ys,la,v1,v2,X,Y')
sage: f=a11*x^2+2*a12*x*y+a22*y^2+2*b1*x+2*b2*y+c
sage: aa11=f.coeff(x^2);aa22=f.coeff(y^2);aa12=f.coeff(x).coeff(y)
sage: A=matrix([[aa11,aa12/2],[aa12/2,aa22]])
sage: laa=solve([la^2-I*la+de],la)
sage: la1=laa[0].right();la2=laa[1].right()
sage: v=vector(SR,[v1,v2]);vv=v.transpose()
sage: m=A*vv-la1*vv;m1=A*vv-la2*vv
sage: ss=solve([m[0][0],m[1][0]],v1,v2);ss
[[v1 == 1/2*r1, v2 == r1]]
sage: f1=vector([ss[0][0].right().subs(r1=2),ss[0][1].right().subs(r1=2)])
sage: g1=f1/f1.norm()
sage: ss1=solve([m1[0][0],m1[1][0]],v1,v2);ss1
[[v1 == -2*r2, v2 == r2]]
sage: f2=vector([ss1[0][0].right().subs(r2=1),ss1[0][1].right().subs(r2=1)])
sage: g2=f2/f2.norm()
sage: M=matrix([g1,g2])
sage: u0=M*vector(SR,[xp,yp])
sage: ff=la1*xp^2+la2*yp^2+2*b1*u0[0]+2*b2*u0[1]+c
sage: ff
5*yp^2 - 11/5*sqrt(5)*xp - 2/5*sqrt(5)*yp - 7
sage: gg=la2*(yp+(ff.coeff(yp)/2)/la2)^2+2*xp*ff.coeff(xp)/2+c-(ff.coeff(yp)/2)^2/la2
sage: h=gg.subs(xp=xs,yp=ys+1/(5*sqrt(5)))
sage: h
5*ys^2 - 11/5*sqrt(5)*xs - 176/25
sage: r=la2*ys^2+h.coeff(xs)*(xs+176/25*sqrt(5)/11);r
5*ys^2 - 11/125*(25*xs + 16*sqrt(5))*sqrt(5)
sage: r.subs(xs=X-176/25*sqrt(5)/11,ys=Y)
5*Y^2 - 11/5*sqrt(5)*X
```

The obtained parabola can be represented not only in the $(Oxy)$ plane, but in the $(Ox''y'')$ plane, too:

```
sage: F=vector([11/(5*sqrt(5)),0]);A=vector([-11/(5*sqrt(5)),0])
sage: Aa=point2d((A[0],A[1]),size=22);t1=text("A",(A[0]-0.1,A[1]+0.3))
sage: Ff=point2d((F[0],F[1]),size=22);t2=text("F",(F[0]-0.2,F[1]+0.2))
sage: O1=point2d((0,0),size=22);t3=text("O",(-0.2,-0.4))
sage: l1=implicit_plot(x, (x,-10,10),(y,-10,10))
sage: l2=implicit_plot(y, (x,-10,10),(y,-10,10))
sage: R.<x,y> = PolynomialRing(RR,2)
sage: I = R.ideal([y^2-11/sqrt(5)*x])
sage: I1=I.plot()
sage: I1+Aa+O1+Ff+l1+l2+t1+t2+t3
```

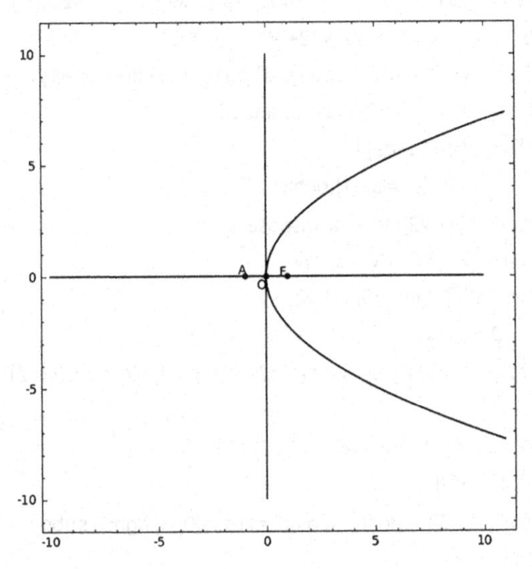


## 7.4 General Equation of a Quadric

**Definition 7.18** (see [7], p. 378). We consider the function $f : \mathbb{R}^3 \to \mathbb{R}$,

$$f(x, y, z) = a_{11}x^2 + a_{22}y^2 + a_{33}z^2 + 2a_{12}xy + 2a_{13}xz + 2a_{23}yz + 2b_1x + 2b_2y + 2b_3z + c. \tag{7.56}$$

An **algebraic surface of the second order** or a **quadric** is the set $\Sigma$ of the points $M(x, y, z)$ from the space, whose coordinates relative to an orthonormal Cartesian reference verify general the equation

$$f(x, y, z) = 0, \tag{7.57}$$

where the coefficients $a_{11}, a_{22}, a_{33}, a_{12}, a_{13}, a_{23}, b_1, b_2, b_3, c$ are some real constants, with $\sum_{i=1}^{3}\sum_{j=1}^{3} a_{ij}^2 \neq 0$; therefore

$$\Sigma = \left\{M(x, y, z) \mid (x, y, z) \in \mathbb{R}^3, f(x, y, z) = 0\right\}. \tag{7.58}$$

**Definition 7.19** (see [2]). The **invariants of a quadric** are those expressions formed with the coefficients of the conic equation, that keep the same value to the changes of an orthonormal reference.

**Proposition 7.20** (see [3], p. 69). We can assign four invariants of the quadric from (7.58):

$$\begin{aligned}
&1)\ I = a_{11} + a_{22} + a_{33} \\
&2)\ J = \begin{vmatrix} a_{11} & a_{12} \\ a_{12} & a_{22} \end{vmatrix} + \begin{vmatrix} a_{11} & a_{13} \\ a_{13} & a_{33} \end{vmatrix} + \begin{vmatrix} a_{22} & a_{23} \\ a_{23} & a_{33} \end{vmatrix} \\
&3)\ \delta = \begin{vmatrix} a_{11} & a_{12} & a_{13} \\ a_{12} & a_{22} & a_{23} \\ a_{13} & a_{23} & a_{33} \end{vmatrix} \\
&4)\ \Delta = \begin{vmatrix} a_{11} & a_{12} & a_{13} & b_1 \\ a_{12} & a_{22} & a_{23} & b_2 \\ a_{13} & a_{23} & a_{33} & b_3 \\ b_1 & b_2 & b_3 & c \end{vmatrix}.
\end{aligned} \tag{7.59}$$

The invariant $\Delta$ determines the nature of the quadric. Thus, if:

- $\Delta \neq 0$, the quadric is called *non- degenerate* (the sphere, the ellipsoid, the hyperboloids and the paraboloids)
- $\Delta = 0$, the quadric is called *degenerate* (the cone, the cylinders).

As in the case of the conics, the center of symmetry for a quadric $\Sigma$, having the Eq. (7.57) is the solution of the linear system:

$$\begin{cases} \frac{\partial f}{\partial x} = 0 \\ \frac{\partial f}{\partial y} = 0 \\ \frac{\partial f}{\partial z} = 0 \end{cases} \Leftrightarrow \begin{cases} a_{11}x + a_{12}y + a_{13}z + b_1 = 0 \\ a_{12}x + a_{22}y + a_{23}z + b_2 = 0 \\ a_{13}x + a_{23}y + a_{33}z + b_3 = 0. \end{cases} \tag{7.60}$$

**Theorem 7.21** (see [4], p. 176). Any quadric has one of the following canonical forms:

(1) imaginary ellipsoid:

$$\frac{x^2}{a^2} + \frac{y^2}{b^2} + \frac{z^2}{c^2} + 1 = 0, a \geq b \geq c > 0 \tag{7.61}$$

(2) real ellipsoid:

$$\frac{x^2}{a^2} + \frac{y^2}{b^2} + \frac{z^2}{c^2} - 1 = 0, a \geq b \geq c > 0 \tag{7.62}$$

(3) hyperboloid of one sheet:

$$\frac{x^2}{a^2} + \frac{y^2}{b^2} - \frac{z^2}{c^2} - 1 = 0, a \geq b > 0, c > 0 \tag{7.63}$$

(4) hyperboloid of two sheets

$$\frac{x^2}{a^2} + \frac{y^2}{b^2} - \frac{z^2}{c^2} + 1 = 0, a \geq b > 0, c > 0 \tag{7.64}$$

(5) imaginary second-order cone

$$\frac{x^2}{a^2} + \frac{y^2}{b^2} + \frac{z^2}{c^2} = 0, a \geq b > 0, c > 0 \tag{7.65}$$

(6) real second-order cone

$$\frac{x^2}{a^2} + \frac{y^2}{b^2} - \frac{z^2}{c^2} = 0, a \geq b > 0, c > 0 \tag{7.66}$$

(7) elliptic paraboloid

$$\frac{x^2}{a^2} + \frac{y^2}{b^2} - 2z = 0, a \geq b > 0 \tag{7.67}$$

(8) hyperbolic paraboloid

$$\frac{x^2}{a^2} - \frac{y^2}{b^2} - 2z = 0, a > 0, b > 0 \tag{7.68}$$

(9) imaginary elliptic cylinder

$$\frac{x^2}{a^2} + \frac{y^2}{b^2} + 1 = 0, a \geq b > 0 \tag{7.69}$$

(10) real elliptic cylinder

$$\frac{x^2}{a^2} + \frac{y^2}{b^2} - 1 = 0, a \geq b > 0 \tag{7.70}$$

(11) hyperbolic cylinder

$$\frac{x^2}{a^2} - \frac{y^2}{b^2} - 1 = 0, a > 0, b > 0 \tag{7.71}$$

(12) pair of imaginary planes, having the intersection a real line

$$\frac{x^2}{a^2} + \frac{y^2}{b^2} = 0, a, b > 0 \tag{7.72}$$

(13) pair of secant planes

$$\frac{x^2}{a^2} - \frac{y^2}{b^2} = 0, a, b > 0 \tag{7.73}$$

(14) parabolic cylinder

$$\frac{x^2}{a^2} - 2z = 0, a > 0 \tag{7.74}$$

(15) pair of parallel imaginary planes

$$\frac{x^2}{a^2} + 1 = 0, a > 0 \tag{7.75}$$

(16) pair of parallel real planes

$$\frac{x^2}{a^2} - 1 = 0, a > 0 \tag{7.76}$$

(17) pair of confounded planes

$$x^2 = 0. \tag{7.77}$$

**The General Table of the Quadric Discussion**

(I) If $\Delta \neq 0$ then [8] for:

(1) $\delta \neq 0$, the quadric has the canonical equation

$$S_1 x^2 + S_2 y^2 + S_3 z^2 + \frac{\Delta}{\delta} = 0, \tag{7.78}$$

where $S_1, S_2, S_3$ are the roots of the *secular equation*

$$S^3 - IS^2 + JS - \delta = 0. \tag{7.79}$$

(2) $\delta = 0$, the quadric has the canonical equation

$$S_1 x^2 + S_2 y^2 = 2\sqrt{-\frac{\Delta}{J}} z. \tag{7.80}$$

If the coefficients of (7.78) have the signs:

- $++++$ the quadric is an imaginary ellipsoid

- $+++-$ the quadric is a real ellipsoid
- $++--$ the quadric is a hyperboloid of one sheet
- $+---$ the quadric is a hyperboloid of two sheets.

If the coefficients of (7.80) have the signs:

- $++$ the quadric is an elliptic paraboloid
- $+-$ the quadric is a hyperbolic paraboloid

(II) If $\Delta = 0$ then for:

(1) $\delta \neq 0$, the quadric has the canonical equation

$$S_1 x^2 + S_2 y^2 + S_3 z^2 = 0. \tag{7.81}$$

(2) $\delta = 0$, the quadric is a cylinder or a pair of planes.

If the coefficients of (7.81) have the signs:

- $+++$ the quadric is an imaginary cone
- $++-$ the quadric is a real cone.

**Example 7.22** (see [8]). Determine the nature of the following quadrics:

(a) $x^2 + y^2 + z^2 + 7xy + yz - 6z = 0$
(b) $36x^2 + y^2 + 4z^2 + 72x + 6y - 40z + 109 = 0$.

**Solution**
(a) We have

$$f(x, y, z) = x^2 + y^2 + 7xy + yz - 6z.$$

We identify the coefficients of the quadric:

$$\begin{cases} a_{11} = 1, a_{22} = 1, a_{33} = 1, a_{12} = \frac{7}{2}, a_{13} = 0, a_{23} = \frac{1}{2} \\ b_1 = 0, b_2 = 0, b_3 = -3 \\ c = 0. \end{cases}$$

We shall obtain:

$$I = 1 + 1 + 1 = 2$$
$$J = \begin{vmatrix} 1 & 7/2 \\ 7/2 & 1 \end{vmatrix} + \begin{vmatrix} 1 & 0 \\ 0 & 1 \end{vmatrix} + \begin{vmatrix} 1 & 1/2 \\ 1/2 & 1 \end{vmatrix} = -\frac{19}{2}$$
$$\delta = \begin{vmatrix} 1 & 7/2 & 0 \\ 7/2 & 1 & 1/2 \\ 0 & 1/2 & 1 \end{vmatrix} = -\frac{23}{2} \neq 0$$

$$\Delta = \begin{vmatrix} 1 & 7/2 & 0 & 0 \\ 7/2 & 1 & 1/2 & 0 \\ 0 & 1/2 & 1 & -3 \\ 0 & 0 & -3 & 0 \end{vmatrix} = \frac{405}{4} \neq 0.$$

As $\delta \neq 0$, the quadratic admits an unique center of symmetry; the center coordinates result solving the system (7.60), which becomes:

$$\begin{cases} 2x + 7y = 0 \\ 7x + 2y + z = 0 \\ y + 2z - 6 = 0 \end{cases} \Rightarrow C\left(-\frac{21}{46}, \frac{3}{23}, \frac{135}{46}\right).$$

We shall solve the secular Eq. (7.79), being:

$$S^3 - 3S^2 - \frac{19}{2}S + \frac{23}{2} = 0 \Rightarrow \begin{cases} S_1 = -2.53548 \\ S_2 = 4.53516 \\ S_3 = 1. \end{cases}$$

The quadric equation will have the canonical form:

$$x^2 + 4.53516y^2 - 2.53548z^2 - 8.80469 = 0,$$

i.e. it is an one-sheeted hyperboloid.

The solution in Sage is:

```
sage: R.<x,y,z> = QQbar[]
sage: f=x^2+y^2+z^2+7*x*y+y*z-6*z
sage: a11=f.coefficient({x:2});a22=f.coefficient({y:2});a33=f.coefficient({z:2})
sage: a12=f.coefficient({x:1,y:1,z:0})/2
sage: a13=f.coefficient({x:1,z:1,y:0})/2
sage: a23=f.coefficient({y:1,z:1,x:0})/2
sage: b1=f.coefficient({x:1,y:0,z:0})/2
sage: b2=f.coefficient({y:1,x:0,z:0})/2
sage: b3=f.coefficient({z:1,x:0,y:0})/2
sage: c=f.coefficient({x:0,y:0,z:0})
sage: I=a11+a22+a33
sage: A1=matrix([[a11,a12],[a12,a22]]);A2=matrix([[a11,a13],[a13,a33]])
sage: A3=matrix([[a22,a23],[a23,a33]]);d1=A1.det();d2=A2.det();d3=A3.det()
sage: J=d1+d2+d3
sage: A=matrix([[a11,a12,a13],[a12,a22,a23],[a13,a23,a33]])
sage: de=A.det()
sage: B=matrix([[a11,a12,a13,b1],[a12,a22,a23,b2],[a13,a23,a33,b3],[b1,b2,b3,c]])
sage: De=B.det()
sage: I;J;de;De
3
-19/2
-23/2
405/4
sage: u=diff(f,x);v=diff(f,y);w=diff(f,z)
sage: u1=u.coefficient(x).n(digits=3);u2=u.coefficient(y).n(digits=3)
sage: u3=u.coefficient(z).n(digits=3);v1=v.coefficient(x).n(digits=3)
sage: v2=v.coefficient(y).n(digits=3);v3=v.coefficient(z).n(digits=3)
```

```
sage: w1=w.coefficient(x).n(digits=3);w2=w.coefficient(y).n(digits=3)
sage: w2=w.coefficient(y).n(digits=3);w3=w.coefficient(z).n(digits=3)
sage: uu=u-(u1*x+u2*y+u3*z);vv=v-(v1*x+v2*y+v3*z);ww=w-(w1*x+w2*y+w3*z)
sage: x0,y0,z0,S=var('x0,y0,z0,S')
sage: solve([u1*x0+u2*y0+u3*z0+uu,v1*x0+v2*y0+v3*z0+vv,w1*x0+w2*y0+w3*z0+ww],x0,y0,z0)
[[x0 == (-21/46), y0 == (3/23), z0 == (135/46)]]
sage: II=I.n(digits=3);JJ=J.n(digits=3);dd=de.n(digits=3);Dd=De.n(digits=3)
sage: s=(S^3-II*S^2+JJ*S-dd).roots()
sage: s0=s[0][0].n(digits=3);s1=s[1][0].n(digits=3);s2=s[2][0].n(digits=3)
sage: F=(s2*x0^2+s1*y0^2+s0*z0^2+Dd/dd).simplify_exp();F
x0^2 + 4.53516*y0^2 - 2.5354*z0^2 - 8.80469
sage: implicit_plot3d(F, (x0, -5, 5), (y0, -5, 5), (z0, -1, 1), opacity=0.5)
```

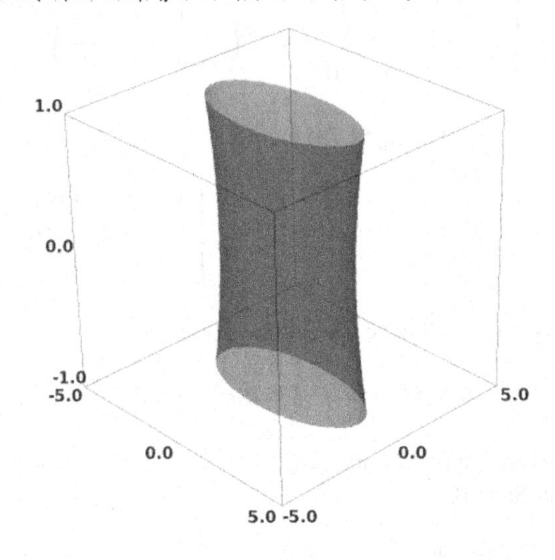


(b) We have

$$f(x, y, z) = 36x^2 + y^2 + 4z^2 + 72x + 6y - 40z + 109.$$

The quadric invariants will be:

$$I = 36 + 1 + 4 = 41$$

$$J = \begin{vmatrix} 36 & 0 \\ 0 & 1 \end{vmatrix} + \begin{vmatrix} 36 & 0 \\ 0 & 4 \end{vmatrix} + \begin{vmatrix} 1 & 0 \\ 0 & 4 \end{vmatrix} = 184$$

$$\delta = \begin{vmatrix} 36 & 0 & 0 \\ 0 & 1 & 1/2 \\ 0 & 0 & 4 \end{vmatrix} = 144 \neq 0$$

$$\Delta = \begin{vmatrix} 36 & 0 & 0 & 36 \\ 0 & 1 & 1/2 & 3 \\ 0 & 0 & 4 & -20 \\ 36 & 3 & -20 & 109 \end{vmatrix} = -5184 \neq 0.$$

The secular equation will be:

$$S^3 - 41S^2 + 184S - 144 = 0 \Rightarrow \begin{cases} S_1 = 1 \\ S_2 = 36 \\ S_3 = 4. \end{cases}$$

The quadric equation will have the canonical form:

$$x^2 + 4y^2 + 36z^2 - 36 = 0$$

or

$$\frac{x^2}{36} + \frac{y^2}{9} + z^2 - 1 = 0$$

i.e. it is a real ellipsoid.

Using Sage, we shall have:

```
sage: R.<x,y,z> = QQbar[]
sage: f=36*x^2+y^2+4*z^2+72*x+6*y-40*z+109
sage: a11=f.coefficient({x:2});a22=f.coefficient({y:2});a33=f.coefficient({z:2})
sage: a12=f.coefficient({x:1,y:1,z:0})/2
sage: a13=f.coefficient({x:1,z:1,y:0})/2
sage: a23=f.coefficient({y:1,z:1,x:0})/2
sage: b1=f.coefficient({x:1,y:0,z:0})/2
sage: b2=f.coefficient({y:1,x:0,z:0})/2
sage: b3=f.coefficient({z:1,x:0,y:0})/2
sage: c=f.coefficient({x:0,y:0,z:0})
sage: I=a11+a22+a33
sage: A1=matrix([[a11,a12],[a12,a22]]);A2=matrix([[a11,a13],[a13,a33]])
sage: A3=matrix([[a22,a23],[a23,a33]]);d1=A1.det();d2=A2.det();d3=A3.det()
sage: J=d1+d2+d3
sage: A=matrix([[a11,a12,a13],[a12,a22,a23],[a13,a23,a33]])
sage: de=A.det()
sage: B=matrix([[a11,a12,a13,b1],[a12,a22,a23,b2],[a13,a23,a33,b3],[b1,b2,b3,c]])
sage: De=B.det()
sage: I;J;de;De
41
184
144
-5184
sage: u=diff(f,x);v=diff(f,y);w=diff(f,z)
sage: u1=u.coefficient(x).n(digits=3);u2=u.coefficient(y).n(digits=3)
sage: u3=u.coefficient(z).n(digits=3);v1=v.coefficient(x).n(digits=3)
sage: v2=v.coefficient(y).n(digits=3);v3=v.coefficient(z).n(digits=3)
sage: w1=w.coefficient(x).n(digits=3);w2=w.coefficient(y).n(digits=3)
sage: w2=w.coefficient(y).n(digits=3);w3=w.coefficient(z).n(digits=3)
sage: uu=u-(u1*x+u2*y+u3*z);vv=v-(v1*x+v2*y+v3*z);ww=w-(w1*x+w2*y+w3*z)
sage: x0,y0,z0,S=var('x0,y0,z0,S')
sage: solve([u1*x0+u2*y0+u3*z0+uu,v1*x0+v2*y0+v3*z0+vv,w1*x0+w2*y0+w3*z0+ww],x0,y0,z0)
[[x0 == -1, y0 == -3, z0 == 5]]
sage: II=I.n(digits=3);JJ=J.n(digits=3);dd=de.n(digits=3);Dd=De.n(digits=3)
sage: s=(S^3-II*S^2+JJ*S-dd).roots()
sage: s0=s[0][0].n(digits=3);s1=s[1][0].n(digits=3).real()
sage: s2=s[2][0].n(digits=3);s0;s1;s2
36.0
4.00
1.00
sage: F=(s2*x0^2+s1*y0^2+s0*z0^2+Dd/dd).simplify_exp();F/(Dd/dd)
-0.0278*x0^2 - 0.111*y0^2 - z0^2 + 1.00
sage: implicit_plot3d(F, (x0,-6, 6), (y0, -4, 4), (z0, -2, 2), opacity=0.5)
```

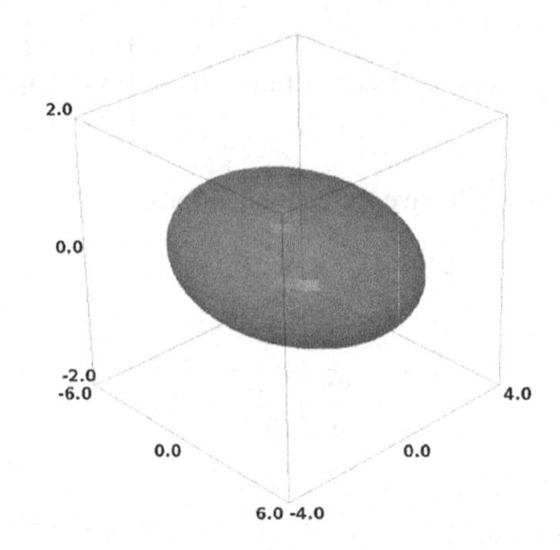


## 7.5 Quadrics on Canonical Equations

If

- $\delta \neq 0$ the quadric has an unique centre of simmetry; ex: sphere, ellipsoid, cone, hyperboloids
- $\delta = 0$ the quadric is called without center; ex: elliptic and hyperbolic paraboloid.

### *7.5.1 Sphere*

**Definition 7.23** (see [1], p. 185). The **sphere** is the set of the points from space equally distant from a fixed point called the *center of the sphere*, the distance from the center to the points of the sphere is called the *radius of the sphere*.

We shall report the sphere plane at an orthonormal Cartesian reference.

Let $C(a, b, c)$ be center of the sphere and $M(x, y, z)$ an arbitrary point of the sphere (see Fig. 7.6).

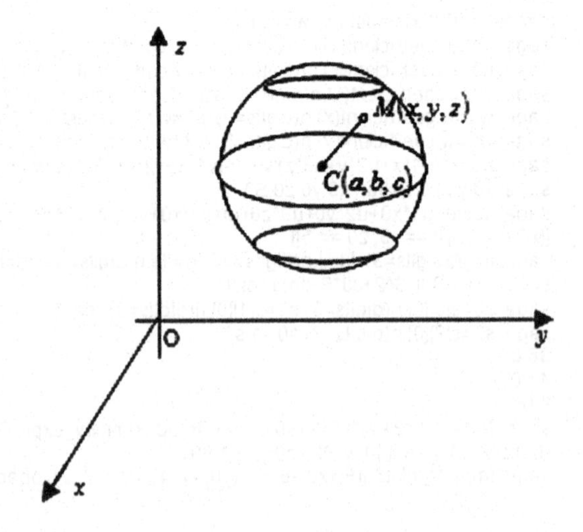


**Fig. 7.6** Sphere

**Remark 7.24** (see [2]). The sphere is a quadric of rotation which is obtained by rotating a circle (semicircle) around one of its diameter.

From the **Definition 7.23** it results that the distance between $C$ and $M$ is constant and equal to the radius $R$ of the sphere:

$$\left\| \overline{CM} \right\| = R,$$

i.e.

$$\sqrt{(x-a)^2 + (y-b)^2 + (z-c)^2} = R \Leftrightarrow$$

$$(x-a)^2 + (y-b)^2 + (z-c)^2 = R^2. \tag{7.82}$$

If we open the squares in (7.82) we get the *general Cartesian equation of the sphere*

$$x^2 + y^2 + z^2 - 2ax - 2by - 2cz + a^2 + b^2 + c^2 - R^2 = 0. \tag{7.83}$$

To obtain other forms of the equation of a sphere we introduce the spherical coordinates: $(\rho, \theta, \varphi)$ (see Fig. 7.7).

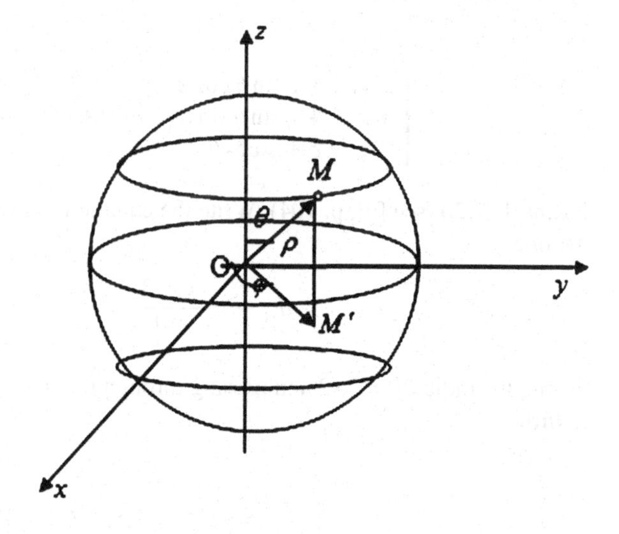


**Fig. 7.7** The spherical coordinates of a point from space

The relations between the Cartesian coordinates $(x, y, z)$ of a point $M$ from space and its spherical coordinates $(\rho, \theta, \varphi)$ are:

$$\begin{cases} x = \rho \sin\theta \cos\varphi \\ y = \rho \sin\theta \sin\varphi \\ z = \rho \cos\theta, \end{cases} \quad (7.84)$$

where:

- $\rho \geq 0$ is the distance from the point $M$ to the origin of axes,
- $\theta, \theta \in [0, \pi]$ is angle made by the position vector of the point $M$ with the O$z$ axis,
- $\varphi, \varphi \in [0, 2\pi]$ means the angle made by the projection of the position vector of the point $M$ on the plane ($x$O$y$) with the O$x$ axis.

**Remark 7.25** (see [6], p. 667). Each triplet of spherical coordinate corresponds to a point, but not any point corresponds to a triplet, as is the case when $M$ is on O$z$ or in origin.

If we make a change of an orthonormal reference, such that $C(a, b, c)$ constitutes the origin of the new reference, then the transition from the coordinates $(x, y, z)$ to the coordinates $(x', y', z')$ in the new reference is achieved by a translation of the vector $\overline{OC}$, characterized by the equations:

$$\begin{cases} x = a + x' \\ y = b + y' \\ z = c + z'. \end{cases}$$

The parametric equations of the sphere with the center $C(a, b, c)$ and the radius $\rho \geq 0$ will be:

$$\begin{cases} x = a + \rho \sin\theta \cos\varphi \\ y = b + \rho \sin\theta \sin\varphi \\ z = c + \rho \cos\theta. \end{cases}, \theta \in [0, \pi], \varphi \in [0, 2\pi]. \quad (7.85)$$

**Example 7.26** (see [9], p. 104). Write the equation of the sphere with the center on the line:

$$d : \frac{x}{1} = \frac{y-1}{-1} = \frac{z+2}{1},$$

having the radius $R = \sqrt{2}$ and passing through the point $A(0, 2, -1)$.

**Solution**

We have

$$\frac{x}{1} = \frac{y-1}{-1} = \frac{z+2}{1} = t, t \in \mathbb{R}.$$

The parametric equations of the straight line $d$ will be:

$$\begin{cases} x = t \\ y - 1 = -t \\ z + 2 = t \end{cases} \Rightarrow \begin{cases} x = t \\ y = 1 - t \\ z = t - 2. \end{cases}$$

As the center of the sphere is situated on the straight line $d$, it results that the point $C\,(t, 1 - t, t - 2)$ is the center of the sphere.

From the condition

$$\left\| \overline{CA} \right\|^2 = R^2$$

we deduce

$$3t^2 = 0 \Rightarrow t = 0.$$

It follows that $C\,(0, -1, 2)$ is the the center of the sphere and the equation of the sphere will be

$$S : x^2 + (y - 1)^2 + (z + 2)^2 = 2.$$

We shall give a solution in Sage, too:

```
sage: x,y,z,t=var('x,y,z,t');R=sqrt(2);r=R^2
sage: s=solve([x==t,y-1==-t,z+2==t],x,y,z)
sage: x0=s[0][0].right();x1=s[0][1].right();x2=s[0][2].right()
sage: A=vector([0,2,-1]);C=vector([x0,x1,x2])
sage: CA=A-C;n=CA.norm().simplify_exp()^2
sage: s=solve([n==r],t);ss=s[0].right()
sage: C1=C.subs(t=0)
sage: f=(x-C1[0])^2+(y-C1[1])^2+(z-C1[2])^2-r
sage: implicit_plot3d(f,(x,-3,3),(y,-3,3),(z,-3,3))
```

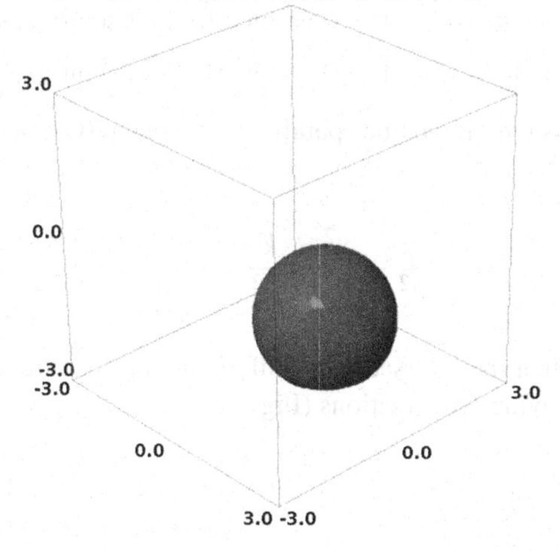

### 7.5.2 Ellipsoid

**Definition 7.27** (see [7], p. 351 and [10], p. 131). The **ellipsoid** is the set of the points in space, whose coordinates relative to an orthonormal reference check the Eq. (7.62), where $a, b, c$ are some strictly positive real numbers called the *semi-axes* of the ellipsoid.

We consider the orthonormal reference, relative to which is given the ellipsoid equation.

To plot the ellipsoid we shall determine its intersections with: the coordinate axes, the coordinate planes, the planes parallel to the coordinate planes.

$$\mathrm{O}x : \begin{cases} y = 0 \\ z = 0 \end{cases} \Rightarrow \frac{x^2}{a^2} - 1 = 0 \Rightarrow x = \pm a;$$

hence the ellipsoid crosses O$x$ in two points: $A\,(a, 0, 0)$ and $A'\,(-a, 0, 0)$.

$$\mathrm{O}y : \begin{cases} x = 0 \\ z = 0 \end{cases} \Rightarrow \frac{y^2}{b^2} - 1 = 0 \Rightarrow y = \pm b;$$

hence the ellipsoid crosses O$y$ in two points: $B\,(0, b, 0)$ and $B'\,(0, -b, 0)$.

$$\mathrm{O}z : \begin{cases} x = 0 \\ y = 0 \end{cases} \Rightarrow \frac{z^2}{c^2} - 1 = 0 \Rightarrow z = \pm c;$$

hence the ellipsoid crosses O$z$ in two points: $C\,(0, 0, c)$ and $C'\,(0, 0, -c)$.

The points $A, A', B, B', C, C'$ are the vertices of the ellipsoid, while the symmetry axes of the ellipsoid are O$x$,O$y$,O$z$.

$(\mathrm{O}xy) : z = 0 \Rightarrow \frac{x^2}{a^2} + \frac{y^2}{b^2} - 1 = 0 \Rightarrow$ an ellipse of semi-axes $a$ and $b$.

$(\mathrm{O}xz) : y = 0 \Rightarrow \frac{x^2}{a^2} + \frac{z^2}{c^2} - 1 = 0 \Rightarrow$ an ellipse of semi-axes $a$ and $c$.

$(\mathrm{O}yz) : x = 0 \Rightarrow \frac{y^2}{b^2} + \frac{z^2}{c^2} - 1 = 0 \Rightarrow$ an ellipse of semi-axes $b$ and $c$.

The intersections with the planes parallel to the plane (O$xy$), by equation $z = k$, can be determined from:

$$\frac{x^2}{a^2} + \frac{y^2}{b^2} + \frac{k^2}{c^2} - 1 = 0.$$

If $k \in (-c, c)$ then the intersections with the planes parallel to the plane (O$xy$) are some ellipses having the equations (Fig. 7.8):

$$\frac{x^2}{\left(\frac{a}{c}\sqrt{c^2-k^2}\right)^2}+\frac{y^2}{\left(\frac{b}{c}\sqrt{c^2-k^2}\right)^2}-1=0.$$

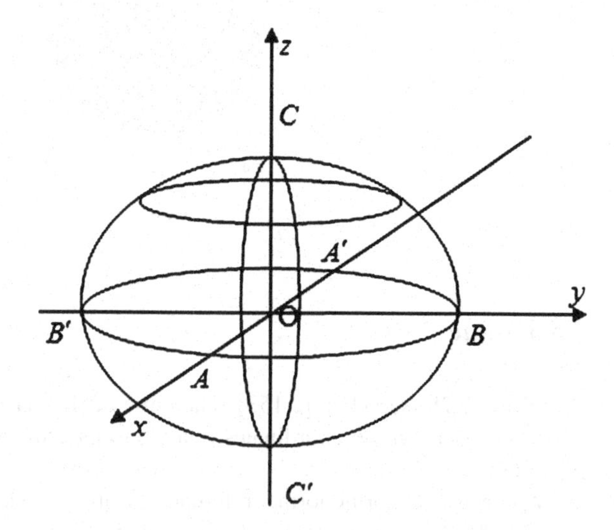


**Fig. 7.8** Ellipsoid

The ellipsoid has: an unique center of symmetry (the origin), symmetry axes (the coordinate axes), planes of symmetry (the coordinate planes).
The parametric equations corresponding to an ellipsoid are:

$$\begin{cases} x = a\sin u\cos v \\ y = b\sin u\sin v \\ z = c\cos u \end{cases}, u\in[0,\pi], v\in[0,2\pi]. \tag{7.86}$$

**Remark 7.28** (see [10], p. 132). The sphere is a special case of ellipsoid, obtained if all the semi-axes of the ellipsoid are equal between themselves. If two semi-axes are equal, then one achieves a rotating ellipsoid, which can be generated by the rotation of an ellipse around of an axis. For example, if $a = b$, then the ellipsoid is of rotation around of $Oz$.

The next figure shows an ellipsoid built in Sage, using (7.86):

```
sage: u,v= var('u, v');a=8;b=2;c=1
sage: f_x(u,v)=a*sin(u)*cos(v)
sage: f_y(u,v)=b*sin(u)*sin(v)
sage: f_z(u,v)=c*cos(u)
sage: parametric_plot3d([f_x,f_y,f_z], (u, 0, pi), (v, 0, 2*pi))
```

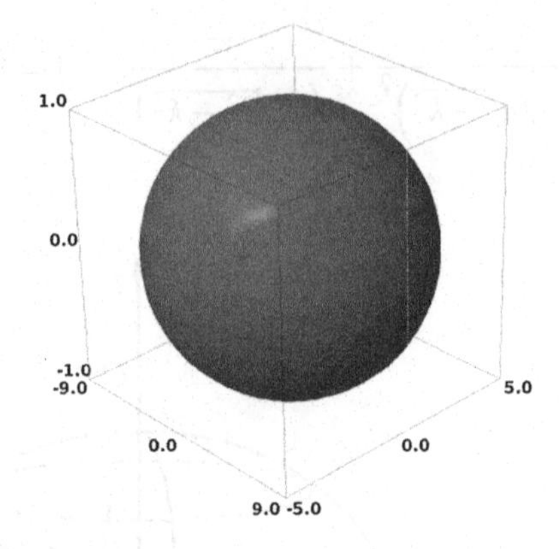


### 7.5.3 Cone

**Definition 7.29** (see [10], p. 153). The **cone of the second order** is the set of the points in space, whose coordinates relative to an orthonormal reference check the Eq. (7.66).

We consider the orthonormal reference, relative to which is given the cone equation. To plot the cone we shall determine its intersections with: the coordinate axes, the coordinate planes, the planes parallel to the coordinate planes.

$$Ox : \begin{cases} y = 0 \\ z = 0 \end{cases} \Rightarrow \frac{x^2}{a^2} = 0 \Rightarrow x = 0$$

hence the cone crosses $Ox$ in origin. Similarly, the cone crosses $Oy$ and $Oz$ in origin, too.

$(Oxy) : z = 0 \Rightarrow \frac{x^2}{a^2} + \frac{y^2}{b^2} = 0 \Rightarrow x = y = 0.$

$(Oxz) : y = 0 \Rightarrow \frac{x^2}{a^2} - \frac{z^2}{c^2} = 0 \Leftrightarrow$

$$\begin{cases} \frac{x}{a} - \frac{z}{c} = 0 \\ \frac{x}{a} + \frac{z}{c} = 0 \end{cases} \Rightarrow \text{two lines concurrent in the origin.}$$

$(Oyz) : x = 0 \Rightarrow \frac{y^2}{b^2} - \frac{z^2}{c^2} = 0 \Leftrightarrow$

$$\begin{cases} \frac{y}{b} - \frac{z}{c} = 0 \\ \frac{y}{b} + \frac{z}{c} = 0 \end{cases} \Rightarrow \text{two lines concurrent in the origin.}$$

The intersections with the planes parallel to the plane $(Oxy)$, by equation $z = k$, can be determined from:

$$\frac{x^2}{a^2} + \frac{y^2}{b^2} - \frac{k^2}{c^2} = 0;$$

therefore the intersections with the planes parallel to the plane $(Oxy)$ are some ellipses having the equations (Fig. 7.9):

$$\frac{x^2}{\left(\frac{a}{c}k\right)^2} + \frac{y^2}{\left(\frac{b}{c}k\right)^2} - 1 = 0.$$

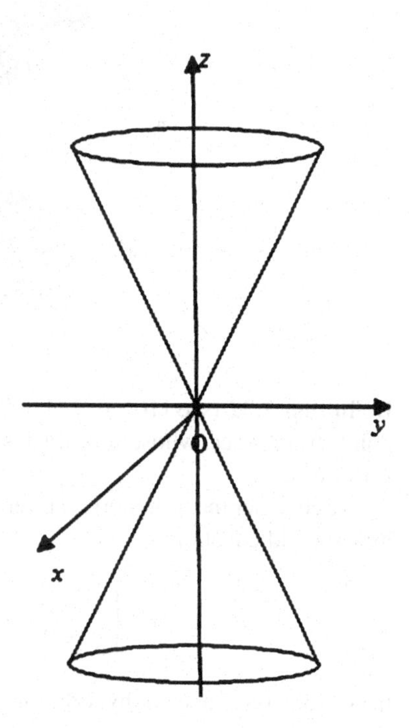


**Fig. 7.9** Cone

The cone has an unique center of symmetry.

**Remark 7.30** (see [10], p. 129). If $a = b$ then one obtains the rotating cone, which can be generated by the rotation of a conic (which represents two concurrent straight lines) by equation

$$\frac{y^2}{a^2} - \frac{z^2}{c^2} = 0 \qquad (7.87)$$

around the $Oz$ axis.

The parametric equations corresponding to the cone of the second order are:

$$\begin{cases} x = av\cos u \\ y = bv\sin u \\ z = \pm c \end{cases}, u \in [0, 2\pi], v \in \mathbb{R}. \quad (7.88)$$

We shall plot in Sage a cone of the second order, defined by the Eq. (7.66):

```
sage: x,y,z=var('x,y,z');a=4;b=2;c=1
sage: implicit_plot3d(x^2/a^2+y^2/b^2-z^2/c^2, (x,-2,2),(y,-1,1),(z,-0.5,0.5))
```

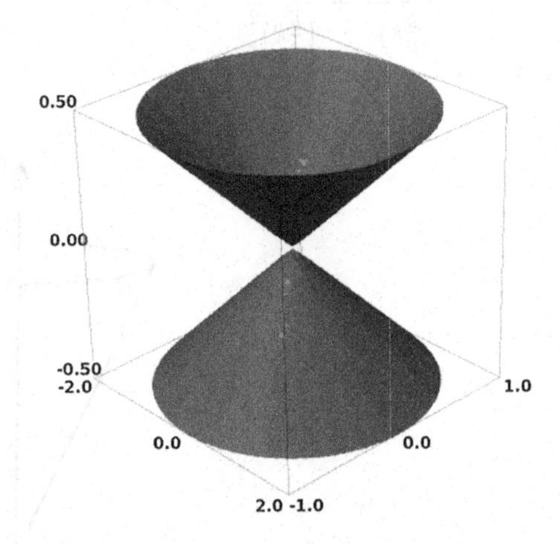


**Definition 7.32** (see [10], p. 141). The **one-sheeted hyperboloid** is the set of the points from space, whose coordinates relative to an orthonormal reference check the Eq. (7.63).

We consider the orthonormal reference, relative to which is given the one-sheeted hyperboloid equation.

$$Ox : \begin{cases} y = 0 \\ z = 0 \end{cases} \Rightarrow \frac{x^2}{a^2} - 1 = 0 \Rightarrow x = \pm a;$$

hence the one-sheeted hyperboloid crosses $Ox$ in two points: $A(a, 0, 0)$ and $A'(-a, 0, 0)$.

$$Oy : \begin{cases} x = 0 \\ z = 0 \end{cases} \Rightarrow \frac{y^2}{b^2} - 1 = 0 \Rightarrow y = \pm b;$$

hence the one-sheeted hyperboloid crosses $Oy$ in two points: $B(0, b, 0)$ and $B'(0, -b, 0)$.

$$Oz : \begin{cases} x = 0 \\ y = 0 \end{cases} \Rightarrow -\frac{z^2}{c^2} - 1 = 0 \Rightarrow z^2 = -c^2;$$

hence the one-sheeted hyperboloid dosn't cross $Oz$ axis.

The points $A, A', B, B'$ are called the vertices of the one-sheeted hyperboloid.

$(Oxy) : z = 0 \Rightarrow \Gamma_1 : \frac{x^2}{a^2} + \frac{y^2}{b^2} - 1 = 0 \Rightarrow$ an ellipse of semi-axes $a$ and $b$.

$(Oyz) : x = 0 \Rightarrow \Gamma_2 : \frac{y^2}{b^2} - \frac{z^2}{c^2} - 1 = 0 \Rightarrow$ a hyperbola.

$(Oxz) : y = 0 \Rightarrow \Gamma_3 : \frac{x^2}{a^2} - \frac{z^2}{c^2} - 1 = 0 \Rightarrow$ a hyperbola.

The intersections with the planes parallel to the plane $(Oxy)$, by equation $z = k$, can be determined from:

$$\frac{x^2}{a^2} + \frac{y^2}{b^2} - \frac{k^2}{c^2} - 1 = 0.$$

The intersections with the planes parallel to the plane $(Oxy)$ are some ellipses having the equations:

$$\frac{x^2}{\left(\frac{a}{c}\sqrt{c^2 + k^2}\right)^2} + \frac{y^2}{\left(\frac{b}{c}\sqrt{c^2 + k^2}\right)^2} - 1 = 0$$

called *clamp ellipse*.

The one-sheeted hyperboloid is an unbounded quadric with an unique center of symmetry (see Fig. 7.10) and that has the following parametric equations:

$$\begin{cases} x = a\sqrt{1 + u^2} \cos v \\ y = b\sqrt{1 + u^2} \sin v \\ z = cu \end{cases}, v \in [0, 2\pi], u \in \mathbb{R}. \quad (7.89)$$

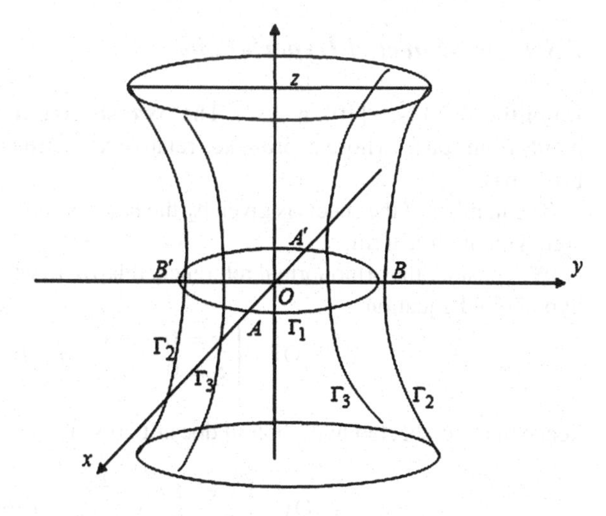


**Fig. 7.10** One-sheeted hyperboloid

**Remark 7.32** (see [10], p. 127) If $a = b$ then the one-sheeted hyperboloid is of rotation around of $Oz$, i.e. it can be generated by the rotation of the hyperbola

$$\frac{y^2}{b^2} - \frac{z^2}{c^2} - 1 = 0 \tag{7.90}$$

around of the $Oz$ axis.

We shall use parametric equation (7.89) to represent the one-sheeted hyperboloid in Sage:

```
sage: u,v= var('u, v');a=4;b=2;c=1
sage: f_x(u,v)=a*sqrt(1+u^2)*cos(v)
sage: f_y(u,v)=b*sqrt(1+u^2)*sin(v)
sage: f_z(u,v)=c*u
sage: parametric_plot3d([f_x,f_y,f_z], (u, -2, 2), (v, 0, 2*pi))
```

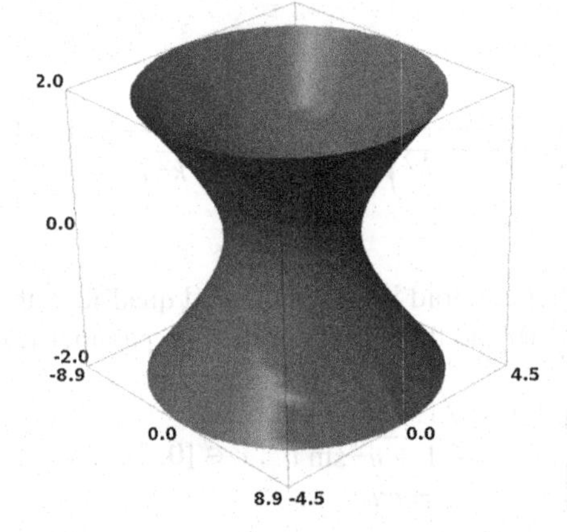


### 7.5.4 Two-Sheeted Hyperboloid

**Definition 7.33** (see [10], p. 137). The **two-sheeted hyperboloid** is the set of the points from space, whose coordinates relative to an orthonormal reference check the Eq. (7.64).

The number of the sheets is given by the number of the squares that have the same sign with the free term.

We consider the orthonormal reference, relative to which is given the two-sheeted hyperboloid equation.

$$Ox : \begin{cases} y = 0 \\ z = 0 \end{cases} \Rightarrow \frac{x^2}{a^2} = -1;$$

hence the two-sheeted hyperboloid doesn't cross the $Ox$ axis.

$$Oy : \begin{cases} x = 0 \\ z = 0 \end{cases} \Rightarrow \frac{y^2}{b^2} = -1 \Rightarrow$$

the two-sheeted hyperboloid doesn't cross the $Oy$ axis.

$$\mathrm{O}z : \begin{cases} x = 0 \\ y = 0 \end{cases} \Rightarrow \frac{z^2}{c^2} = 1 \Rightarrow z = \pm c;$$

hence the one-sheeted hyperboloid crosses in two points: $C\ (0, 0, c)$ and $C'\ (0, 0, -c)$.

$(\mathrm{O}yz) : x = 0 \Rightarrow \frac{y^2}{b^2} - \frac{z^2}{c^2} + 1 = 0 \Rightarrow \Gamma_1 : \frac{z^2}{c^2} - \frac{y^2}{b^2} = 1$ a hyperbola.

$(\mathrm{O}xz) : y = 0 \Rightarrow \frac{x^2}{a^2} - \frac{z^2}{c^2} + 1 = 0 \Rightarrow \Gamma_2 : \frac{z^2}{c^2} - \frac{x^2}{a^2} = 1$ a hyperbola.

The intersections with the planes parallel to the plane $(\mathrm{O}xy)$, by equation $z = k$, can be determined from:

$$\frac{x^2}{a^2} + \frac{y^2}{b^2} = \frac{k^2}{c^2} - 1.$$

If $k \in (-\infty, -c) \cup (c, \infty)$ then the intersections with the planes parallel to the plane $(\mathrm{O}xy)$ are some ellipses having the equations:

$$\frac{x^2}{\left(\frac{a}{c}\sqrt{c^2 - k^2}\right)^2} + \frac{y^2}{\left(\frac{b}{c}\sqrt{c^2 - k^2}\right)^2} - 1 = 0.$$

The two-sheeted hyperboloid is an unbounded quadric with an unique center of symmetry (see Fig. 7.11) and that has the following parameter equations:

$$\begin{cases} x = a \sinh u \cos v \\ y = b \sinh u \sin v \\ z = \pm c \cosh u \end{cases}, v \in [0, 2\pi], u \in \mathbb{R}. \tag{7.91}$$

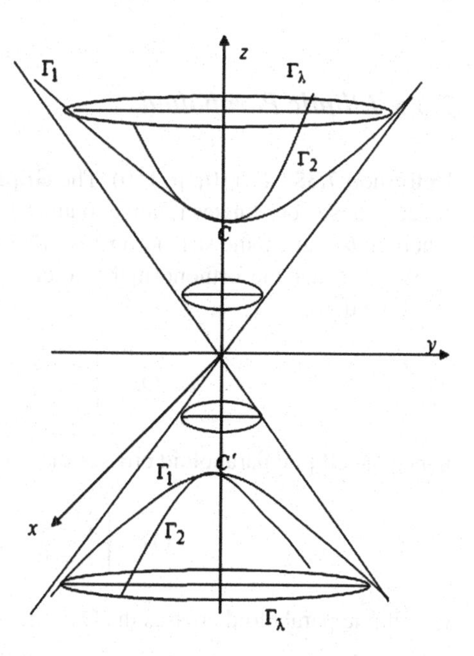


**Fig. 7.11** Two-sheeted hyperboloid

**Remark 7.34** (see [10], p. 128) If $a = b$ then the two-sheeted hyperboloid is of rotation around of O$z$, i.e. it can be generated by the rotation of the hyperbola

$$\frac{y^2}{b^2} - \frac{z^2}{c^2} + 1 = 0 \tag{7.92}$$

around of the O$z$ axis.

The following Sage code allows us to plot a two-sheeted hyperboloid:

```
sage: x,y,z=var('x,y,z');a=7;b=1;c=0.5
sage: implicit_plot3d(x^2/a^2+y^2/b^2-z^2/c^2+1, (x,-3,3),(y,-3,3),(z,-1.5,1.5))
```

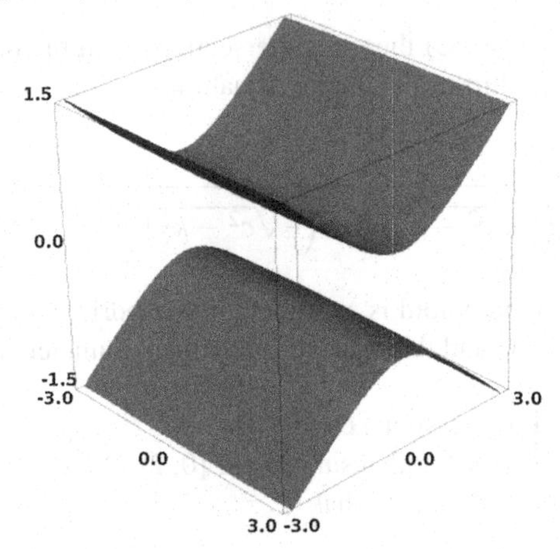


## 7.5.5 Elliptic Paraboloid

**Definition 7.35** (see [10], p. 150). The **elliptic paraboloid** is the set of the points in space, whose coordinates relative to an orthonormal reference check the Eq. (7.67), where $a, b, z$ are some strictly positive real numbers.

We consider the orthonormal reference, relative to which is given the elliptic paraboloid.

$$\text{O}x : \begin{cases} y = 0 \\ z = 0 \end{cases} \Rightarrow \frac{x^2}{a^2} = 0;$$

hence the elliptic paraboloid crosses the O$x$ axis in the origin.

$$\text{O}y : \begin{cases} x = 0 \\ z = 0 \end{cases} \Rightarrow \frac{y^2}{b^2} = 0 \Rightarrow$$

the elliptic paraboloid crosses the O$y$ axis in the origin.

$$\text{O}z : \begin{cases} x = 0 \\ y = 0 \end{cases} \Rightarrow \frac{z^2}{c^2} = 0 \Rightarrow$$

the elliptic paraboloid crosses the O$z$ axis in the origin.

(O$xy$) : $z = 0 \Rightarrow \frac{x^2}{a^2} + \frac{y^2}{b^2} = 0 \Rightarrow x = y = 0 \Rightarrow$ the intersection is the origin.

(O$yz$) : $x = 0 \Rightarrow \frac{y^2}{b^2} - 2z = 0 \Rightarrow \Gamma_1 : y^2 = 2b^2 z$ a parabola.

(O$xz$) : $y = 0 \Rightarrow \frac{x^2}{a^2} - 2z = 0 \Rightarrow \Gamma_2 : x^2 = 2a^2 z$ a parabola.

The intersections with the planes parallel to the plane (O$xy$), by equation $z = k$, can be determined from:

$$\frac{x^2}{a^2} + \frac{y^2}{b^2} = 2k.$$

If $k > 0$ then the intersections with the planes parallel to the plane (O$xy$) are some ellipses having the equations:

$$\Gamma_\lambda : \frac{x^2}{\left(a\sqrt{2k}\right)^2} + \frac{y^2}{\left(b\sqrt{2k}\right)^2} = 1.$$

The the elliptic paraboloid is an unbounded quadric without a center of symmetry (see Fig. 7.12) and that has the following parameter equations:

$$\begin{cases} x = a\sqrt{2v}\cos u \\ y = b\sqrt{2v}\sin u \\ z = v \end{cases}, u \in [0, 2\pi], v > 0. \tag{7.93}$$

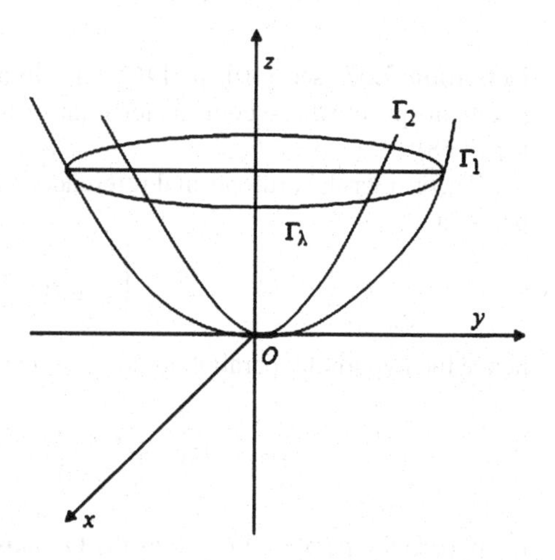


**Fig. 7.12** Elliptic paraboloid

**Remark 7.36** (see [10], p. 128) If $a = b$ then elliptic paraboloid is of rotation around of O$z$, i.e. it can be generated by the rotation of the parabola

$$y^2 = 2a^2 z \tag{7.94}$$

around of the O$z$ axis.

We need the following Sage code to represent an elliptic paraboloid:

```
sage: u,v= var('u, v');a=4;b=2
sage: f_x(u,v)=a*sqrt(2*v)*cos(u)
sage: f_y(u,v)=b*sqrt(2*v)*sin(u)
sage: f_z(u,v)=v
sage: parametric_plot3d([f_x,f_y,f_z], (u, 0, 2*pi), (v, 0, 2))
```

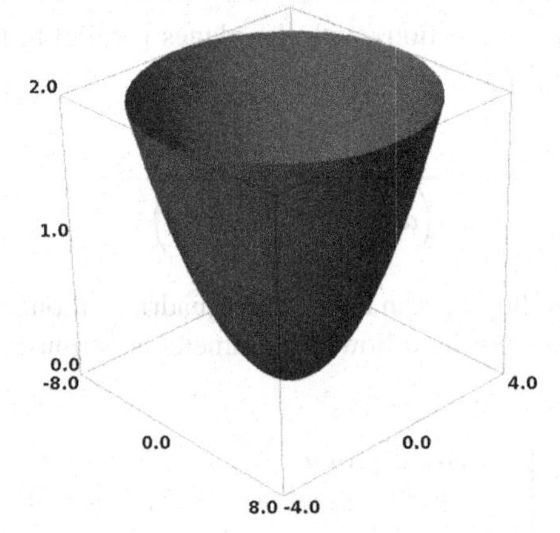


### 7.5.6 Hyperbolic Paraboloid

**Definition 7.37** (see [10], p. 146). The **hyperbolic paraboloid** is the set of the points in space, whose coordinates relative to an orthonormal reference satisfy the Eq. (7.68).

We consider the orthonormal reference, relative to which is given the hyperbolic paraboloid.

$$\text{O}x : \begin{cases} y = 0 \\ z = 0 \end{cases} \Rightarrow \frac{x^2}{a^2} = 0;$$

hence the hyperbolic paraboloid crosses the O$x$ axis in the origin.

$$\text{O}y : \begin{cases} x = 0 \\ z = 0 \end{cases} \Rightarrow \frac{y^2}{b^2} = 0 \Rightarrow$$

the hyperbolic paraboloid crosses the O$y$ axis in the origin.

$$\mathrm{O}z : \begin{cases} x = 0 \\ y = 0 \end{cases} \Rightarrow \frac{z^2}{c^2} = 0 \Rightarrow$$

the hyperbolic paraboloid crosses the O$z$ axis in the origin.

$(\mathrm{O}xy) : z = 0 \Rightarrow \frac{x^2}{a^2} - \frac{y^2}{b^2} = 0 \Leftrightarrow \begin{cases} \frac{x}{a} - \frac{y}{b} = 0 \\ \frac{x}{a} + \frac{y}{b} = 0 \end{cases} \Rightarrow$ two line concurrent in the origin.

$(\mathrm{O}yz) : x = 0 \Rightarrow -\frac{y^2}{b^2} - 2z = 0 \Rightarrow \Gamma_1 : y^2 = -2b^2 z$ a parabola with O$z$ as axis of symmetry, pointing in the negative direction of the straight line O$z$.

$(\mathrm{O}xz) : y = 0 \Rightarrow \frac{x^2}{a^2} - 2z = 0 \Rightarrow \Gamma_2 : x^2 = 2a^2 z$ a parabola with O$z$ as axis of symmetry, pointing in the positive direction of the straight line O$z$.

The intersections with the planes parallel to the plane (O$xy$), by equation $z = k$, can be determined from:

$$\frac{x^2}{a^2} - \frac{y^2}{b^2} = 2k.$$

If $k > 0$ then the intersections with the planes parallel to the plane (O$xy$) are some ellipses having the equations:

$$\Gamma_\lambda : \frac{x^2}{\left(a\sqrt{2k}\right)^2} - \frac{y^2}{\left(b\sqrt{2k}\right)^2} = 1.$$

The the elliptic paraboloid is an unbounded quadric without a center of symmetry (see Fig. 7.12) and that has the following parameter equations (Fig. 7.13):

$$\begin{cases} x = a\sqrt{2v}\cos u \\ y = b\sqrt{2v}\sin u \\ z = v\cos 2u \end{cases}, u \in [0, 2\pi], v > 0. \tag{7.95}$$

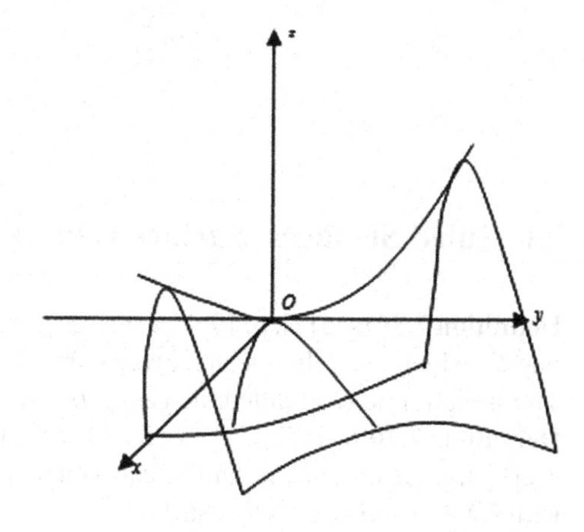


**Fig. 7.13** Hyperbolic paraboloid

**Remark 7.38** (see [6], p. 686). There is not a hyperbolic paraboloid of rotation; the hyperbolic paraboloid is the only surface of second degree, which is not a surface of rotation (because any section through a hyperbolic parabolid is not an ellipse).

The hyperbolic paraboloid is a surface of translation, this being obtained by the translation of a parabola (which has the opening in the bottom)

$$y^2 = -2b^2 z \tag{7.96}$$

on a parabola (which has the opening upward)

$$x^2 = 2a^2 z. \tag{7.97}$$

We shall illustrate a hyperbolic paraboloid, made in Sage:

```
sage: u,v= var('u, v');a=3;b=2
sage: f_x(u,v)=a*sqrt(2*v)*cos(u)
sage: f_y(u,v)=b*sqrt(2*v)*sin(u)
sage: f_z(u,v)=v*cos(2*u)
sage: parametric_plot3d([f_x,f_y,f_z], (u, 0, 2*pi), (v, 0, 2))
```

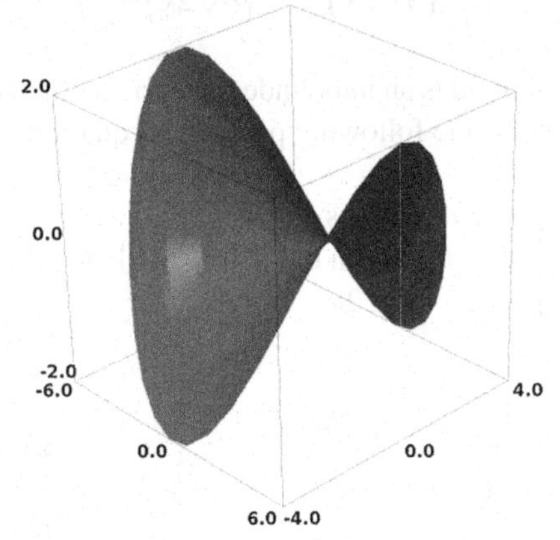


## 7.6 Ruled Surfaces. Surface Generation

**Definition 7.39** (see [1], p. 196). A surface that can be generated by moving a straight line $G$, which one relieses on a given curve $\Gamma$ from space is called a **ruled surface**. The straight line $G$ is called the *generatrix* (or *generator*) *of the surface*.

**Definition 7.40** (see [2]). A family of straight lines, which has the property that every straight line of the family can generate a surface represents the **rectilinear generators** of the respective surface.

**Proposition 7.41** (see [4], p. 174). The straight line $d$ passing through the point $A(x_0, y_0, z_0)$ and which has the director vector $\overline{v} = (l, m, n)$ constitutes the rectilinear generators of the quadric from (7.58) if and only if the following conditions are satisfied:

$$\begin{cases} f(x_0, y_0, z_0) = 0 \\ a_{11}l^2 + a_{22}m^2 + a_{33}n^2 + 2a_{12}lm + 2a_{13}ln + 2a_{23}mn = 0 \\ l\frac{\partial f}{\partial x}(x_0, y_0, z_0) + m\frac{\partial f}{\partial y}(x_0, y_0, z_0) + n\frac{\partial f}{\partial z}(x_0, y_0, z_0) = 0. \end{cases} \quad (7.98)$$

The only ones ruler surfaces are the two non-degenerate quadrics: the one-sheeted hyperboloid and the hyperbolic paraboloid.

**Example 7.42** (see [4], p. 239). Determine the rectilinear generators of the quadric by equation:

$$\Sigma : x^2 + y^2 + z^2 + 2xy - 2xz - yz + 4x + 3y - 5z + 4 = 0$$

passing through the point $A(-1, -1, 1)$.

**Solution**

We check firstly the first condition from (7.99), i.e. if $f(-1, -1, 1) = 0$, where:

$$f(x, y, z) = x^2 + y^2 + z^2 + 2xy - 2xz - yz + 4x + 3y - 5z + 4.$$

We want to determine the vector $\overline{v} = (l, m, n)$ so that, it should be the director vector of the rectilinear generatrix.

We shall check the the second condition of (7.98):

$$l^2 + m^2 + n^2 + 2lm - 2ln - mn = 0. \quad (7.80)$$

As

$$\begin{cases} \frac{\partial f}{\partial x}(x, y, z) = 2x + 2y - 2z + 4 \\ \frac{\partial f}{\partial y}(x, y, z) = 2y + 2x - z + 3 \\ \frac{\partial f}{\partial z}(x, y, z) = 2z - 2x - y - 5 \end{cases} \Rightarrow \begin{cases} \frac{\partial f}{\partial x}(-1, -1, 1) = -2 \\ \frac{\partial f}{\partial y}(-1, -1, 1) = -2 \\ \frac{\partial f}{\partial z}(-1, -1, 1) = 0, \end{cases}$$

the third condition of (7.98) becomes: $l = -m$. Substituting it in (7.80) we shall achieve:

$$n(m + n) = 0 \Leftrightarrow \begin{cases} n = 0 \\ l = -m \end{cases} \text{ or } \begin{cases} m = -n \\ l = n. \end{cases}$$

We determined the two director vectors:

$$\begin{cases} \overline{v}_1 = (-m, m, 0) = m\,(-1, 1, 0) \\ \overline{v}_2 = (n, -n, n) = n\,(1, -1, 1) \end{cases} \overset{m=1, n=1}{\Rightarrow} \begin{cases} \overline{v}_1 = (-1, 1, 0) \\ \overline{v}_2 = (1, -1, 1); \end{cases}$$

therefore, the two rectilinear generators pass through the point $A\,(-1, -1, 1)$. Their equations are:

$$d_1 : \begin{cases} \frac{x+1}{-1} = \frac{y+1}{1} \\ z = 1 \end{cases}$$

and, respectively:

$$d_2 : \frac{x+1}{1} = \frac{y+1}{-1} = \frac{z-1}{1}.$$

The solution in Sage will be given, too:

```
sage: x,y,z,l,m,n=var('x,y,z,l,m,n')
sage: f=x^2+y^2+z^2+2*x*y-2*x*z-y*z+4*x+3*y-5*z+4
sage: A=vector([-1,-1,1])
sage: f.subs(x=A[0],y=A[1],z=A[2])
0
sage: d1=diff(f,x).subs(x=A[0],y=A[1],z=A[2])
sage: d2=diff(f,y).subs(x=A[0],y=A[1],z=A[2])
sage: d3=diff(f,z).subs(x=A[0],y=A[1],z=A[2])
sage: solve(l*d1+m*d2+n*d3,l)
[l == -m]
sage: a11=f.coeff(x^2);a22=f.coeff(y^2);a33=f.coeff(z^2)
sage: a12=f.coeff(x).coeff(y);a13=f.coeff(x).coeff(z)
sage: a23=f.coeff(y).coeff(z)
sage: f2=a11*l^2+a22*m^2+a33*n^2+a12*l*m+a13*l*n+a23*m*n
sage: solve(f2.subs(l=-m),n)
[n == -m, n == 0]
sage: v1=vector([l,m,n]).subs(l=-m,n=0)
sage: vv1=v1.subs(m=1)
sage: v2=vector([l,m,n]).subs(l=n,m=-n)
sage: vv2=v2.subs(n=1)
sage: A1=point3d((A[0],A[1],A[2]),size=15,color='red')
sage: i=implicit_plot3d(f,(x,-33,33),(y,-13,13),(z,-10,10),opacity=0.5)
sage: l2=line3d([(-8,6,-6),(8,-10,10)],thickness=7,color='magenta')
sage: l1=line3d([(-8,6,1),(8,-10,1)],thickness=7,color='magenta')
sage: (l1+i+l2+A1).show()
```

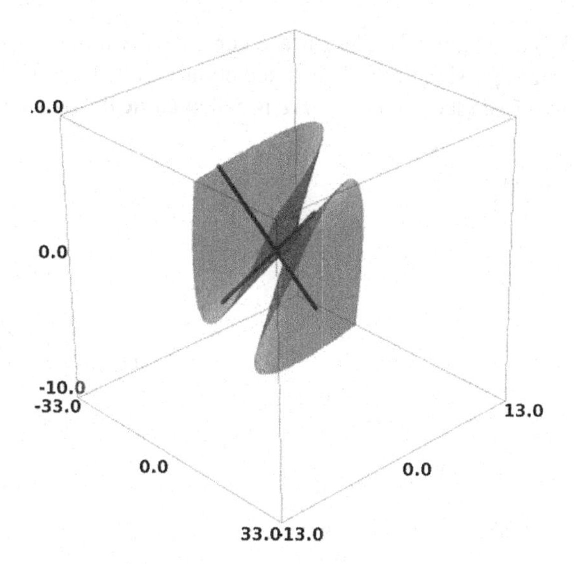


**Proposition 7.43** (see [4], p. 176). The one-sheeted hyperboloid has two families of rectilinear generators:

$$\begin{cases} \frac{x}{a} - \frac{z}{c} = \lambda \left(1 - \frac{y}{b}\right) \\ \lambda \left(\frac{x}{a} + \frac{z}{c}\right) = 1 + \frac{y}{b} \end{cases} \tag{7.81}$$

and

$$\begin{cases} \frac{x}{a} - \frac{z}{c} = \lambda \left(1 + \frac{y}{b}\right) \\ \lambda \left(\frac{x}{a} + \frac{z}{c}\right) = 1 - \frac{y}{b} \end{cases}, (\forall)\, \lambda \in \overline{\mathbb{R}}. \tag{7.81}$$

**Proposition 7.44** (see [4], p. 176). The hyperbolic paraboloid has two families of rectilinear generators:

$$\begin{cases} \frac{x}{a} - \frac{y}{b} = 2\lambda z \\ \lambda \left(\frac{x}{a} + \frac{y}{b}\right) = 1 \end{cases} \tag{7.82}$$

and

$$\begin{cases} \frac{x}{a} - \frac{y}{b} = \lambda \\ \lambda \left(\frac{x}{a} + \frac{y}{b}\right) = 2z \end{cases}, (\forall)\, \lambda \in \overline{\mathbb{R}}. \tag{7.83}$$

All of the degenerate quadrics have generators. Therefore, the cylindrical surfaces and the conical surfaces (including the quadrics by the cone type and the cylinder type) are ruled surfaces.

Let $V\,(a, b, c)$ be a fixed point and

$$\Gamma : \begin{cases} f\,(x, y, z) = 0 \\ g\,(x, y, z) = 0 \end{cases} \tag{7.84}$$

be a given curve.

**Definition 7.45** (see [11], p. 87). The surface generated by moving a straight line $D$ called the generatrix, passing through the fixed point $V$ called *vertix* and one relieses on the given curve $\Gamma$, called *director curve* is called **conical surface** (Fig. 7.14).

**Fig. 7.14** A conical surface

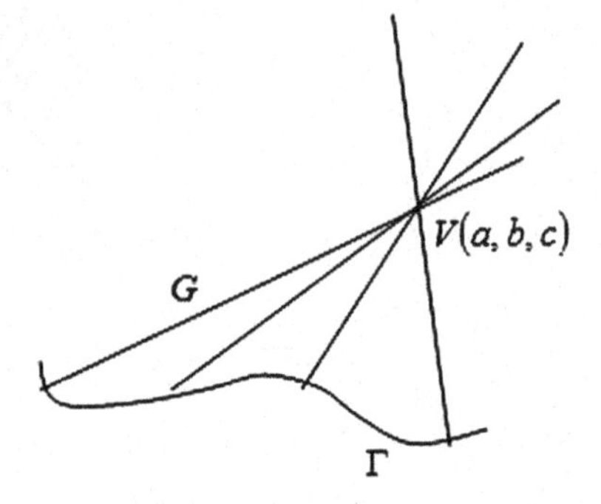


We want to find the equation of the conical surface. An arbitrary straight line, which passes through the point $V$ has the equations:

$$\frac{x-a}{l}=\frac{y-b}{m}=\frac{z-c}{n},\ l^2+m^2+n^2\neq 0.$$

We denote

$$\begin{cases} \frac{l}{n}=\alpha \\ \frac{m}{n}=\beta \end{cases},\ n\neq 0. \tag{7.85}$$

The straight line, which generates the conical surface (the generatrix) will be

$$\frac{x-a}{\alpha}=\frac{y-b}{\beta}=\frac{z-c}{1}. \tag{7.86}$$

The condition of supporting the generatrix on the director curve returns to the algebraic condition of compatibility of the system:

$$\begin{cases} f(x,y,z)=0 \\ g(x,y,z)=0 \\ \frac{x-a}{\alpha}=\frac{y-b}{\beta}=\frac{z-c}{1}. \end{cases} \tag{7.87}$$

Eliminating $x, y, z$ from the equations of the system, the compatibility condition become

$$\varphi(\alpha,\beta)=0. \tag{7.88}$$

Eliminating now $\alpha$ and $\beta$ from (7.86) and (7.88) it results that the requested conical surface has the equation:

$$\varphi\left(\frac{x-a}{z-c}, \frac{y-b}{z-c}\right) = 0. \tag{7.89}$$

**Example 7.46**. Find the Cartesian equation of the conical surface, which has the vertex $V(2, 1, 3)$ and the director curve

$$\Gamma : \begin{cases} z = 0 \\ x^2 + y^2 = 9. \end{cases}$$

**Solution**

The generatrix of the conical surface has the equations:

$$G : \frac{x-2}{l} = \frac{y-1}{m} = \frac{z-3}{n}, l^2 + m^2 + n^2 \neq 0.$$

Imposing the condition that the straight line $G$ to support on the curve $\Gamma$ we deduce that the system

$$\begin{cases} \frac{x-2}{l} = \frac{y-1}{m} = \frac{z-3}{n} \\ z = 0 \\ x^2 + y^2 = 9 \end{cases} \tag{7.90}$$

is compatible.

Using the notation (7.85) we achieve:

$$G : \frac{x-2}{\alpha} = \frac{y-1}{\beta} = \frac{z-3}{1}. \tag{7.91}$$

The system (7.90) becomes

$$\begin{cases} \beta x - \alpha y = 2\beta - \alpha \\ y = 1 - 3\beta \\ x = 2 - 3\alpha \\ x^2 + y^2 = 9 \end{cases} \Leftrightarrow (2 - 3\alpha)^2 + (1 - 3\beta)^2 = 9;$$

therefore, the condition of compatibility is reduced to the relation

$$9\alpha^2 + 9\beta^2 - 12\alpha - 6\beta - 4 = 0. \tag{7.92}$$

From (7.91) we deduce

$$\begin{cases} \alpha = \frac{x-2}{z-3} \\ \beta = \frac{y-1}{z-3}. \end{cases} \tag{7.93}$$

Substituting (7.93) in (7.92) it results that

$$9\left(\frac{x-2}{z-3}\right)^2+9\left(\frac{y-1}{z-3}\right)^2-12\frac{x-2}{z-3}-6\frac{y-1}{z-3}-4=0.$$

The Cartesian equation of the conical surface will be

$$9\,(x-2)^2+9\,(y-1)^2-12\,(x-2)\,(z-3)-6\,(y-1)\,(z-3)-4\,(z-3)^2=0.$$

The solution in Sage is:

```
sage: x,y,z,al,be,t=var('x,y,z,al,be,t')
sage: V=vector([2,1,3])
sage: s1=solve([(x-V[0])/al==(0-3)/1],x)
sage: s2=solve([(y-V[1])/be==(0-3)/1],y)
sage: f=x^2+y^2-9
sage: f1=expand(f.subs(x=s1[0].right(),y=s2[0].right()))
sage: s3=solve([(x-V[0])/al==(z-3)/1],al)
sage: s4=solve([(y-V[1])/be==(z-3)/1],be)
sage: F=f1.subs(al=s3[0].right(),be=s4[0].right()).factor()
sage: i=implicit_plot3d(F,(x,-20,20),(y,-20,20),(z,-20,20),opacity=0.5)
sage: Vv=point3d((V[0],V[1],V[2]),size=15,color='green')
sage: c=parametric_plot([3*cos(t), 3*sin(t), 0], (t, 0, 2*pi),thickness=5, color='red')
sage: i+c+Vv
```

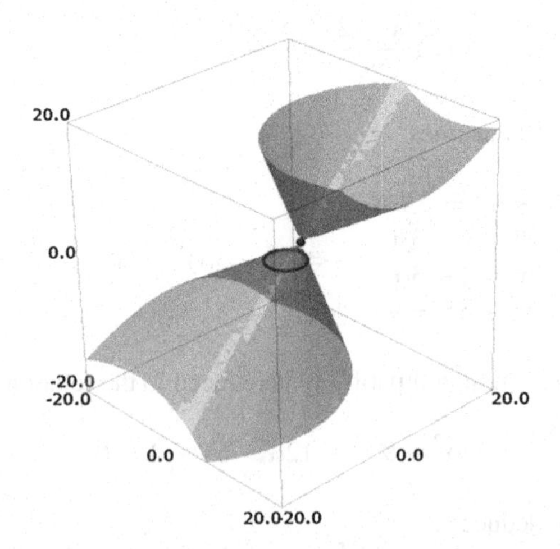


**Definition 7.47** (see [11], p. 89). We call a **cylindrical surface**, the surface generated by a straight line which remains parallel to a given direction and which supports on a given curve, called the *director curve* (Fig. 7.15).

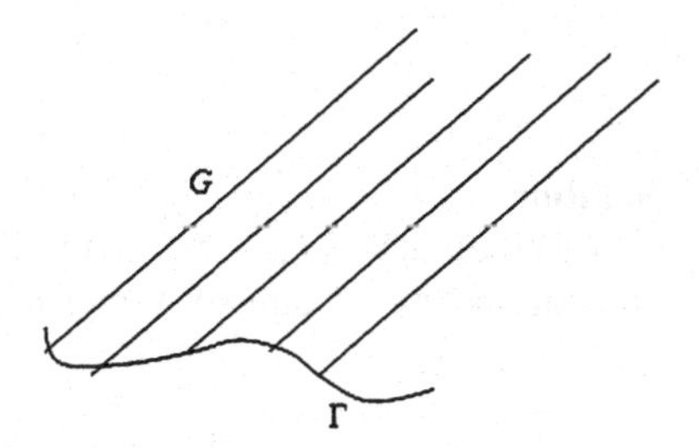


**Fig. 7.15** A cylindrical surface

Let $\overline{v} = (l, m, n)$ be the given direction and $\Gamma$ from (7.84) be the director curve. An arbitrary straight line, by the director vector $\overline{v}$ has the equations:

$$\begin{cases} nx - lz = \lambda \\ ny - mz = \mu. \end{cases} \tag{7.94}$$

The condition of supporting the generator on the director curve returns to the algebraic condition of compatibility of the system:

$$\begin{cases} f(x, y, z) = 0 \\ g(x, y, z) = 0 \\ nx - lz = \lambda \\ ny - mz = \mu. \end{cases}$$

Eliminating $x, y, z$ from the equations of the system, the compatibility condition becomes:

$$\varphi(\lambda, \mu) = 0. \tag{7.95}$$

Eliminating now $\lambda$ and $\mu$ from (7.94) and (7.95) it results that the requested cylindrical surface has the equation:

$$\varphi(nx - lz, ny - mz) = 0. \tag{7.96}$$

**Example 7.48** (see [4], p. 215). Write the equation of the cylindrical surface, which has the director curve

$$\Gamma : \begin{cases} z = 0 \\ x^2 + y^2 + 2x - y = 0 \end{cases}$$

and the generators parallel to the straight line by equation

$$d: \frac{x-1}{1} = \frac{y-2}{-1} = \frac{z}{2}.$$

**Solution**

We note that $\overline{v} = (1, -1, 2)$ is the director vector of the straight line $d$. An arbitrary straight line, by the director vector $\overline{v}$ has the equations:

$$\begin{cases} x + y - 3 = \lambda \\ 2x - z - 2 = \mu \end{cases}, (\forall)\, \lambda, \mu \in \mathbb{R}. \tag{7.97}$$

The condition of supporting the generatrix on the director curve returns to the algebraic condition of compatibility of the system:

$$\begin{cases} x^2 + y^2 + 2x - y = 0 \\ z = 0 \\ x + y - 3 = \lambda \\ 2x - z - 2 = \mu \end{cases} \Leftrightarrow \begin{cases} x^2 + y^2 + 2x - y = 0 \\ x + y - 3 = \lambda \\ 2x - 2 = \mu. \end{cases}$$

From the third equation of the system we deduce

$$x = \frac{\mu + 2}{2}; \tag{7.98}$$

substituting this expression of $x$ in the second equation of the system we get:

$$y = \lambda + 3 - \frac{\mu + 2}{2} = \frac{2\lambda - \mu + 4}{2}. \tag{7.99}$$

If in the first equation of the system, we take into account of (7.98) and (7.99) we achieve:

$$\left(\frac{\mu + 2}{2}\right)^2 + \left(\frac{2\lambda - \mu + 4}{2}\right)^2 + 2 \cdot \frac{\mu + 2}{2} - \frac{2\lambda - \mu + 4}{2} = 0,$$

i.e

$$(\mu + 2)^2 + (2\lambda - \mu + 4)^2 + 6\mu - 4\lambda = 0. \tag{7.100}$$

Eliminating $\mu$ and $\lambda$ between (7.100) and (7.97) it follows that the cylindrical surface equation is

$$(2x - z)^2 + (2y + z)^2 + 8x - 4y - 6z = 0.$$

Solving with Sage this problem, we have:

```
sage: x,y,z,la,miu=var('x,y,z,la,miu')
sage: v=vector([1,-1,2])
sage: s1=solve([(x-1)/v[0]-(y-2)/v[1]==la],la)
sage: s2=solve([(x-1)/v[0]-z/v[2]==miu],miu)
sage: f=x^2+y^2+2*x-y
sage: s3=solve(s2[0].subs(z=0),x)
sage: s4=solve(s1[0].right().subs(x=s3[0].right())==s1[0].left(),y)
sage: f1=f.subs(x=s3[0].right(),y=s4[0].right())
sage: F=f1.subs(la=s1[0].right(),miu=s2[0].right()).factor()
sage: F
x^2 - x*z + y^2 + y*z + 1/2*z^2 + 2*x - y - 3/2*z
sage: i=implicit_plot3d(F,(x,-7,7),(y,-4,4),(z,-4,4),opacity=0.5)
sage: l1=line3d([(-3-2,6-2,-8-2),(3-2,0-2,4-2)],thickness=5,color='green')
sage: l2=line3d([(-3-1.2,6-1.2,-8-1.2),(3-1.2,0-1.2,4-1.2)],thickness=5,color='green')
sage: c=parametric_plot([sqrt(5)/2*cos(t)-1, sqrt(5)/2*sin(t)+1/2, 0], (t, 0, 2*pi),thickness=5, color='red')
sage: i+l1+l2+c
```

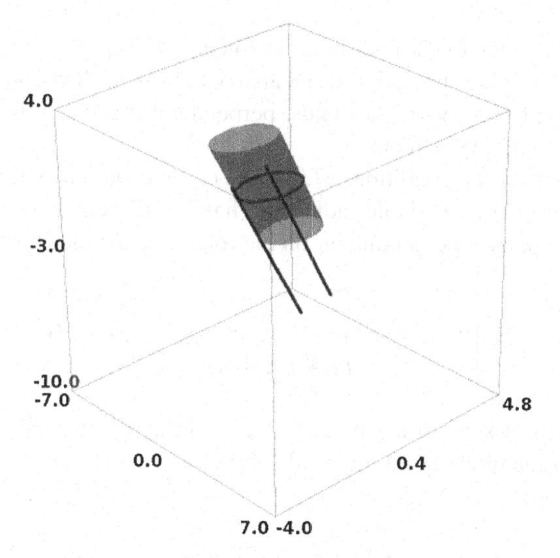


**Definition 7.49** (see [11], p. 91). We call a **surface of rotation**, a surface generated by rotating a curve around of a straight line called the *axis of rotation* (Fig. 7.16).

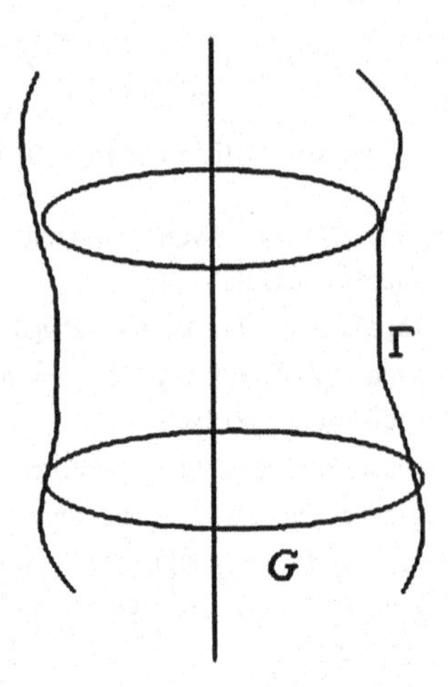


**Fig. 7.16** A surface of rotation

The axis of rotation has the equations:

$$\frac{x - x_0}{l} = \frac{y - y_0}{m} = \frac{z - z_0}{n}, l^2 + m^2 + n^2 \neq 0$$

and the curve Γ, which one rotates being defined in (7.84).

The surface from the above figure can also result by the displacement of a circle, parallel with itself, with variable radius, perpendicular to the axis of rotation and which one relies on the given curve.

In order to satisfy the conditions (7.95), the circle equations will be achieved by crossing a sphere with a variable radius and having the center on the rotation axis with a family of planes, perpendicular on the rotation axis; therefore

$$G : \begin{cases} (x - x_0)^2 + (y - y_0)^2 + (z - z_0)^2 = \alpha^2 \\ lx + my + nz = \beta. \end{cases} \tag{7.101}$$

The condition of supporting the generatrix on the director curve returns to the algebraic condition of compatibility of the system:

$$\begin{cases} f(x, y, z) = 0 \\ g(x, y, z) = 0 \\ (x - x_0)^2 + (y - y_0)^2 + (z - z_0)^2 = \alpha^2 \\ lx + my + nz = \beta. \end{cases}$$

Eliminating $x, y, z$ from the equations of the system, the compatibility condition becomes (7.88).

Eliminating now $\alpha$ and $\beta$ from (7.101) and (7.88) it results that the requested cylindrical surface has the equation:

$$\varphi\left((x-x_0)^2+(y-y_0)^2+(z-z_0)^2, lx+my+nz\right)=0. \qquad (7.102)$$

## 7.7 Problems

1. Let be the points: $A(-1,4)$, $B(3,-2)$. Write the equation of the circle, which has $AB$ as a diameter.

**Solution**

Solving this problem in Sage, we have:

```
sage: A=vector([-1,4]);B=vector([3,-2])
sage: AB=B-A;r=AB.norm()/2
sage: xc=(A[0]+B[0])/2;yc=(A[1]+B[1])/2
sage: x,y=var('x,y')
sage: (x-xc)^2+(y-yc)^2==r^2
(y - 1)^2 + (x - 1)^2 == 13
sage: C=circle((xc,yc),r)
sage: A1=point2d((A[0],A[1]),size=22);t1=text("A",(A[0]-0.1,A[1]+0.1))
sage: B1=point2d((B[0],B[1]),size=22);t2=text("B",(B[0]+0.1,B[1]-0.2))
sage: C1=point2d((xc,yc),size=22);t3=text("C",(xc+0.1,yc+0.1))
sage: l=line([(A[0],A[1]), (B[0],B[1])])
sage: A1+B1+C+t1+t2+l+C1+t3
```

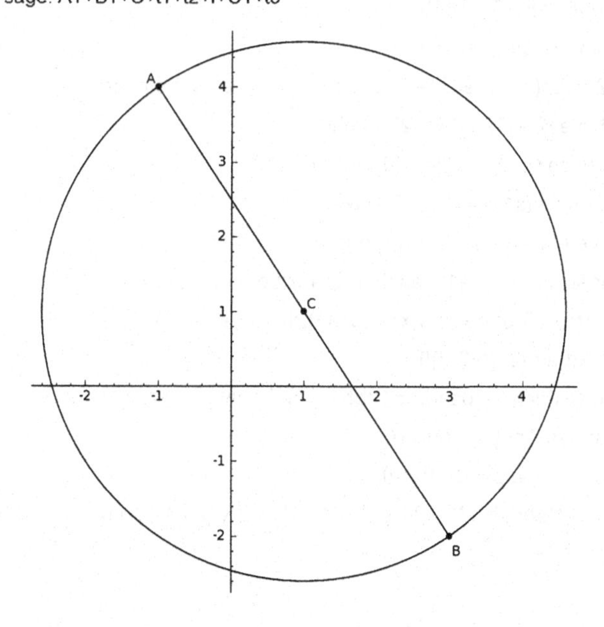

2. Determine the vertices and the semi-axes of the ellipse

$$2x^2 + 4y^2 - 5 = 0.$$

3. Let be the conic

$$\Gamma : 4x^2 - 12xy + 9y^2 - 2x + 3y - 2 = 0.$$

Bring it to the canonical form, indicating the required reference changes and recognize the achieved conic.

4. Write the equation of the parabola which passes through the points:

$$O\,(0, 0)\,,\, A\,(1, 0)\,,\, B\,(0, 1)\,,\, C\,(2, 3)\,.$$

**Solution**

The solution in Sage is:

```
sage: a11,a12,a22,b1,b2,c,x,y=var('a11,a12,a22,b1,b2,c,x,y')
sage: A=vector([1,0]);B=vector([0,1])
sage: C=vector([2,3]);O=vector([0,0])
sage: f=a11*x^2+2*a12*x*y+a22*y^2+2*b1*x+2*b2*y+c
sage: e1=f.subs(x=O[0],y=O[1])
sage: s1=solve(e1,c);cc=s1[0].right()
sage: e2=f.subs(x=A[0],y=A[1])
sage: e3=f.subs(x=B[0],y=B[1])
sage: e4=f.subs(x=C[0],y=C[1])
sage: ee2=e2.subs(c=cc);ee3=e3.subs(c=cc);ee4=e4.subs(c=cc)
sage: s2=solve(ee2,a11);aa11=s2[0].right()
sage: s3=solve(ee3,a22);aa22=s3[0].right()
sage: ef4=ee4.subs(a11=aa11,a22=aa22)
sage: s4=solve(ef4,a12);aa12=s4[0].right()
sage: g=f.subs(a11=aa11,a12=aa12,a22=aa22,c=cc).expand()
sage: a=g.coeff(x^2);b=g.coeff(x).coeff(y);d=g.coeff(y^2)
sage: A=matrix([[a,b/2],[b/2,d]])
sage: s5=solve(A.det().expand(),b1);bb1=s5[0].right();bc1=s5[1].right()
sage: gg=g.subs(b1=bb1).factor()
sage: h=gg/b2;ff=h.expand();ff==0
12*sqrt(6)*x^2 - 4*sqrt(6)*x*y - 30*x^2 + 12*x*y - 2*y^2 - 12*sqrt(6)*x + 30*x + 2*y == 0
```

We shall also use Sage to represent the achieved parabola:

```
sage: l=implicit_plot(ff,(-3,3),(-3,3),linewidth=2)
sage: A1=point2d((A[0],A[1]),size=22);t1=text("A",(A[0]+0.05,A[1]-0.1))
sage: B1=point2d((B[0],B[1]),size=22);t2=text("B",(B[0]-0.05,B[1]+0.1))
sage: O1=point2d((O[0],O[1]),size=22);t3=text("O",(O[0]-0.1,O[1]-0.1))
sage: C1=point2d((C[0],C[1]),size=22);t4=text("C",(C[0]+0.05,C[1]+0.1))
sage: l1=implicit_plot(y,(x,-3,3),(y,-3,3),linewidth=1)
sage: l2=implicit_plot(x,(x,-3,3),(y,-3,3),linewidth=1)
sage: l+A1+B1+t1+t2+O1+t3+l1+l2+C1+t4
```

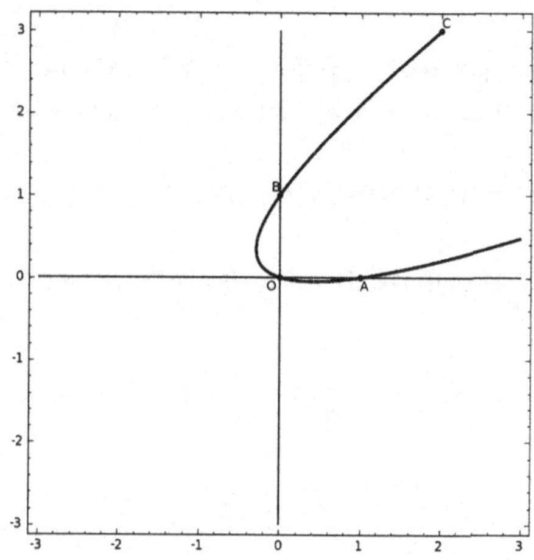


5. Let be the points $A(-1,1,2)$, $B(1,3,3)$. Write the equation of the sphere with center in the point and which passes through the point $B$.
6. Write the equations of the rectilinear generators

$$\frac{x^2}{16} - \frac{y^2}{4} = z,$$

that are parallel to the plane $\pi : 3x + 2y - 4z = 0$.

7. Find the rectilinear generators of the quadrics :

$$\Sigma : x^2 + 3y^2 + 4yz - 6x + 8y + 8 = 0.$$

**Solution**

we need the following Sage code to solve this problem:

```
sage: R.<x,y,z> = QQbar[]
sage: f=x^2+3*y^2+4*y*z-6*x+8*y+8
sage: a11=f.coefficient({x:2});a22=f.coefficient({y:2});a33=f.coefficient({z:2})
sage: a12=f.coefficient({x:1,y:1,z:0})/2
sage: a13=f.coefficient({x:1,z:1,y:0})/2
sage: a23=f.coefficient({y:1,z:1,x:0})/2
sage: b1=f.coefficient({x:1,y:0,z:0})/2
sage: b2=f.coefficient({y:1,x:0,z:0})/2
sage: b3=f.coefficient({z:1,x:0,y:0})/2
sage: c=f.coefficient({x:0,y:0,z:0})
sage: I=a11+a22+a33
sage: A1=matrix([[a11,a12],[a12,a22]]);A2=matrix([[a11,a13],[a13,a33]])
sage: A3=matrix([[a22,a23],[a23,a33]]);d1=A1.det();d2=A2.det();d3=A3.det()
sage: J=d1+d2+d3
sage: A=matrix([[a11,a12,a13],[a12,a22,a23],[a13,a23,a33]])
sage: de=A.det()
sage: B=matrix([[a11,a12,a13,b1],[a12,a22,a23,b2],[a13,a23,a33,b3],[b1,b2,b3,c]])
sage: De=B.det()
sage: I;J;de;De
4
-1
-4
4
sage: u=diff(f,x);v=diff(f,y);w=diff(f,z)
sage: u1=u.coefficient(x).n(digits=3);u2=u.coefficient(y).n(digits=3)
sage: u3=u.coefficient(z).n(digits=3);v1=v.coefficient(x).n(digits=3)
sage: v2=v.coefficient(y).n(digits=3);v3=v.coefficient(z).n(digits=3)
sage: w1=w.coefficient(x).n(digits=3);w2=w.coefficient(y).n(digits=3)
sage: w2=w.coefficient(y).n(digits=3);w3=w.coefficient(z).n(digits=3)
sage: uu=u-(u1*x+u2*y+u3*z);vv=v-(v1*x+v2*y+v3*z);ww=w-(w1*x+w2*y+w3*z)
sage: x0,y0,z0,S,t=var('x0,y0,z0,S,t')
sage: solve([u1*x0+u2*y0+u3*z0+uu,v1*x0+v2*y0+v3*z0+vv,w1*x0+w2*y0+w3*z0+ww],x0,y0,z0)
[[x0 == 3, y0 == 0, z0 == -2]]
```

```
sage: II=I.n(digits=3);JJ=J.n(digits=3);dd=de.n(digits=3);Dd=De.n(digits=3)
sage: s=(S^3-II*S^2+JJ*S-dd).roots()
sage: s0=s[0][0].n(digits=3);s1=s[1][0].n(digits=3);s2=s[2][0].n(digits=3)
sage: F=(s2*x0^2+s0*y0^2+s1*z0^2+Dd/dd).simplify_exp();F
x0^2 + 4.0*y0^2 - z0^2 - 1.0
```

```
sage: i=implicit_plot3d(F, (x0, -5, 5), (y0, -5, 5), (z0, -4, 4), opacity=0.5)
sage: a=F.coeff(x0^2);b=F.coeff(y0^2);c=F.coeff(z0^2)
sage: f1=x0/a-z0/c-t*(1-y0/b).factor()
sage: f2=t*(x0/a+z0/c)-(1+y0/b).factor()
sage: f3=x0/a-z0/c-t*(1+y0/b).factor()
sage: f4=t*(x0/a+z0/c)-(1-y0/b).factor()
sage: f1==0;f2==0;f3==0;f4==0
1/4*(y0 - 4)*t + x0 + z0 == 0
(x0 - z0)*t - 1/4*y0 - 1 == 0
-1/4*(y0 + 4)*t + x0 + z0 == 0
(x0 - z0)*t + 1/4*y0 - 1 == 0
sage: f1.subs(t=1);f2.subs(t=1);f3.subs(t=1);f4.subs(t=1)
x0 + 1/4*y0 + z0 - 1
x0 - 1/4*y0 - z0 - 1
x0 - 1/4*y0 + z0 - 1
x0 + 1/4*y0 - z0 - 1
sage: l1=line3d([(-4,4,1),(4,-4,1)],thickness=6,color='orange')
sage: l2=line3d([(-0.7,-0.7,-0.7),(0.7,0.7,-0.7)],thickness=6,color='orange')
sage: l3=line3d([(-0.7,-0.7,0.7),(0.7,0.7,0.7)],thickness=6,color='orange')
sage: l4=line3d([(-4,4,-1),(4,-4,-1)],thickness=6,color='orange')
sage: l1+l2+l3+l4+i
```

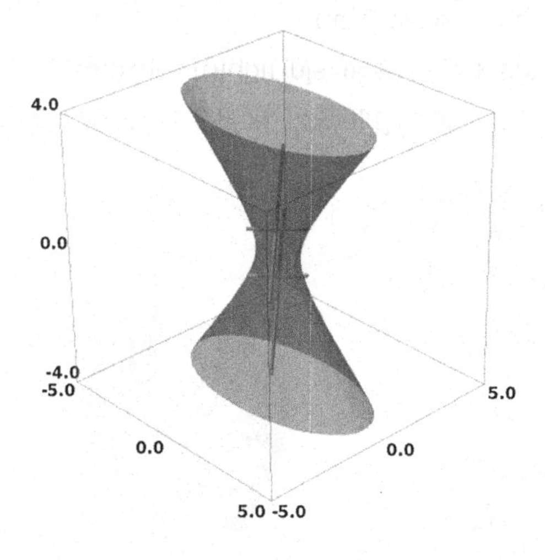

8. Find the equation of the conical surface having as vertex the point $V(-3,0,0)$ and the director curve:

$$\Gamma : \begin{cases} 3x^2 + 6y^2 - z = 0 \\ x + y + z - 1 = 0. \end{cases}$$

9. Determine the equation of the cylindrical surface that has the director curve:

$$\Gamma : \begin{cases} x^2 + y^2 - 1 = 0 \\ z = 0 \end{cases}$$

and the generators parallel to the straight line

$$d : x = y = z.$$

10. Write the equation of the surface generated by rotating the parabola

$$p : \begin{cases} y^2 = 2px \\ z = 0 \end{cases}$$

around the O$x$ axis.

**Solution**

A solution in Sage of this problem is:

```
sage: x,y,z,al,be=var('x,y,z,al,be')
sage: f=x^2+y^2+z^2-al^2
sage: f1=f.subs(x=be,y=sqrt(2*be),z=0)
sage: s=solve(f1,al)
sage: F=f.subs(al=s[0].right()).subs(be=x)
sage: implicit_plot3d(F,(x,-3,3),(y,-3,3),(z,-3,3))
```

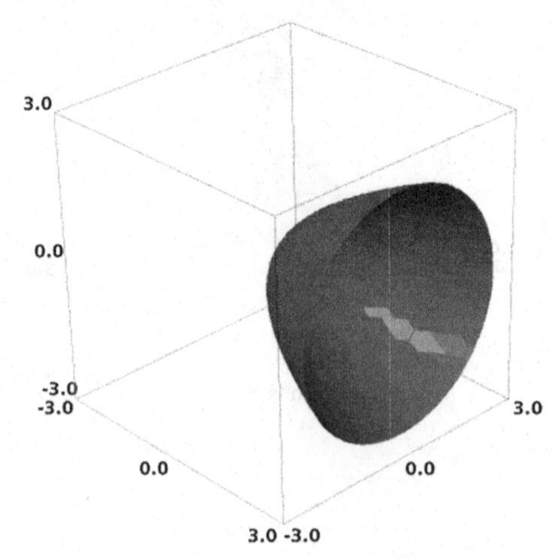

# References

1. V. Balan, *Algebră liniară, geometrie analitică, ed* (Fair Partners, Bucureşti, 1999)
2. I. Vladimirescu, M. Popescu, M. Sterpu, *Algebră liniară şi geometrie analitică* (Universitatea din Craiova, Note de curs şi aplicaţii, 1993)
3. C. Udrişte, *Aplicaţii de algebră, geometrie şi ecuaţii diferenţiale, ed* (Didactică şi Pedagogică R.A, Bucureşti, 1993)
4. I. Vladimirescu, M. Popescu, *Algebră liniară şi geometrie analitică, ed* (Universitaria, Craiova, 1993)
5. Gh. D. Simionescu, *Geometrie analitică, ed* (Didactică şi pedagogică, Bucureşti, 1968)
6. V. Postelnicu, S. Coatu, *Mică enciclopedie matematică, ed* (Tehnică, Bucureşti, 1980)
7. E. Murgulescu, S. Flexi, O. Kreindler, O. Sacter, M. Tîrnoveanu, *Geometrie analitică şi diferenţială, ed* (Didactică şi pedagogică, Bucureşti, 1965)
8. V.T. Postelnicu, I.M. Stoka, Gh Vrânceanu, *Culegere de probleme de geometrie analitică şi proiectivă, ed* (Tehnică, Bucureşti, 1962)
9. Gh Atanasiu, Gh Munteanu, M. Postolache, *Algebr ă liniară, geometrie analitică şi diferenţială, ecua ţii diferenţiale, ed* (ALL, Bucureşti, 1998)
10. G. Mărgulescu, P. Papadapol, *Curs de geometrie analitică, diferenţială şi algebră liniară* (Catedra de Matematici, 1976)
11. T. Didenco, *Geometrie analitică şi diferenţială* (Academia Militară, Bucureşti, 1977)

# Index



G. A. Anastassiou and I. F. Iatan, *Intelligent Routines II*,
Intelligent Systems Reference Library 58, DOI: 10.1007/978-3-319-01967-3,

Printed in the United States
By Bookmasters